SAUDI ARAMCO
AND ITS WORLD

SAUDI ARAMCO AND ITS WORLD

Arabia And The Middle East

A revised edition of *Aramco And Its World*, 1981

Edited by

Ismail I. Nawwab   Peter C. Speers   Paul F. Hoye

With principal research and writing by
Paul Lunde, Lyn Maby and John A. Sabini

Revised edition, 1995

Edited and with additional material by

Arthur P. Clark   James P. Mandaville   William Tracy   Jane Waldron Grutz

The Saudi Arabian Oil Company (Saudi Aramco)
Dhahran, Saudi Arabia

ISBN 0-9601164-3-5
Copyright © 1995 Aramco Services Company
Library of Congress Catalogue Number: 95-70752

Aramco Services Company
Houston, Texas
1995

Printed in the Netherlands by Boekhoven-Bosch B.V.
Part of the Roto Smeets de Boer Group

Preface

Saudi Aramco and Its World is a revised and updated edition of *Aramco and Its World*, published in 1981. This present edition of the book notes some of the momentous social, political and economic changes that have taken place in the Middle East and the world during the past 15 years. It adds considerable new material, including maps and photographs, on the impressive developments which have occurred within Saudi Arabia and the company during the same period.

Saudi Aramco and Its World had its origins in the late 1940s. The Arabian American Oil Company (Aramco), which became Saudi Aramco (the Saudi Arabian Oil Company) in 1988, was moving through a period of tremendous growth in the years following World War II. This expansion involved the rapid enlargement of the company's work force of many different nationalities. To provide these newcomers with basic information about the company and its place in the worldwide oil industry, and to help them adjust to work and life in a new land among a people whose language they did not understand and whose culture was very different from what they had known, Aramco in January 1950 issued a series of five spiral-bound booklets with the general title *Handbooks for American Employees*.

These booklets attracted interest both inside and outside the company and the demand for copies became so great that two years later, in January 1952, a revised hardbound edition was published that consisted of two volumes entitled *Aramco and World Oil* and *The Arabia of Ibn Saud*. A single-volume edition containing the same material, further revised, appeared in 1960 with the title *Aramco Handbook*. When the fourth edition was published in 1968, it was apparent that the book had become much more than a "handbook for employees" and accordingly it carried the additional subtitle *Oil and the Middle East*.

In 1980, when the fifth edition was published under the name *Aramco and Its World*, the book assumed its present format, richly illustrated with color photographs, drawings, and extensive maps. Immediately successful, it was reprinted the following year.

Saudi Aramco and Its World may thus be considered the sixth edition of a work first published some 45 years ago. It continues the trend away from a narrowly circumscribed audience of company employees and toward a more general readership. Its content has also evolved. In the early 1950s most Westerners considered Saudi Arabia and the rest of the Middle East to be exotic lands, and comprehensive, accurate, up-to-date books about the area hardly existed. Today, although global communications and broadened education have helped make more people aware of the Arab world's proud history and its contributions to the world's spiritual and intellectual heritage, and even though the role of oil in the global economy has pushed the Arabian Peninsula on to center stage, the region remains a relatively little-known and surely an insufficiently understood part of the world.

Saudi Aramco and Its World aims to contribute to increased understanding of Saudi Arabia, the Arab world in general, and the rest of the Middle East on the part of a readership which now includes many with no direct connection with the region, with Saudi Arabia, or Saudi Aramco. We hope it will also reach others with little or no knowledge of Islam, the Arab world, or the oil industry, but who have an interest in some or all of those things and a desire to learn something about them.

An Introduction

Some of the most crucial events in the history of mankind have taken place in the region that this volume attempts to survey. Agriculture began in Mesopotamia and later, in this "land between the rivers," the first cities grew up, with the attendant development of laws, crafts, kingship, and writing. From the broader area we call the Middle East came the basic elements of mathematics, the wheel and the arch, sciences such as astronomy and medicine, and the framework of organized trade and commerce. From this region, too, came the three great monotheistic religions: Judaism, Christianity, and Islam. Indeed, civilization itself — the triumph of human experience over time — began in the lands that lie between the valleys of the Nile and the Indus.

Over the past 6,000 years great cultures have sprung from this source and ebbed and flowed across the map of the Middle East. Some of them rivaled or surpassed the levels of culture later reached by the Greeks and Romans of classical times, who rightly acknowledged their debt to the civilizations of Mesopotamia and the Nile. And when the Greek and Roman civilizations dimmed and sank into the shadows of the Dark Ages, their afterglow lingered in the Byzantine lands — which included Syria and Egypt — and much of their philosophy and their scientific and practical knowledge was preserved in the new classical language, Arabic, by the new torchbearers of civilization, the Muslims. The great flowering of Islamic culture between the ninth and thirteenth centuries owed much to this heritage. When the Golden Age of Islamic civilization ended, it was passed on again, vastly enriched, to the renascent West, which in turn came once more to play an important role in the political, economic, and cultural history of the region.

No one volume can do full justice to such an immense sweep of history. Nor can it completely satisfy the often conflicting demands of scholars, technical experts, and ordinary readers. To compress the developments of several thousand years of history into a readable and reasonably brief narrative has inevitably required painful omissions, and occasionally a narrow focus on particular details to the exclusion of other significant material. *Saudi Aramco and Its World* strives for a balance that we hope will appeal to the general reader and, at the same time, succeed in presenting accurate information that is based on sound scholarship and, wherever possible, on firsthand knowledge of the events and places discussed.

A few comments about the Arabic language are necessary for readers of this book. Arabic personal names and place names are spelled according to a system used by Saudi Aramco which closely follows a generally accepted system of transliteration from Arabic to English. The system does not always represent the pronunciation of the Arabic original with complete accuracy, as the Arabic language contains letters and sounds for which no equivalent exists in English.

One Arabic consonant which has no counterpart in English is the letter 'ayn, which is represented generally by an inverted apostrophe ('). It often appears in personal or place names, such as Al Sa'ud or Ka'bah. But when an Arabic word has acquired a common English usage, the popular form is used, such as Saudi Arabia (instead of Sa'udi 'Arabia) or Dhahran (instead of az-Zahran).

It may be helpful to explain four common Arabic words that the reader will come across repeatedly in this book. The word *al-* (joined to the following word and with a small *a* unless it comes at the beginning of a sentence) is the definite article and corresponds to the English "the." The similar *Al* (always with a capital *A* and never joined to the following word) means "House (or family) of," as in Al Sa'ud, the name of the ruling family of Saudi Arabia. *'Abd* means "servant of," and in conjunction with one of the attributes of God it is very commonly used to form Arabic personal names such as 'Abd Allah ("Servant of God") or 'Abd al-Rahman ("Servant of the Merciful"). The word *ibn*, sometimes pronounced *bin*, means "son of" or "descendant of the House (or family) of," as in Ibn Sa'ud or Ibn Khaldun. Its plural is *banu* or *bani* ("sons of"), as in Banu Musa or Bani Hilal.

Metric measurements, standard in Saudi Arabia and most other countries, are used throughout *Saudi Aramco and Its World*. Readers who would like to convert from the metric to the English system of measurement are asked to consult the tables below.

Like its predecessors, this new edition of *Saudi Aramco and Its World* is dedicated to bringing to its readers a greater understanding of a history and culture that have contributed so much to the artistic, scientific and philosophical evolution of both East and West.

Metric-to-English Conversion Table

Length:	1 meter = 1.09 yards = 3.28 feet = 39.37 inches 1 kilometer (1,000 meters) = .62 mile = 3,280.8 feet
Area:	1 square meter = 1.196 square yards = 10.76 square feet 1 hectare (10,000 square meters) = 2.47 acres = 107,593 square feet 1 square kilometer (100 hectares) = .386 square mile = 247 acres
Volume:	1 liter = .264 U.S. gallon 1 barrel = 159 liters = 42 U.S. gallons
Mass:	1 kilogram = 2.2 pounds 1 metric ton (1,000 kilograms) = 1.1 short tons = 2,204.6 pounds

Acknowledgements

Saudi Aramco and Its World is the work of many hands, and the editors wish that space allowed them to acknowledge fully the contributions of all who participated. Each can take credit for the success of the book; none should be held responsible for any errors that may be found in it.

The 1981 edition of *Aramco and Its World* would not have been possible without the support of three members of the company's senior management of the time, John J. Kelberer, Majed Elass and Faysal M. al-Bassam. This 1995 revised edition grew out of that same tradition, and the editors wish to thank Saudi Aramco's Ali I. Naimi, Ahmad S. al-Nassar, and again Faysal M. al-Bassam for their personal as well as corporate support. At Aramco Services Company in Houston, Ibrahim S. Mishari and Shafiq W. Kombargi deserve mention for their support.

This book reflects the work of the many scholars, writers, researchers, and reviewers who worked on earlier volumes. They include past and present employees of Saudi Aramco and others with little or no direct connection with the company. Among them have been Thomas C. Barger, Brainerd S. Bates, S. D. Bowers, John M. Curry, Fred H. Drucker, D. R. Fate, Marny Golding, George F. Hourani, Cathy Hoye, Yusuf Ibish, Libby Dudley, Abdallah S. Jum'ah, Frank Jungers, Leslie G. Lewis, Robert L. Maby, Elizabeth Monroe, William E. Mulligan, John M. Munro, Nawal Naoum, Mary Norton, Peter P. Pease, George S. Rentz, Ellen Speers, Caroline Stone, F. S. Vidal, and William A. Ward. Individuals who assisted in the preparation of the 1995 edition include Robert Arndt, Mohammad I. al-Bassam, David D. Bosch, Hassan I. al-Husseini, Mohammad A. Mashat, Muhammed R. al-Mughamis, Muhammad A. Tahlawi, and Elaine Thomason.

The 1981 edition was designed by Don Thompson of Aramco's production staff, who has also provided the paintings which conclude the present revised edition. The design of the revised edition incorporates the work of Herring Design in Houston.

The majority of new photographs in the 1995 edition connected with Islam, Saudi Arabia, and Saudi Aramco are by Abdulaziz Abdullatif, S. M. Amin, Ian Bennett, John Champney, Abdullah al-Dobais, and Adrian Waine—all of them company photographers now or in the past. Former and current staff photographers who contributed to earlier editions and whose work is represented on the following pages include A. G. Daghish, David Denning, Sa'id Ghamdi, Geoffrey Hunter, Michael Isaac, A. A. al-Khalifa, A. M. al-Khalifa, A. A. al-Mentakh, Burnett H. Moody, Khalil Abou el-Nasr, E. E. Seal, M. S. al-Shabeeb, Thomas F. Walters, and A. L. Yousif. Most of the drawings, and all of the maps and charts in the 1981 edition were by Don Thompson, while the 1995 revision contains additional maps and charts by Herring Design. Mohamed Zakariya crafted the astrolabe pictured on the dust cover.

Additional photographs and drawings are by Robert Azzi, Rick Bloom, John E. Burchard, Dick Doughty, Tor Eigeland, Randy Faris, Harris Foster, Steve and Claudine Furman, Marny Golding, Christine Kokesh Hill, Lawson Hitt, Peter Keen, Sydney King, Dickran Kouymjian, Chris Kutschera, James P. Mandaville, Roland and Sabrina Michaud, Dorothy Miller, Joe Mountain, Ed Mullis, Robert Yarnall Richie, Don Rokke, Brian Smith, Max Steineke, Katrina Thomas, William Tracy, Nik Wheeler, Penny Williams Yaqub, Adam Woolfitt and Chad Wyatt. Other illustrations have been taken from *Aramco World* and *Dimensions* magazines, *Facts about Oil*, Humble Oil Company, International Publication Services (Beirut), Iraq Petroleum Company, the Metropolitan Museum of Art (New York), the National Museum (Damascus), National Photo Company (al-Khobar), Petroleum Information Bureau (London), the Royal Commission for Jubail and Yanbu', and SIPA Press (Paris).

Contents

Contents

Contents

SAUDI ARAMCO
AND ITS WORLD

BEFORE ISLAM

In the Middle East, as elsewhere, geography and ecology have been among the important architects of history.

During the last Ice Age — which ended about 15,000 years ago — parts of Europe lay under a sheet of ice. The mass of cold air this generated blocked off the Atlantic rainstorms and sent them across North Africa and the Middle East, creating a climate that supported park and grassland and fed lakes, streams, and great river systems — some still visible today as dry wadis in the Arabian Peninsula and the Sahara. The climate also nourished a wide variety of animal life, now vanished from the area, which Paleolithic man hunted and memorialized in rock carvings in the midst of now nearly lifeless desert.

Later, as the European ice cap melted and the Atlantic rain path moved northward again, progressive desiccation set in, turning the grasslands and forests of the Middle East into steppe and desert. The process was gradual, however, and for long periods the climate of early historical times remained more temperate than it is today. Up to the second millennium B.C., Libya was rich in olives, wine, and livestock. Egypt was a land of marshes teeming with wildlife, and lower Mesopotamia a marshland covered with extensive reed forests.

As desiccation progressed, however, areas with favorable conditions were reduced considerably. By the dawn of history the habitable area of Egypt was already confined to the cliff valley of the Nile in the south, spreading across the Delta in the north. To the north and east of Egypt, the area of favorable conditions narrowed down to a great arc pressed between the Anatolian plateau and the Arabian Desert. Stretching from the head of the Arabian Gulf northward to include the land watered by the Tigris and Euphrates, west across present-day Syria, then south to include the whole eastern shore of the Mediterranean and the lands behind it as far south as the Gulf of Aqaba, this region has been aptly called the Fertile Crescent.

As the highlands turned into steppe and desert, man was forced to descend into the river valleys in search of game and edible vegetation. By Neolithic times mass migrations were common and agriculture, perhaps the most revolutionary development in man's history, had made its first appearance. Agriculture is believed to have originated in the Tigris-Euphrates Valley, where very old

In the marshy river valleys of the Middle East man first attempted to live in settled communities.

The earliest agricultural sites in the world have been found in the foothills of Kurdistan in the Tigris Valley.

agricultural sites have been found in the foothills of what is now Kurdistan. Excavations at Jericho in the Jordan River Valley have also revealed evidence of settled farming going back to 8000 B.C. Irrigation, a separate and later development, probably originated in Mesopotamia too and did more than simple agriculture — which in earliest times, before the discovery of crop rotation, probably involved a constant search for new and fertile soil—to tie human groups to specific localities and so foster the growth of permanent communities. Both agriculture and irrigation developed in the Nile Valley as well, probably independently but a little later than in Mesopotamia.

Another development closely connected to agriculture was the introduction of pottery — possibly an outgrowth of the need to store food surpluses and seed for planting. Peoples of what is known as the 'Ubayd culture (from a site in southern Iraq) were, despite the lack of the potter's wheel, among the world's first master potters. They left a trail of distinctive sherds running up the Tigris and Euphrates Valley and as far south as Tarut Island in what is now Saudi Arabia, which probably indicates the spread of the new science of agriculture as it radiated slowly outwards.

Prehistoric peoples developed distinctive cultures and, chiefly through trade, cultural contacts with one another. Peoples speaking Semitic languages, whose origins are unknown but who first appear in the lands of the Fertile Crescent and the Arabian Peninsula, undoubtedly had cultural and commercial contacts with the peoples of the Nile, for Semitic elements are found in the earliest written records of the Egyptians. Strictly speaking, the terms "Semitic" and "Semite" refer respectively to a group of related languages which includes Akkadian, Aramaic, Syriac, Hebrew, Arabic, and others and to a group of peoples speaking these languages.

The culture of prehistoric peoples eventually flowered into what we call civilization—at which point they possessed writing, monumental architecture, religious art, legal codes, and complex political systems. This efflorescence first occurred in the fourth millennium B.C. in the southern reaches of the Tigris-Euphrates Valley and, a little later, on the banks of the Nile.

Today Westerners regard the birth of Jesus as occurring in very ancient times. But Jesus was born into a civilization whose past was twice as long as the period since his birth. Abraham was born about 1,900 years before Jesus, in the Mesopotamian city of Ur, which was already about a thousand years old. When Moses lived in Egypt, the Pyramids were already monuments of antiquity, the earliest of them some 16 centuries old. And when Joshua fought the battle of Jericho in about 1210 B.C., Canaanite settlements were already some 1,800 years old.

During those centuries, the history of the Middle East was as dynamic as it was prolonged. Empires rose and fell, civilizations appeared, rose to their zenith, and sank into eclipse or were absorbed by rivals or engulfed in new waves of conquest. As recently as 1975, for example, Italian archaeologists, after digging for more than 10 years in northern Syria, unearthed the state archives of an ancient city called Ebla, previously known only as a name mentioned in Sumerian and Akkadian records. Discovery of the archives — some 16,500 clay tablets so far — disclosed that Ebla was one of the major urban centers of northern Syria about 2300 B.C. It dominated its neighbors and traded with the kingdoms of Mesopotamia. But by the time the Patriarch Abraham was born Ebla had vanished, leaving no more to history than occasional references in the annals of its neighbors.

Yet there was continuity as well. The conquerors learned the arts of civilization from the conquered, each civilization borrowing technology, law, and sometimes elements of language and religion from its predecessor. Even today, social patterns, customs and beliefs may go back thousands of years. The past is always present in the Middle East.

In this long and complex history, geography remains a constant, and it is geography that can supply a thread to guide us through the historical maze, or rather two threads — the two river valleys where the seeds of civilization first germinated.

The first thread is the Tigris-Euphrates Valley, whose rivers rise in the Anatolian plateau, now in Turkey, and flow southeastward through Syria and what is now Iraq, creating the land of Mesopotamia ("between the rivers" in Greek) before meeting near the head of the Arabian Gulf. The second thread is the Nile, rising in two sources in East Africa and flowing in two streams northward to Khartoum in the Sudan, where the streams meet and roll in a mighty flood between cliffs and deserts northward through Egypt to spread into the Delta before emptying into the Mediterranean.

Like the Nile itself, the history of Egypt, despite interruptions and upheavals, appears to flow with an impressive unity, whereas in Mesopotamia the historical chronology, as well as the hydrography, is more irregular, with nation succeeding nation, empire swallowing empire, and people with new names and of diverse origins taking turns at the center of the stage.

These two cultural areas, Egypt and Mesopotamia, were at times brought together by conquest, first by one, then by the other, again by outsiders. Cultural contacts and borrowings multiplied and deepened; yet each area retained a distinctive character, just as each of the rivers maintained a recognizable physical thread throughout the centuries.

EARLY MESOPOTAMIA: The first civilization on earth, as far as we know, was that of Sumer — a name, incidentally, unknown to archaeologists as recently as 150 years ago. A people speaking a non-Semitic language, the Sumerians first appear in the southern part of the Tigris-Euphrates Valley in the fourth millennium B.C. There are several theories as to their origin: that they came from the north, or from the Indus Valley, or that they were the descendants of the indigenous 'Ubayd culture. Most probably they came from Central Asia, in small groups, across the Iranian Plateau. Although they all spoke the same language and had a common culture, the Sumerians did not form a unified state but lived in a cluster of independent city-states often at war with one another. The Sumerian cities best known to us are Ur, Erech (Uruk), Lagash, and Nippur, all originally on the riverbanks but now left, literally, high and dry by the gradual diversion of the rivers. Eridu, for example, a Sumerian seaport for the Arabian Gulf, is now 210 kilometers inland.

The Sumerians seem to have been the first people to use the potter's wheel, a technological innovation of great importance. They also built elaborate irrigation systems, traded with their neighbors and other peoples,

Gray stone statue represents a respected Sumerian scribe of the period of Ur-Nina, king of Lagash in the twenty-fifth century B.C.

MEDITERRANEAN · SEA

BLACK SEA

CASPIAN SEA

Byblos

Euphrates River

Tigris

AKKAD
Sippar

Heliopolis
Memphis

SINAI

EGYPT

Nile River

Abydos

Thebes

FIRST CATARACT   Aswan

Nile River

RED

SEA

THE MIDDLE EAST IN THE TWENTY-FIFTH CENTURY B.C.

Cities — Ur, Eridu, Lagash, and others — and a system of writing have been developed by a non-Semitic people in southern Mesopotamia. Akkad to the north is inhabited by a Semitic people. Elam to the east largely controls the far-flung maritime trade of the time, which includes Dilmun as a flourishing entrepôt. Northern and southern Egypt as far as the First Cataract have been unified in the Old Kingdom, with its capital at Memphis in the north. The Great Pyramids of Giza are already in existence and the basic forms of Egyptian religion, art, and writing have been established. Byblos and other cities in the Levant maintain trade and cultural links with both Mesopotamia and Egypt.

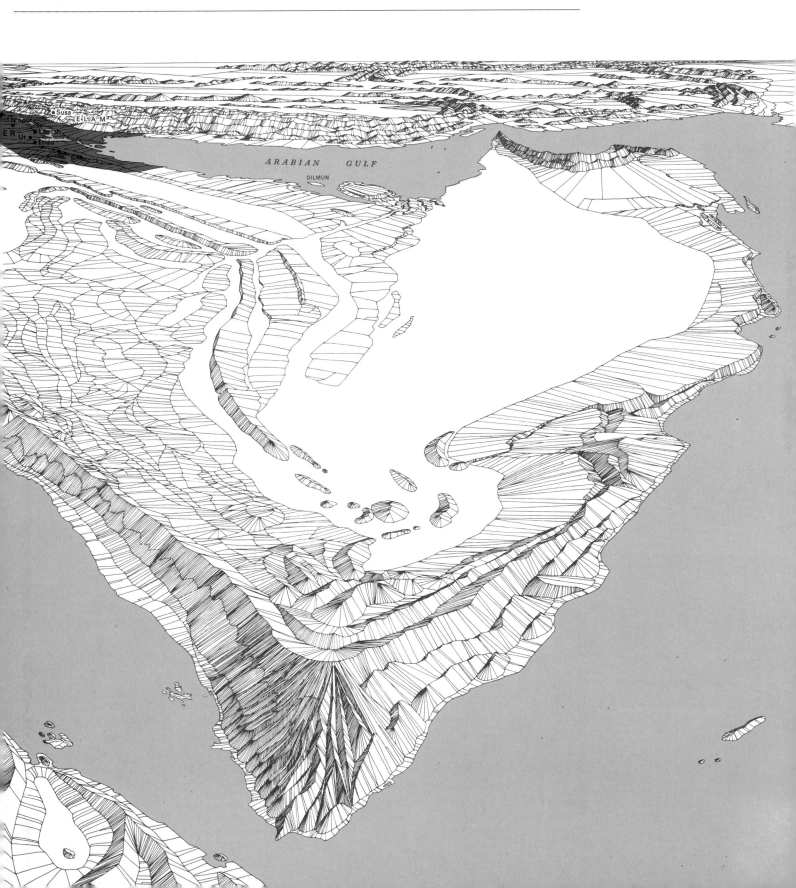

drew up codes of law, introduced a monetary system of weighted metal rings, and used wheeled vehicles and the architectural arch. Through their interest in astrology, they discovered many of the laws of astronomy. They could solve simple equations and calculate the square and cube roots of numbers and the areas of squares and rectangles. To this day their sexagesimal system of numbers is reflected in the division of the hour into 60 minutes and the circle into 360 degrees.

Brick set in bitumen lines the stairway of the ziggurat of Nanna, the moon god, built at Ur at the end of the third millennium B.C.

Despite the lack of building stone in the river valley, the Sumerians constructed impressive buildings of sun-dried brick, notably the imposing tower temples known as ziggurats — artificial heights on the floodplain reflecting, perhaps, ancient memories of the mountains in their original homeland. Ziggurats, whose construction continued long after the Sumerians, are thought to have inspired the biblical story of the Tower of Babel. A large part of one ziggurat survives at Ur today.

But the Sumerians' most important innovation was the development of the art of writing. From their original pictorial writing they evolved a cuneiform script — arrangements of wedge-shaped strokes impressed on clay tablets which were then dried in the sun or baked. Many of these hard clay tablets have survived, most of them commercial documents (contracts, receipts, deeds, and wills) but others containing medical prescriptions, proverbs, poetry, and chronicles.

The splendor of Sumerian culture burst upon the modern world when Sir Leonard Woolley opened the royal tombs at Ur in the 1920s and exhumed the possessions of the rulers along with those of courtiers, wives, and guards who apparently were killed to accompany the rulers into the afterlife. The graves have yielded superb examples of sculpture, furniture, clothing, jewels, musical instruments, and eating and drinking vessels, all of skillful workmanship and a high level of art. Among the objects found have been a harp inlaid with gold, silver, and lapis lazuli; a gameboard of shell, bone, and stone squares set in bitumen; and a ceremonial helmet of beaten gold in the form of a wig. Despite their many cultural achievements, however, the Sumerians were unable to unite politically and were eventually overshadowed and then absorbed by such peoples as the Akkadians and

A gold ceremonial helmet found at the royal cemetery of Ur reflects the splendor of Sumerian art.

the Amorites, both part of the larger group known collectively as the Semites.

The first of these peoples to attain political dominance in Mesopotamia were the Akkadians. About 2360 B.C., Sargon I of Akkad (which lay somewhere north of Sumer) extended his power beyond a single city-state and created the first Semitic kingdom, a development that was to lead to the later Semitic empires. The Sumerian cities along with their culture were absorbed into a blend of Sumerian and Semitic culture which was to enlighten most of the civilized world for centuries to come and which continued to flourish even after the Akkadian kingdom was overrun by barbarians from the Taurus Mountains and the Elamites of western Iran captured the city of Ur.

In the third millennium B.C. a wave of immigrants from the west, the Amorites, had entered Mesopotamia and settled on the site of the already ancient city of Babylon. They subsequently absorbed the culture of the Sumerians and Akkadians, gradually pushed back the Elamites, and founded the Old Babylonian Empire. Their greatest king, Hammurabi or Hammurapi, inaugurated a reform and unification of the Sumerian-Semitic culture, which henceforth may be called simply Babylonian.

Hammurabi's reforms were religious, social, and political, but he is best known today for his legal code, which sought to harmonize the laws that had evolved through the centuries in Mesopotamia. This code has been found carved in 3,600 lines of cuneiform writing on a diorite column at Susa — modern Dizful in Iran. Punishment under the code was harsh, but in many ways the code was also humanitarian. It provided, for example, state compensation to the victims of crimes. During Hammurabi's rule, and under the rule of his successors, Babylon enjoyed a state of security that permitted two centuries of intellectual development, particularly in mathematics and astronomy, up to the sixteenth century B.C. New forces now enter the history of the Fertile Crescent with other peoples, bringing the hitherto self-sufficient and relatively isolated river valleys of Mesopotamia into the larger stage of the outside world. At the dawn of the second millennium B.C. an Indo-European people entered Anatolia and founded what was to become the Hittite kingdom. At about the same time, the Hurrians, a people speaking neither a Semitic nor an Indo-European tongue, moved from the area of Armenia into northern Mesopotamia; and the Kassites, Indo-Europeans long settled in Iran, moved west into Babylonia. The Hittites, one of the first peoples to successfully smelt iron, raided Mesopotamia and added northern Syria and the Cilician plain to their empire, whose center remained in Anatolia. They borrowed the cuneiform script and adopted much of Babylonian culture, especially its literature and religion. The Hurrians, under an Indo-European warrior class, learned the arts of horsemanship and carved out the Kingdom of Mitanni in northern Mesopotamia, with their principal city at Nuzu near modern Kirkuk. So strong was the attraction of the ancient civilization of Mesopotamia that the newcomers adopted its language and culture. The Kassites, meanwhile, occupied Babylonia, so that about 1400 B.C. the Fertile Crescent was divided among the Kassites in the southeast, the Mitanni in the north-central region, and the Hittites in the northwest. A fourth contender for power was Egypt, which had pushed up the Mediterranean coast into Palestine, Lebanon, and southern Syria, and whose parallel, but very different, civilization had been developing since the fourth millennium.

EGYPT: In contrast to Mesopotamia, Egyptian civilization has a perennial quality, an aura of unity and permanence that resembles the flow of the Nile itself. As Herodotus observed 2,500 years ago, the land of Egypt is literally the gift of the river. Life itself depended on its waters and on the silt brought by the annual flood to fertilize the Delta and the narrow strip of river valley lying between two vast deserts. Civilization flowered here a little later than in the Tigris-Euphrates Valley and, in the soil of the Nile, grew a distinctive form and character of its own.

The Nile flows some 1,350 kilometers through Egypt, from the border of Sudan to the Delta without tributaries and with no great natural physical barriers — a geographical unity that also encouraged political unity and resulted in a strong central government that could guarantee the efficient use of the life-giving waters for the entire country. From this need emerged the office of the pharaoh, who, with a retinue of priests, attempted to guarantee the annual rising of the Nile by religious observances. As strict adherence to the prescribed observances was thought essential, a rigid conservatism in religion and politics evolved. Art was similarly formalized, with its conventions remaining essentially unchanged for 3,000 years, most of it devoted to religious ends, particularly the glorification of the pharaoh.

During the centuries, as pharaoh-worship became more intense, a great part of the wealth and labor of the kingdom was diverted to his glorification. Massive stone buildings were erected in his honor, and his exploits were depicted in painting and sculpture, or written in hieroglyphs on the walls. As his powers were believed to continue after death, his body was carefully preserved and his possessions — or, later, artistic representations of them — were buried with him. Above all, his tombs were built to last forever and, in the early form of the pyramids, they practically have done so. Eventually, as this mortuary cult penetrated other levels of Egyptian society, anyone who could afford it arranged to be buried in an elaborate tomb with evidence of his wealth around him, even including the bodies of sacred animals embalmed as mummies. These burial customs have preserved for us an extraordinarily vivid picture of life in ancient Egypt.

This preservation of the past was in itself a cause of traditionalism. As building stone was readily available, Egyptian structures (unlike those of Mesopotamia) endured, and each successive generation lived, literally, in the shadow of its own past. Even the climate fostered conservatism: the absence of rain and the dry desert air preserved the more perishable materials of art such as wood, paint, and cloth, and the written records on papyrus.

Historians traditionally divide the history of Egypt into four periods, interrupted by episodes of internal turbulence, foreign rule, or both. The four periods are called the Old Kingdom (from about 3000 to 2270 B.C.), the Middle Kingdom (about 2100 to 1700 B.C.), the New Kingdom, or Empire (1550 to 1090 B.C.), and the Late Period (712 to 525 B.C.). After the fourth century B.C. Egypt, while still important in world affairs, was rarely independent and usually subservient to other empires.

The Old Kingdom was created by the unification of the Kingdom of the North and

Late Hittite statue of a king reveals the strong influence of Assyrian art.

At Giza near modern Cairo, the enigmatic Sphinx (a royal head on a lion's body) stands at the beginning of Egyptian art and history.

Egyptian artists excelled in free-standing statuary such as this figure of Tutankhamen harpooning, made of wood overlaid with gold.

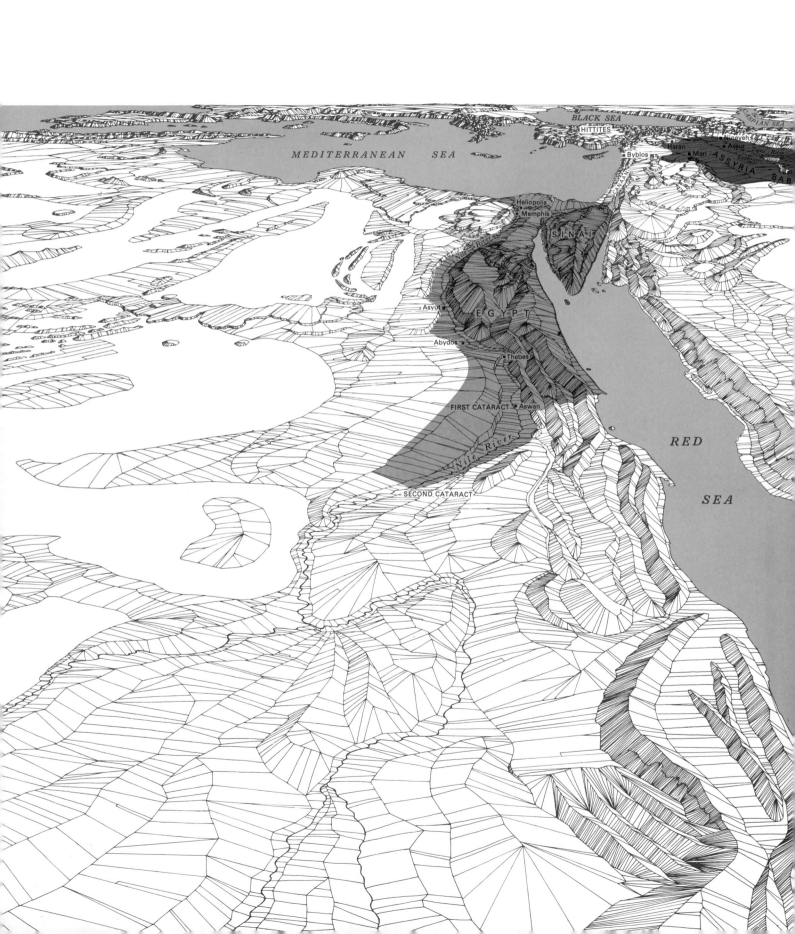

BLACK SEA

CASPIAN SEA

HITTITES

MEDITERRANEAN SEA

Byblos

Haran

Nineveh

Assur

Mari

Tigris River

Euphrates River

ASSYRIA

BAB

Heliopolis

Memphis

SINAI

Nile River

CANAAN

EGYPT

Asyut

Abydos

Thebes

FIRST CATARACT Aswan

RED

SEA

Nile River

SECOND CATARACT

THE MIDDLE EAST IN THE EIGHTEENTH CENTURY B.C.

Northern Mesopotamia is controlled by a new element, the Assyrians, with cities at Assur and Tell al-Rimah (near Nineveh). In southern Mesopotamia, Babylonia flourishes during the distingusihed reign of Hammurabi. This is a period of great prosperity for Canaan, the commercial center through which pass much of the trade and cultural influences of the civilized world. In Egypt the Theban dynasties of the Middle Kingdom push the boundary southward to the Second Cataract.

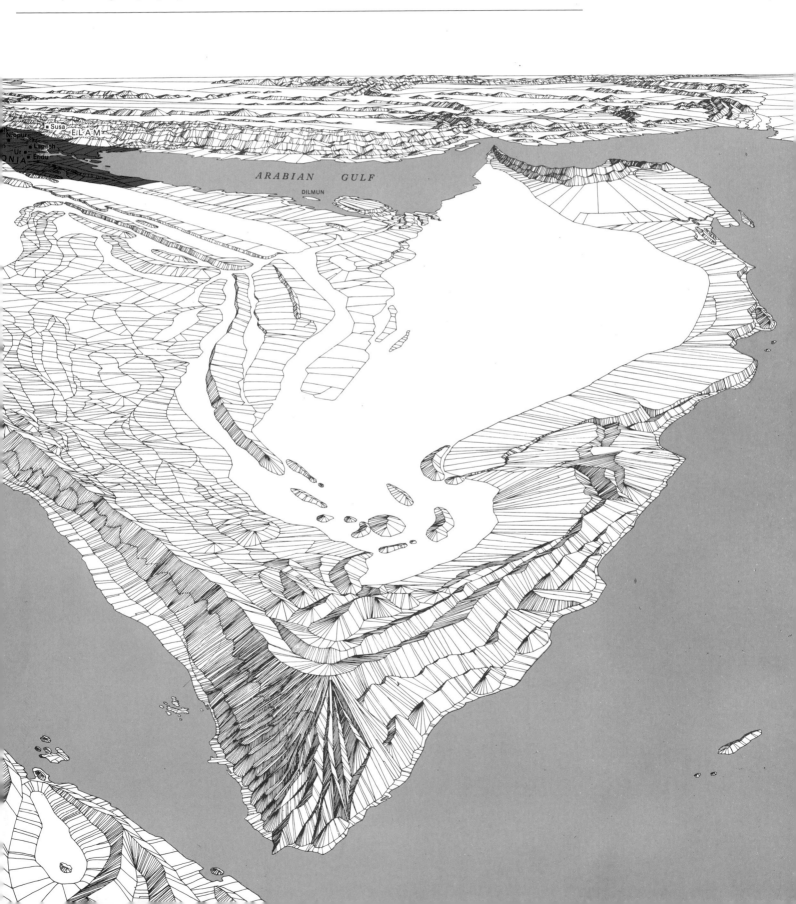

the Kingdom of the South by the Pharaoh Menes of the First Dynasty. The capital was in the north at Memphis, near modern Cairo. During this period the most important pyramids were built, and the basic forms of Egyptian religion, art, and writing took shape. As Egypt had to import wood and certain metals, its merchants carried on trade with neighbors, exporting gold, ivory, ebony, and manufactured objects, and importing such things as cedar wood from Lebanon. After about eight centuries of relative peace and prosperity, the Old Kingdom collapsed from internal strains, and the dynasties of various cities competed in civil war.

When Moses lived in Egypt, the Pyramids of Giza were already ancient monuments, some of them 1,600 years old.

About 2100 B.C. a dynasty of Thebes in the south seized power and founded the Middle Kingdom, the second period of Egypt's greatness. The Theban rulers pushed the boundaries southward as far as the Second Cataract of the Nile, now in northern Sudan, conducted trade with Syria and Palestine, and drained and cultivated the Fayyum, a large fertile depression in the desert west of the Nile. In 1700 B.C. the Semitic Hyksos from Asia — the so-called Shepherd Kings — conquered the Delta and ruled there for 150 years.

Another Theban dynasty ended Hyksos rule in 1555 B.C. and created the New Kingdom or Empire, the zenith of ancient Egypt's power and splendor. In this period the northern borders of the empire reached into Syria, gifts were exchanged with the kingdoms of Mesopotamia, the great temples of Karnak and Luxor were built, and pharaohs were buried with pomp and luxury in the cliff-tombs of the Valley of the Kings. In this period too, the Pharaoh Akhenaten tried to reform the art, literature, and religion of his times. He moved his capital to Tell Amarna, where he encouraged a naturalistic style in art, one of the glories of which was the famous bust of his wife, Nefertiti. He introduced colloquial language into literature and substituted the worship of a form of the primeval sun-god, Aten, for that of the pantheon of gods current at that time. His Hymn to Aten is often compared to the 23rd Psalm. Yet on his death Egyptian religion, literature, and art reverted to the old fixed forms.

Akhenaten's son-in-law Tutankhamen, a warrior king who died young, is known today because of the wealth of magnificent art

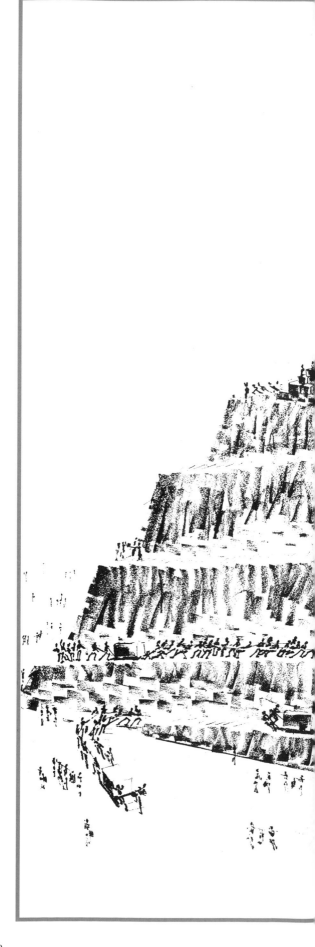

The building of the pyramid tomb of an Egyptian pharaoh was a tremendous engineering feat involving 100,000 to 200,000 workers — laborers, overseers, masons, architects, scribes, paymasters, interpreters — many tons of sand and stone, and up to 20 years of work. The core of the pyramid was built in horizontal courses rising like steps, each level a bit smaller than the step below. As the courses rose higher, inclines of sand and rubble were constructed along each side and fitted crosswise with logs to facilitate the movement of wooden sledges on which building stones were dragged up the ever-rising surface. Once the core of the pyramid was completed, the outer casing of stones was laid, beginning at the peak. As the casing proceeded downwards, the inclines were gradually removed until the lowest course was finished and the monument stood as a true pyramid — its smooth sides sloping down in an unbroken surface from peak to base. The tomb of the pharaoh, often with false doors and misleading tunnels, was secreted in the base.

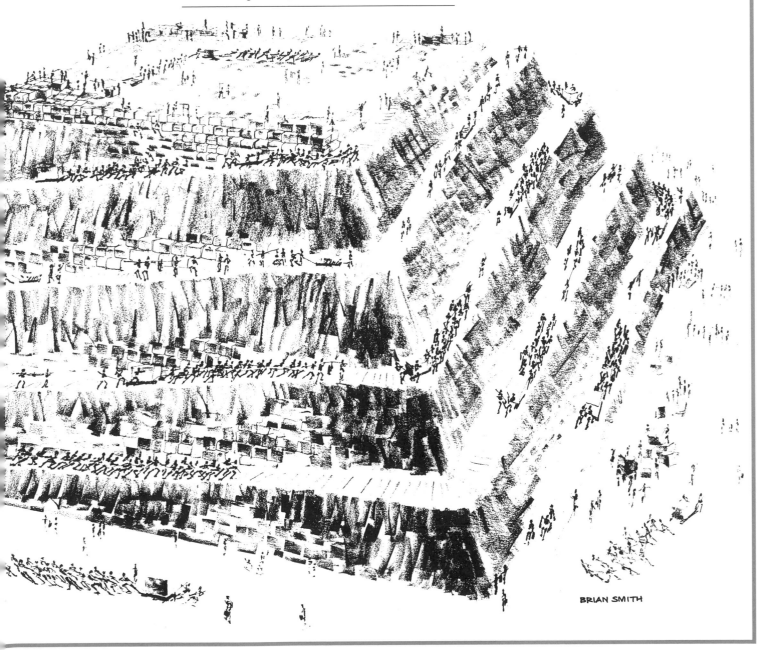

BRIAN SMITH

The splendor of Egyptian art supported religion and the state for three millennia. Opposite page: The pharaoh was considered to be a reincarnation of the falcon-headed god Horus, shown here on the Temple of Isis at Philae. This page, top left: A giant head of Ramses II bears symbols of the goddesses of Upper and Lower Egypt. Top right: The funerary mask of Tutankhamen was intended to preserve some of the godlike power of the departed pharaoh. Above: Two of the four colossal statues of Ramses II in front of the temple of Abu Simbel. At right: The temple of the Roman emperor Trajan at Philae.

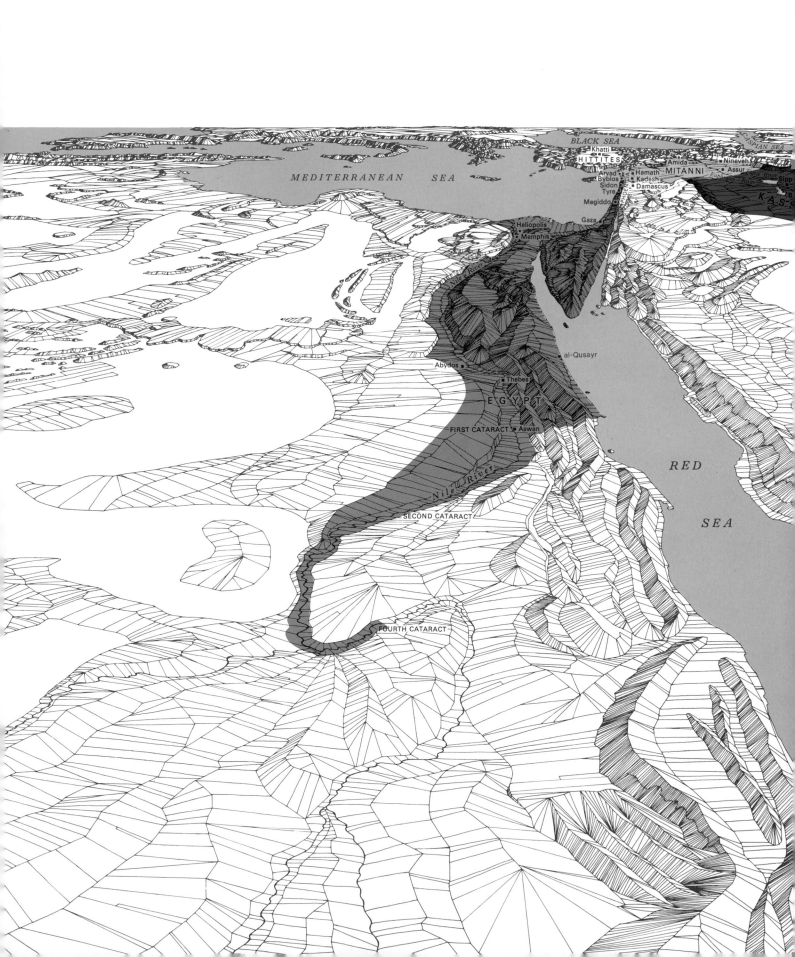

THE MIDDLE EAST IN THE FIFTEENTH CENTURY B.C.

Egypt reaches its zenith in the New Kingdom, and the capital at Thebes is embellished with the monumental temples of Karnak and Luxor. The empire extends from the Fourth Cataract in the south to the upper Euphrates in Syria. The expanding Hittite Empire is encroaching upon both the Egyptian Empire and the Kingdom of Mitanni which, with Assyria in a period of decline, is dominant in northern Mesopotamia. A Hittite raid brings down the First Dynasty of Babylon, but it is the Kassites who inherit political power in southern Mesopotamia.

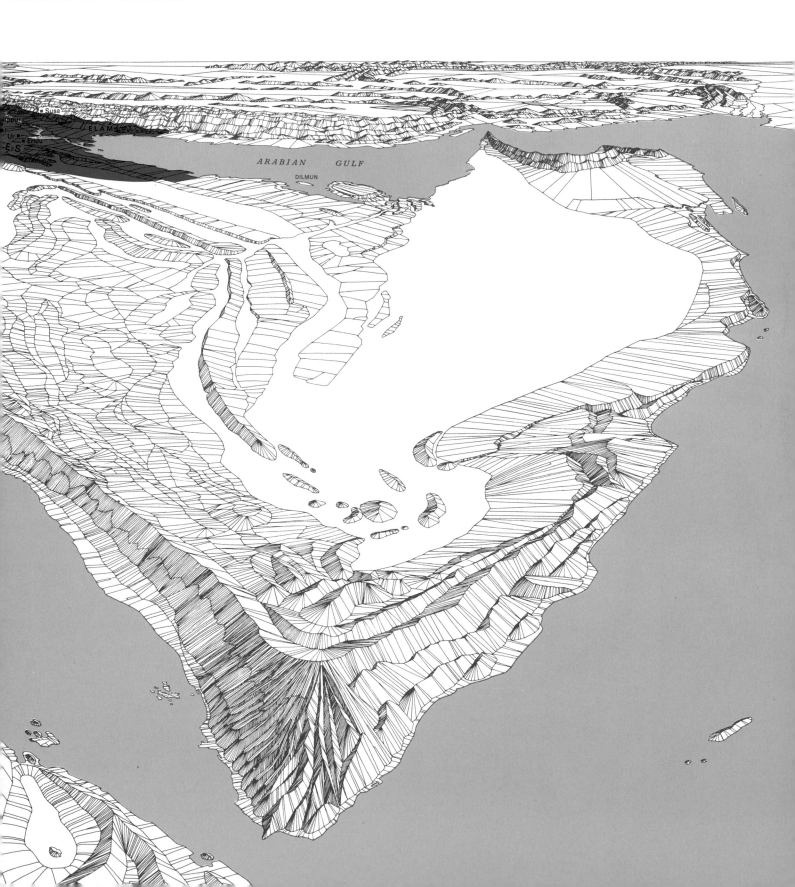

objects found in his tomb in 1922. The strongest of the pharaohs of the period were Ramses II, called the Great, who fought and concluded a treaty of peace with the Hittites, and Ramses III, who had to contend with raiders from the sea, possibly the Philistines who eventually settled in Palestine. During the third period of eclipse, starting about 1090 B.C., Egypt was ruled by foreigners, including Libyans and Ethiopians, and in 712 B.C. was invaded by a people called the Assyrians, the new ruling power in Mesopotamia.

During the Late Period, beginning in the eighth century, native rule was reestablished within Egypt, but efforts to restore its previous power and cultural predominance were challenged by younger rivals in the eastern Mediterranean, and attempts to reconquer the former empire were blocked by Nebuchadnezzar of Babylon. In this period too, Greek merchants flourished in the Delta and the Pharaoh Necho reportedly commissioned a Phoenician admiral to sail around the continent of Africa. In its final centuries ancient Egypt was more often than not ruled by others: the Persians, who in 525 B.C. annexed the country for 200 years; the all-conquering Alexander the Great in the fourth century; and his successors, the Ptolemies, a Greek-Macedonian family. During the two and a half centuries that the Ptolemies ruled, the city of Alexandria developed into the center of Greek culture. In 31 B.C. Cleopatra, the last of the Ptolemies, in alliance with Mark Antony, lost the Battle of Actium to the forces of Caesar Augustus and Egypt became a Roman colony.

The culture of ancient Egypt may seem alien to Westerners today, but it fertilized Western civilization, first through interchanges with the cultures of Mesopotamia, and then through the Mycenaeans of Greece, who borrowed greatly from Egypt and passed on much to the later Greeks. Classical Greece and ancient Rome looked on Egypt as a treasure house of wisdom. Imhotep, a genius of the Old Kingdom, proficient in architecture and astronomy, was revered in the ancient world as the father of medicine, and Egyptian medicine reached a level not achieved in Europe till the end of the Middle Ages. The Egyptians were the first people to build monumental stone buildings and their architecture, sculpture, and painting still rank among the

After accepting the surrender of the Egyptian kingdom, Alexander founded the capital that bears his name.

When Cleopatra ruled in the first century B.C., Alexandria was a city of culture and grandeur, second only to Rome.

Quest For The Past

Until 200 years ago, sources for the history and culture of the ancient Middle East were few and limited: the Bible, Greek and Roman authors (particularly Herodotus), accounts by such Muslim travelers as 'Abd al-Latif al-Baghdadi and Ibn Jubayr, and vague accounts passed on by nomads who had glimpsed traces of abandoned civilizations beneath shifting sands.

It was not until 1748, when the ruins of Pompeii were dug from the ancient ash of Mount Vesuvius, that man began to search systematically beneath the ground for knowledge of his vanished past and, through the science called archaeology, assess and add to those sources. And it was not until 1871 — when Heinrich Schliemann, stripping away the earth in layers, century by century, discovered classical Troy — that archaeology started to become a true science. In the Middle East, however, the quest for the past had been under way since 1798, when Napoleon invaded Egypt and, beneath the Pyramids, said to his troops, "Soldiers, 40 centuries are looking down on you."

This was more than rhetoric: The future emperor had brought with him nearly 200 scientists and scholars to study the Egyptian past. The result was the famous *Description de l'Egypte* — a meticulous account of the antiquities, geography, and monuments of ancient and modern Egypt. Painstakingly researched and lavishly illustrated, this work opened the eyes of the Western world to the splendors of ancient Egypt and established Egyptology as an important discipline.

The most important challenge facing the early Egyptologists was how to decipher the baffling hieroglyphic system of writing. It was not until 1821 that the brilliant French linguist Jean-François Champollion unlocked its mysteries. The key, one of the treasures brought back by Napoleon's expedition, was a slab of fine-grained basalt found near the town of Rosetta on the banks of the Nile — and hence known as the Rosetta Stone. As it was engraved with three blocks of scripts — hieroglyphic, hieratic (a cursive version of hieroglyphic), and Greek — the Rosetta Stone offered the possibility of deciphering the language

18

of ancient Egypt by comparing it to a known language: Greek.

Examining the Rosetta Stone, Champollion realized that the hieroglyphic system of writing consisted of a combination of alphabetic, syllabic, and ideographic signs. His first success in determining the value of some of the alphabetic signs in this complex system came when he succeeded in reading names of Ptolemy and Cleopatra on a bilingual Greekhieroglyphic inscription on an obelisk found at Philae. Then, assuming that the signs enclosed in "cartouches" — oval rings — were personal names, he compared the relative positions of the cartouches in the hieroglyphic text with the personal names in the Greek and thus determined the values of P, T, O, C, L, E, M, R, and A in the alphabetic system of hieroglyphic writing. As the names of Ptolemy and Cleopatra also occurred on the Rosetta Stone, Champollion was able to vindicate his method and, moreover, provide scholars with a fundamental technique for deciphering other unknown languages.

In Egypt, meanwhile, the soaring interest in Egyptology had resulted in a serious problem: Egypt was being looted of its treasures by an antique-hungry Europe. The seriousness of the problem was first noted by a French archaeologist named Auguste Mariette who, in the 1850s, excavated the great Avenue of Sphinxes at Luxor and the tomb of Tiy, a noble of the Egyptian Fifth Dynasty.

Excavations such as this were a valuable contribution, but in noting the looting of Egypt — and acting to prevent further looting — Mariette contributed much more. He helped establish, in 1858, the Egyptian Museum. He drafted strict restrictions on the export of antiquities. Perhaps more importantly, he also trained the first generation of Egyptian archaeologists.

Unfortunately, Mariette was too late by centuries. Looters had been plundering Egypt since the pharaohs were first buried. In 1875, for example, investigations disclosed that one family of professional tomb robbers had found and hidden the mummies and treasures of no less than 40 ancient Egyptian kings. It is easy, therefore, to understand the excitement of Howard Carter — and the world — when, on November 26,1922, Carter opened the undisturbed tomb of Tutankhamen and gazed upon the treasures of the young pharaoh who had died some 3,270 years before.

The King Tut treasures were the most famous archaeological discovery ever made — as well as the most dramatic. In 1906, Carter, a young and gifted archaeologist, persuaded George Herbert, Earl of Carnarvon, to finance years of work in the Valley of Kings near Luxor. The site he chose had been thoroughly picked over by several generations of archaeologists, but in 1922, just as he was about to abandon hope, Carter discovered a single stairway leading down into the earth, cabled Carnarvon, and, in an agony of impatience, waited three weeks for his patron to arrive. Finally, the door of the tomb was slowly opened and they beheld the untouched gold and other goods of the long-dead pharaoh.

Elsewhere in the Middle East the search for the past was less dramatic, but was certainly more difficult. This was particularly true in the Tigris and Euphrates Valley, where the great civilizations of Sumer, Akkad, Nineveh, and Babylon had arisen, flourished, and vanished.

Some of these civilizations had been known to historians from biblical references, but it was not until 1840 that archaeology began to confirm such references. In that year Paul-Émile Botta, the new French consul at Mosul, noticed the strange mounds that dotted the flat, muddy landscape between the Tigris and Euphrates rivers and decided to explore them. Botta started digging at a site called Kuyunjik — where he very nearly found Nineveh — but then learned that in the nearby village of Khorsabad there were a large number of ancient bricks marked with traces "like a bird's-footprints." Excited by this news, Botta dispatched two workmen who, almost with the first shovelful of earth, came on massive walls carved with the reliefs of mythical animals. Summoned by the workmen, Botta galloped to the site and found himself gazing on the walls of the summer palace of Sargon II, King of Assyria.

(Continued on page 20)

Quest For The Past

Botta's discovery of the palace was, of course, important. But his subsequent discovery of a hundred tablets bearing the "bird's footprints" was immeasurably more valuable. For those "footprints" were cuneiform inscriptions and the tablets would provide the clues needed to break this ancient code, one of archaeology's greatest achievements.

Unlike Champollion, the men who deciphered cuneiform had no convenient Greek inscriptions to help them and, furthermore, could only guess at what language, or languages, lay concealed by the enigmatic script. Nevertheless, as early as 1802, a young German schoolmaster named George Grotefend had, in response to a beer-hall bet, made an astounding breakthrough. Working from no more than some poor copies of ancient Persian inscriptions, he deduced that the inscriptions were trilingual. As he did not know what three languages were represented, he decided, like Champollion, to begin with names. Since the names of the Kings of Achaemenian Persia were known from Herodotus and other sources, Grotefend reasoned that he might be able to match some of the recurring arrangements of signs in the inscriptions with the names of known rulers. Grotefend eventually was able to read the names of three rulers — Darius, Hystaspes, and Xerxes — and to identify correctly nine signs of the Old Persian cuneiform syllabary.

The next step was not taken until 1837, the year that Grotefend finally published his observations. In that year an Englishman, Major Henry Creswicke Rawlinson, dangling at the end of a rope on the face of a cliff near Bisitun in Persia, copied 14 columns of cuneiform writing inscribed on the rock. Later that year he also published translations of the first two paragraphs.

Finally, a few years later, Botta found his hundred tablets and supplied the final clues; by a fortunate chance the tablets turned out to be Sumerian-Assyrian lexicographical texts. With these, and with the values discovered for the Old Persian texts at Bisitun and Persepolis, Rawlinson and others were soon able to read Assyrian cuneiform tablets and, within a decade, understand the principles of the cuneiform writing system.

In 1845 another discovery shook the scholarly world: Austen Henry Layard, perhaps the most famous Assyriologist of all, had, with the help of a Bedouin shaykh, uncovered the royal palace of King Assurnasirpal II, who reigned in the city of Nimrud from 883 to 859 B.C.

Layard was even luckier with his subsequent excavations. Returning to Kuyunjik, which Botta had abandoned, he found the great palace of Sennacherib, with its walls of glazed bricks. Later he found the library of Assurbanipal, the grandson of Sennacherib, which contained the oldest major work of literature in the world, the Epic of Gilgamesh, cut into hundreds of tablets buried in the palace rubble.

The Epic of Gilgamesh was translated by a man originally trained as a bank-note engraver whose secret passion was Assyriology. His name, too often unremembered, was George Smith. Gripped by the story related in the epic he was translating, George Smith was dismayed to discover that the end was missing — so dismayed that when *The Daily Telegraph* offered 1,000 guineas to anyone who could find the missing tablets, George Smith decided to undertake the search himself. By a stroke of the extraordinary luck that blessed those early Assyriologists, in just five days of searching in the vast heap of rubble at Kuyunjik he found the missing tablets — 384 of them. Back in England, he claimed his prize, deciphered the tablets, and — in another spectacular discovery — found that the story of Gilgamesh also told the story of the Flood.

To some, the golden age of archaeology may seem to have occurred in the nineteenth century. In fact, exciting discoveries have been made regularly since then. At the turn of the twentieth century Robert Koldeway began to excavate Babylon and struck the great wall of the city in just 14 days. Not long after, he discovered the ruins

of the Hanging Gardens of Babylon and then a ziggurat he identified as the Tower of Babel. In 1928 Leonard Woolley excavated Ur and went on to find, below Ur, a 2½-meter deposit of clay that tended to corroborate both the biblical account of the flood and the story of the deluge in the Gilgamesh epic.

More recently, Kathleen Kenyon's important excavations at Jericho and Jerusalem have thrown much light on the prehistory of man outside the great river valleys of Egypt and Mesopotamia; Mortimer Wheeler's discoveries in what is now Pakistan have revealed a flourishing urban civilization contemporary with Ur; and excavations in Bahrain have shown that this small island had an advanced culture contemporary with the oldest evidence of civilization from Mesopotamia and India. In 1975 Italian archaeologists in northern Syria unearthed thousands of cuneiform tablets from the state archives of Ebla, until then virtually unknown except for rare references in Akkadian and Sumerian tablets. Ebla, however, apparently was strong enough to withstand the better-known states of Mesopotamia, and had a far-flung commercial empire.

Egypt, too, continues to yield up secrets. In the spring of 1995 archaeologists announced the discovery of a huge T-shaped tomb in the Valley of the Kings. Containing at least 62 chambers — with the possibility of more at a lower level — it is believed to have been built by the thirteenth-century B.C. Pharaoh Ramses II for his many sons, three of whom are named in hieroglyphics inside. One Egyptologist called the discovery "absolutely amazing."

The tomb, the largest of some 60 in the Valley of the Kings, is unique in construction. Robbed in ancient times and buried under debris, it was all but forgotten — until Kent Weeks, an archaeologist with the American University in Cairo, began to probe the site in 1988. His find offers opportunities for new insights into the rule of Ramses II, Egypt's longest-reigning pharaoh, and poses new riddles for Egyptologists. It is also a reminder that, likely as not, much more remains to be uncovered in the Middle East to help point the way in mankind's continuing quest for the past.

world's supreme works of art. Above all, the Egyptians developed and preserved a civilization of the highest order while most of the world was still submerged in barbarism.

LATER MESOPOTAMIA: In Mesopotamia, meanwhile, a people called the Assyrians had ended the period of anarchy following the breakup of the Hittite Empire. One of those Semitic peoples who had been settled in Mesopotamia since the third millennium, the Assyrians took their name from Assur, the name of their national god and of their first capital on the upper Tigris. Later they moved their capital to Nineveh, and in the ninth century B.C. they welded a new empire out of the neighboring states of Mesopotamia. The Assyrians later extended their boundaries to embrace all of Mesopotamia and parts of Egypt, Armenia, and Arabia.

The Assyrians were famous for their military proficiency. They fought with iron weapons, perfected the technique of the siege, and developed a deadly combination of archers, heavy infantry, and cavalry. Even their art glorified war; their famous manheaded winged bulls conveyed vivid images of power and terror.

Assyrian achievements were not exclusively militaristic: they also built roads and introduced a government postal system and the use of silver currency. The Assyrians also compiled in royal libraries a wide range of written documents. Indeed, the 22,000 clay tablets found at Nineveh were one of the most important archaeological discoveries of all time and included the Epic of Gilgamesh, a poem of 3,000 lines which relates the Sumerian legend of a hero-king.

Nevertheless, it was militarism that eventually brought about the downfall of the empire. It drained the treasury, decimated the empire's agricultural labor force, and alienated its subjects. In 612 B.C. the ruler of Babylon and the king of the Medes joined forces to destroy Nineveh. The Assyrian Empire collapsed and Nineveh became a byword for vanished glory. With the Assyrians routed, the city of Babylon experienced a brilliant cultural revival as the capital of the Neo-Babylonian or Chaldean-Babylonian Empire, Chaldea being the name given to the southernmost part of the valley of the Tigris and Euphrates. Under Nebuchadnezzar in the

The winged manheaded bull, a benevolent being often seen in pairs as gateway guardians, had a long life in the successive civilizations of the ancient Middle East.

The central figure of a literary epic written more than a thousand years before the *Iliad* or the *Odyssey*, Gilgamesh was the hero par excellence of the ancient world, symbolizing man's vain drive for glory and immortality.

Objects from Iran to the western edge of the Fertile Crescent reflect three millennia of art. From Ugarit, a cosmopolitan Canaanite center on the Syrian coast, comes the ivory head seen at top left. Directly above is a tablet from the fifteenth century B.C. bearing the Ugaritic cuneiform alphabet. Massive winged bulls with human heads (left) guard the gateway built in the time of Xerxes I, fifth century B.C., at the Persian capital of Persepolis. At nearby Pasargadae, the winged guardian of the bas-relief above wears a tiara of Egyptian inspiration. Facing page, top: Parade helmet of iron and silver is an example of Palmyrene art of the first century A.D. Gypsum statuette of Urananshe, leader of the temple choir at Mari on the upper Euphrates, dates from about 2500 B.C. Directly opposite is a ceramic-brick bull from the Ishtar Gate in Nebuchadnezzar's Babylon.

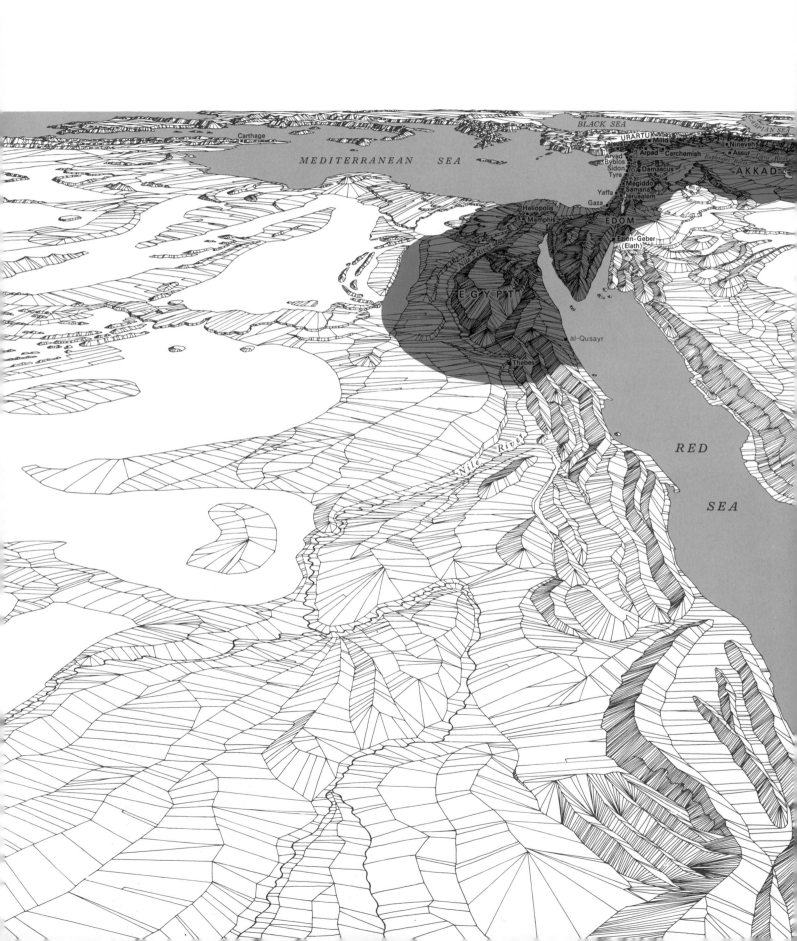

BLACK SEA

CASPIAN SEA

MEDITERRANEAN SEA

Carthage

URARTU
Tarsus
Milid
Ninveh
Arvad
Arpad
Carchemish
Byblos
Sidon
Tyre
Damascus
Assur
AKKAD
Megiddo
Yaffa
Samaria
Jerusalem
Gaza

Heliopolis
Memphis
EDOM
Ezion-Geber
(Elath)

Euphrates River
Tigris River

EGYPT

Nile River

al-Qusayr

Thebes

Nile River

RED

SEA

THE MIDDLE EAST IN THE SEVENTH CENTURY B.C.

Assyria, at the height of its power, controls all of Mesopotamia including Elamite Susa, the Levant, and much of Egypt and Armenia. Detente with Assyria allows Urartu to recover from Assyrian assaults of the previous century. The Phoenician cities of Tyre and Sidon pay tribute to Assyria but preserve their commercial freedom, while their colonies as far away as Carthage flourish. Caravan routes and the spice trade provide continuing prosperity for the kingdoms of southern Arabia. Overextended, the Assyrian Empire soon comes under pressure, and Egypt regains its independence in midcentury. The Chaldeans of Babylon combine with the Medes and the Persians to destroy Nineveh in 612 B.C.

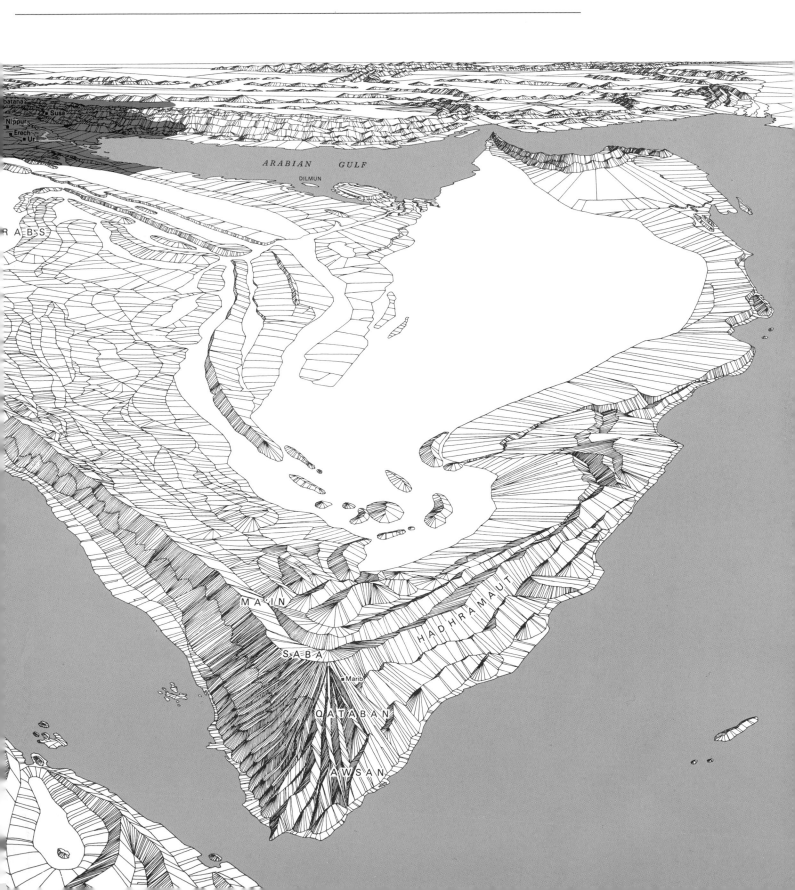

sixth century B.C., Babylon became synonymous with splendor and luxury. The Hanging Gardens, a late form of the ziggurat, were one of the Seven Wonders of the World and the royal palaces, the city walls, and the gorgeously decorated Ishtar Gate were equally magnificent. Herodotus, visiting Babylon, said that no one could believe the fertility of the surrounding plain if he had not been there. Because of the great advances in astronomy and astrology made during this period, astrologers for many centuries thereafter were simply called Chaldeans.

The ruins at Babylon bear mute testimony to the greatness of one of the mightiest cities of antiquity.

Although humane compared to the Assyrians, the Babylonians continued the practice of deporting subject peoples. The best known of these deportations was the Babylonian Captivity of the Hebrews during which, the Bible says, Daniel went into the lions' den and Belshazzar, at a feast, stared in terror at the writing on the wall.

But the glory of the Neo-Babylonian Empire was short-lived. In 539 B.C. it surrendered to the Persian conqueror Cyrus the Great, thus ending Mesopotamian independence for more than a thousand years.

Monumental basalt lion stands over a fallen enemy at Babylon.

OTHER ANCIENT PEOPLES: There were also, during these periods, many ancient peoples who did not rule over extensive empires, but who nevertheless played major roles in the history of what is now called the Middle East — and also made important contributions to Western culture. Among the best known were the Canaanites, the Hebrews, the Philistines, the Phoenicians, and the Aramaeans.

Canaan is a geographic term designating the region of southern Syria, Lebanon, and Palestine, and including such towns as Jericho, Lachish, Byblos, and Ugarit. Famous later as "a land flowing with milk and honey," Canaan attracted diverse peoples from Paleolithic times onwards — Semites, Hurrians, Hittites, and others. By the third millennium B.C. this amalgam of peoples had reached a high stage of development. From one of the Semitic peoples in Sinai, for example, the world probably received the alphabet.

Through trade, the Canaanites absorbed Egyptian and Minoan influences from the west and Assyrian-Sumerian influences from the east. They worshiped a fertility goddess, Ashtoreth, and a god, Baal, the "giver of life,"

whose name appears in place-names where his temples once stood — such as Baalbek in northern Lebanon. The Canaanite civilization reached its peak from the fifteenth to the thirteenth centuries B.C. and its history then merges with that of the northern branch, the Phoenicians, and other immigrant peoples such as the Hebrews and Philistines.

Because they left an indelible record of their history and ideas in the Old Testament, the Hebrews are better known to posterity than many other peoples of the Middle East. In fact, however, they were politically of secondary importance compared to their more powerful neighbors; their fame rests primarily on their outstanding contribution to literature and religion.

The Hebrews seem to have developed from a group of related semi-nomadic tribes with a common patriarch, Abraham — a situation still paralleled among Bedouin tribes today. As told in the Book of Genesis, their early history is closely intertwined with that of Mesopotamia, where civilization was thousands of years old when the Hebrews emerged as a people, settled in Canaan, and eventually, under King David in the tenth century B.C., conquered it. Earlier some Hebrew tribes had settled in the Delta of the Nile until, oppressed by the pharaohs, they were led out of Egypt by Moses.

Living at the crossroads of the powerful empires of Egypt, Assyria, Persia, Greece, and Rome, the Hebrews were able to rule this territory only at intervals between the thirteenth century B.C. and the Roman destruction of the Temple in Jerusalem in A.D. 70. Despite this, the Hebrews left a rich legacy of poetry, history, and religious thought.

The Philistines, a people from somewhere in the west — possibly Crete or Mycenae — disputed control over western Canaan with the Hebrews for centuries, a contest that was never finally settled in favor of one or the other. With control of iron as one source of their strength, the Philistines welded the cities of Philistia — Gaza, Ashkelon, Ashdod, Ekron, and Gath — into a confederacy that dominated the major commercial route between Egypt and Syria. They worshiped gods similar to those of the earlier Canaanites and at some period in their history abandoned their original tongue in favor of a Semitic one.

The Phoenicians never formed a unified

The Art Of Writing

Aleph, the first letter of the Canaanite alphabet, was gradually transformed into the letter "A" of the European languages.

To the best of present knowledge, it appears that the world's first system of writing was evolved by the Sumerians toward the end of the fourth millennium B.C. in the southern reaches of the Tigris-Euphrates Valley. Not long after that, the Egyptians developed a system of their own.

The Sumerian script is called cuneiform — an arrangement of wedge-shaped marks impressed with a stylus on wet clay tablets, which were then dried or baked until they were nearly indestructible. Cuneiform was adapted by later Mesopotamian peoples to their own quite different languages and was in use by the Persians as late as the first century A.D., over 30 centuries after its earliest development. The Egyptian script — hieroglyphic (meaning "sacred carving" in Greek) and a later cursive form called hieratic — lasted as long but was confined to Egypt.

The Sumerian-Assyrian and Egyptian scripts were not true alphabets — that is, systems of writing in which one symbol stands for a single sound. They were transitional systems representing a stage beyond simple picture-writing or ideographic writing (a somewhat more advanced system in which pictures can be used to represent not only objects but also some abstract concepts) but not yet a true alphabet. Cuneiform writing was pictographic in its origins, and Egyptian hieroglyphics remained strongly pictorial throughout their very long history, but both systems of writing developed a number of additional signs which had no meaning in themselves but were a guide to the pronunciation of the symbols.

The next stage in the development of writing was the use of syllabaries (sets of signs representing syllables rather than whole words or concepts). Both the Egyptians and the Assyrians developed syllabaries which they used as a further help in reading their hieroglyphic and cuneiform writings.

The final step was the creation of a true alphabet, and this probably took place in Sinai. The Egyptians had penetrated the Sinai Peninsula in search of the turquoise mined there by the Semitic inhabitants. Hieroglyphic inscriptions have been found in Sinai and it can be assumed that the local people were familiar with Egyptian writing. From it they took the Egyptian syllabary and reduced it to some 30 letters, which they applied to their own Semitic language. It was at about this stage that they or, according to some scholars, Semitic peoples further to the north in Palestine took the crucially important step of converting these from a syllabary to a system in which each sign or letter represented a single sound.

The Semitic invention of the alphabet had great advantages over the mixed and complex writing systems of other peoples. It required the memorization of only a score or so of characters and even people without a long and arduous education in sacred and historical matters could easily master it. This took writing out of the exclusive hands of priests and scribes and put it at the disposal of all who took the trouble to memorize the alphabet.

Soon the Phoenicians — the Greek name given to the seafaring Canaanites of the Syrian-Lebanese coast — were spreading the alphabet throughout the Mediterranean by leaving examples of this priceless tool in the form of orders, invoices, bills of lading, and receipts wherever they went in search of trade.

The Aramaeans, a closely related Semitic people who were also traders, contributed to the spread of the alphabet too. They adapted it to their needs and, as their language became the lingua franca of the Middle East from Egypt to Persia for more than a thousand years, introduced the alphabet wherever Aramaic was spoken. As a result, the Semitic alphabet became the remote basis of virtually every alphabet used in Asia: Arabic, Hebrew, the many Indian scripts, Burmese, Thai, Malayan, Tibetan, Manchu, Korean, and scores of others.

In Greece, a regular port of call for the Phoenicians, the alphabet was modified. Because the Indo-European languages of the West put more emphasis on vowels than the Semitic tongues, the Greeks assigned certain of the borrowed letters to vowel sounds and invented and adapted others to accommodate the sounds of Greek. But even then the Semitic alphabet remained the basis; as written in the sixth century B.C., Greek retained 19 of the original Semitic symbols.

From Greece the alphabet spread to the Etruscans, who introduced still more modifications, and then, as altered, to the Romans. Thus the alphabet entered Europe and the New World where, today, it still bears a name directly derived from the first two letters of the Semitic alphabet, for "alpha," the name of the first letter in the Greek alphabet, comes from "aleph" and "beta" is from the Semitic "beit" or "beth"—thus "alphabet," the key element in the art of writing.

Bronze figures covered with gold leaf were among the Bronze Age finds at Byblos, considered by the Phoenicians the most ancient city on their coast.

The political, administrative, and religious center of the Persian Empire was at Persepolis, a city of palaces and ceremonial staircases built primarily by Darius I and his son, Xerxes I.

state and the boundaries of Phoenicia were vague. But they were the greatest seafarers of the ancient world; from the independent ports of Tyre, Sidon, Byblos, and Tripoli on what is now the coast of Lebanon, and Aradus in Syria, the Phoenicians, beginning about the twelfth century B.C., ranged throughout the Mediterranean and into the Atlantic, founding colonies — essentially trading posts — in Cyprus, North Africa, France, and Spain. They also sailed along the west coast of Africa and may have reached Britain, where they are thought to have traded for tin in Cornwall.

The Phoenicians carried many items of trade, especially cloth in rich shades of Tyrian purple, a dye made from the murex, a shellfish found on their coast. They also disseminated the Canaanite alphabet, especially to the Greeks, who passed it on to Europe.

Of all Phoenician colonies, Carthage, in what is now Tunisia, played the most spectacular role. Founded, according to tradition, by Elissa, or Dido, a fugitive princess of Tyre, Carthage outstripped her mother-city and became strong enough to challenge the Roman Empire. In the three Punic (Phoenician) Wars of the third and second centuries B.C., Carthage almost defeated Rome in a long contest for control of the Mediterranean world. In 217 B.C., under Hannibal, a Carthaginian army accompanied by elephants crossed the Alps into Italy and threatened Rome itself. But the Romans won the final battle. In 146 B.C. they took Carthage by siege, destroyed the city and — according to legend — sowed salt on its site. The Phoenician cities of the eastern Mediterranean, meanwhile, had fallen to the army of Alexander the Great and were eventually absorbed into the Roman Empire.

The Aramaeans are chiefly remembered for the extraordinary diffusion and longevity of their language. They appeared in Syria sometime before 1200 B.C. and made the already ancient city of Damascus their capital. A commercial people like the Phoenicians, they penetrated other lands in pursuit of trade, spreading their language — which was closely related to other Semitic tongues in the region — until it became the lingua franca of the Middle East for more than a thousand years. The Egyptians and Persians used it in their diplomatic correspondence and the people of Palestine spoke it during the lifetime of Jesus.

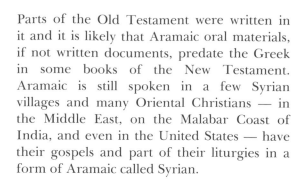

Parts of the Old Testament were written in it and it is likely that Aramaic oral materials, if not written documents, predate the Greek in some books of the New Testament. Aramaic is still spoken in a few Syrian villages and many Oriental Christians — in the Middle East, on the Malabar Coast of India, and even in the United States — have their gospels and part of their liturgies in a form of Aramaic called Syrian.

THE OUTSIDERS: From the sixth century B.C. to the seventh century A.D., a period of more than a thousand years, outsiders ruled in the Fertile Crescent and the Nile Valley. Non-Semitic peoples from the east and the west — Persians, Macedonian Greeks, Romans, Byzantines, Parthians, and Sassanians — governed part or all of the Middle East and each made a contribution to the intricate mosaic of Middle Eastern culture.

The conquest of Babylon by Cyrus the Great of Persia in 539 B.C. was an early step in the formation of the greatest empire the world had yet seen. Persian rule eventually extended from areas now in Pakistan and Afghanistan through the Fertile Crescent into Egypt, and northward through Asia Minor to the banks of the Danube. Only the mainland Greeks, at the battles of Marathon and Salamis, were able to withstand the Persian advance. Despite its size, the empire was highly centralized, with a remarkably efficient administration and excellent communications based on the use of fast horses. The Persians showed unusual tolerance toward their subject peoples, allowing them to retain local customs and religions as long as these did not conflict with Persian hegemony, and they permitted peoples deported by the Assyrians and Babylonians to return to their homelands. The Persian religion, Zoroastrianism, had a high ethical content and Persian art was eclectic: the great complex of palaces at Persepolis shows elements borrowed from Mesopotamia, Egypt, and Greece, but assimilated into a distinctive Persian style.

After a little more than two centuries, the old Persian or Achaemenid Empire was overwhelmed by the conquests of Alexander the Great in 334-323 B.C. It was in Babylon that Alexander died in 323 B.C., after wresting the ancient city from Persian rule and after completing his conquest of the whole civilized

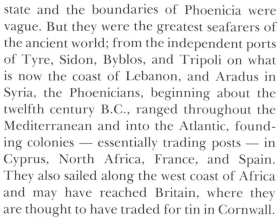

world from Macedonia to India. Greek culture, which already had been spreading over the Middle East, continued there in Hellenistic form through the succeeding era of the Seleucids in Syria and the Ptolemies in Egypt, as well as the later Roman and Byzantine empires, down to the age of Islam.

In the first century B.C. the tide of Roman conquests swept away such personal dynasties and absorbed their fiefs into the Roman Empire. Rome itself gradually became orientalized by the older and richer cultures of the Middle East — a change reflected in the Roman poet Juvenal's complaint that Syria's Orontes River was flowing into the Tiber. But as the tendency of the Roman Empire was toward uniformity, Roman art and architecture eventually blanketed the civilized world — as witnessed by the remains of Roman cities at such places as Baalbek in Lebanon, Palmyra and Bosra in Syria, and Jerash in Jordan.

The vitality of Persia was shown in its revival in the Parthian Empire (250 B.C. — A.D. 226) and in the Sassanian Empire (A.D. 225-650). The Parthians kept the power of Rome from going east of the Euphrates. The frequent occurrence on Bahrain Island, as well as on the western shore of the Arabian Gulf, of certain types of Parthian pottery indicates a close relationship between Mesopotamia and the Arabian Gulf, at least during the second half of the Parthian Empire. The Sassanians were often more than a match for Byzantium in war and made significant achievements in architecture, art, and textiles. The historic and fertile region of Mesopotamia attracted both the later Persian empires. Ctesiphon, on the Tigris southeast of Baghdad, was the Persian capital when the Sassanians were conquered by the Muslim Arab armies in 637-650.

Another of the great outside powers that ruled and influenced the Middle East was Byzantium, heir to the Roman provinces in Syria, Asia Minor, and Egypt. In A.D. 330 the Roman Emperor Constantine I moved his capital from Rome to Constantinople — on the site of an old Greek town called Byzantium. At the end of the fourth century, when the Roman Empire broke in two, Constantinople became the capital of the eastern, and more brilliant, half. Byzantium was Greek in language, Orthodox Christian in religion, and Roman in political institutions. In 636 the

Byzantine armies were defeated by the Arabs at the Battle of Yarmuk and Byzantine power withdrew into Asia Minor and the Balkans, where it continued to flourish although in a steadily diminishing area until the Ottoman Turks took Constantinople in 1453. The Battle of Yarmuk was not the first appearance of the Arabs in history, but until then they were peripheral to it. Now they step into its full light.

THE ARABS: The Arabs themselves call their homeland the "Island of the Arabs," a phrase that is effectively, if not literally, true. Cut off from Africa and Asia by water on three sides — the Red Sea, the Gulf of Aden, the Arabian Sea, the Gulf of Oman, and the Arabian Gulf —Arabia was for centuries also sealed off from the Fertile Crescent to the north by a sea of sand and stone as forbidding as the coral reefs and shoals of the coast. Invaders could and did breach this barrier but it was a formidable challenge; as a result the peninsular Arabs enjoyed long uninterrupted periods of separate development.

But as the Island of the Arabs was also located athwart the ancient trade routes that linked the Mediterranean world with the Far East by sea and land, the Arabs almost inevitably became the middlemen of East-West commerce. Although the West could rarely reach them, they could reach the West; as a consequence they found themselves in contact not only with material products of both worlds, but also with their inventions and discoveries. They were, in effect, middlemen for the passage of ideas as well as things between the Orient and the Occident.

Other geographical influences were climate and terrain. The Arabian Peninsula — a veritable subcontinent almost the size of Western Europe or one-third the size of the continental United States — is sharply divided into two different regions by the amount of rainfall each receives. The southern mountainous fringe of the peninsula receives abundant rains from the monsoons blowing off the Indian Ocean, and these regions—in ancient times part of Arabia Felix (Happy or Fortunate Arabia) and today comprising the 'Asir province of Saudi Arabia, Yemen and Oman — are relatively green and fertile. The northern and central parts of the peninsula — that is, the coastal area north from about Jiddah and the vast

The Seleucids, Greek successors to Alexander the Great, ruled Syria and Mesopotamia, introducing Hellenistic styles of art as in this bust found at Hatra on the Tigris.

A great elliptical, barrel-shaped vault — more than 36 meters high at the crown — spans what was once a palace hall at Ctesiphon, the Sassanian capital on the Tigris.

interior — receive little rain and are mostly desert except for occasional green oases where groundwater is near the surface.

Corresponding to these differences in climate and terrain is the clear distinction in Arab tradition between the Arabs of the north and the Arabs of the south. The Arabs of the north had a heterogenous social fabric; the southern Arabs formed a homogenous society. The northerners were divided into camel-herding nomads and sedentary oasis-dwellers. The nomads, organized on a tribal basis, depended for their livelihood on their herds and hence moved constantly in search of pasture over a vast and forbidding area, while most of the sedentary people dwelt in far-flung oases, hospitable dots of greenery surrounded by the harsh desert. The southern Arabs, on the other hand, lived in a densely populated, rain-fed area rich in vegetation, and resided in towns and villages. These population centers developed an urban culture that had a vigorous material and political growth, flourishing for almost a millennium and interrelating and interacting not only with the rest of the Arabian Peninsula but also with the whole of the ancient world.

The Arabs of the north, furthermore, spoke Thamudic, Lihyanite, Safaitic, and Arabic; those of the south spoke several related Semitic tongues but, in ancient times, not Arabic. There were even differences in physical appearance (probably reflecting a difference in ancestry), as there still are today. These differences were not, of course, absolute; in early times there were Arabic-speaking nomads in the south and colonies of southern merchants in the north (at al-'Ula, for example).

One thing stands out clearly: Despite the diversity of the Arabs there was an essential unity. Regardless of the political, cultural, social, and economic differences arising from the complex interaction of man and nature, all the inhabitants of the peninsula were recognizably one people: the Arabs.

THE SOUTH ARABIANS: It was in the south, during the millennium preceding the birth of Christ, that the first advanced civilizations in the Arabian Peninsula developed. The earliest of these was the Sabaean, named for the Kingdom of Saba or Sheba, whose queen (according to the Bible — and the Quran)

For centuries the number of a man's camels was a measure of his wealth among the nomadic tribes of Arabia.

An ancient village in southern Arabia is sheltered by walls of a canyon cut deeply into the Hadhramaut plateau.

Early Times

For many years historians regarded the ancient Arabian Gulf as no more than an artery of seafaring and trade between Mesopotamia and the lands to the south and east. In the 1950s, however, an archaeological expedition reported that the islands and coast on the western side of the Gulf had not only played a vital role in the economy of the ancient Middle East, but had developed an autonomous civilization going back to the third millennium B.C.

In the 1970s, moreover, the Department of Antiquities of Saudi Arabia excavated evidence confirming that soon after 5000 B.C. peoples on the western coast and islands of the Gulf had contact with the 'Ubayd culture of Mesopotamia. An Aramco geologist had already reported that during an early rainy period — some 20,000 to 30,000 years ago — the Arabian Peninsula was inhabited by Stone Age man.

The Rub' al-Khali, now the largest sand desert in the world, was quite different then. Because the climate as a whole was milder, there were numerous lakes and streams in the Rub' al-Khali. Large crude stone axes have been found on what were the shores of those lakes,

In The Arabian Gulf

Thousands of years ago, ships that were the ancestors of today's dhow plied the waters of the Arabian Gulf.

and similar axes have been found elsewhere in the peninsula.

In a later rainy age — between 6,000 and 9,000 years ago — the peoples of the time left hundreds of thousands of beautifully worked stone tools by the now vanished lakes and streams. And during the later centuries, people living beside a spring in what is now the northern edge of the al-Hasa oasis were using 'Ubayd pottery.

The discovery of 'Ubayd pottery in eastern Arabia was a surprise. This pottery, with its characteristic stripes, zigzags, and loops, is identical to pottery found primarily in Iraq, Iran, and Syria, and such early contacts between those areas and the Gulf had never been suspected. The subsequent excavation of three 'Ubayd sites in Saudi Arabia (two of them near what are now Jubail and Hofuf), as well as two in Qatar and one in Bahrain, showed that 'Ubayd culture extended nearly 650 kilometers farther south than archaeologists had believed. At some of these sites archaeologists also found sharp arrowheads and tools made from volcanic glass imported from the Lake Van region of Turkey nearly 1,500 kilometers to the north. An 'Ubayd site near Dhahran yielded pottery and other objects coated with bitumen that was probably imported from southern Iraq. This late fifth-millennium site provided some of the earliest known evidence for man's use of petroleum.

By about 3000 B.C., a thousand years after the rise of Sumer, the earliest known civilization, eastern Arabia was widely settled and, as the Gulf by then had become an artery of ancient trade, the area was a prosperous one. Tarut Island, for example, was in direct or indirect contact with civilizations in what are now parts of Russia, Pakistan, Iran, Iraq, Syria, Kuwait, and Oman. Among the goods traded were copper, stone, wood, gold, ivory, carnelian, and lapis lazuli — which were exchanged for woven goods and foodstuffs. Designs on bowls found at Tarut Island — snakes, leopards, bulls, and various geometric patterns — are identical to those on bowls found in the other civilizations, but more different designs have been reported from Tarut than from any other place. One particularly attractive find on Tarut Island was a figure, about four centimeters tall, which was carved in lapis lazuli, a stone mined in northern Afghanistan.

(Continued on page 32)

This tomb excavated at Jawan on the Saudi Arabian mainland contained an alabaster statuette and jewelry of gold and precious stones dating from the first century A.D.

Early Times

The most famous Gulf civilization of this period — and therefore the focus of archaeological investigation — was Dilmun. Cuneiform texts found at Ur refer to a land called Ni-tuk-ki in Sumerian and Dilmun in Akkadian. In Sumerian myths it was a pure, bright place where "the lion kills not, the wolf snatches not the lamb." It was also the scene of certain incidents in the legend of Gilgamesh, the epic which fascinated Mesopotamian peoples for thousands of years.

Dilmun was also an actual country that may have developed in the east Arabian mainland, but later centered around Bahrain. Much of Dilmun's economy was based on the export and transshipment of goods into the Mesopotamian economy from such other lands as Oman and India. "Ships of Dilmun from the foreign lands brought me wood as a tribute," boasts a ruler of Lagash in the middle of the third millennium B.C. The expansionist empires of Mesopotamia often tried to draw Dilmun into their political and economic orbit, but it was not until about 600 B.C. that the Assyrians brought Dilmun's independence to an end.

To amend and amplify what was known about the Gulf in ancient times, a Danish archaeological expedition, sent by the Museum of Aarhus, began working in the Arabian Gulf in the 1950s. Most of the digging was in Bahrain, where, with the active support of the Bahrain Government, archaeologists unearthed a large palace, and a town site going back to the third millennium. Within the palace complex they found the skeletons of snakes buried in covered pots — a practice that perhaps may be related to the "snake of immortality" referred to in the Epic of Gilgamesh. They also found weights suggesting a possible commercial tie with the Indus Valley, where weights of similar value were used. The expedition also unearthed a Dilmun town on Failaka Island in Kuwait, the remains of a people who lived in Abu Dhabi during the same period, and copper works in Oman — suggesting that it might be the lost land of Makan — which produced the copper on which the Bronze Age civilization of Mesopotamia was based.

The most notable archaeological feature of the western Gulf is the existence of large fields of burial mounds on Bahrain Island and on the mainland south of Dhahran. Over the years travelers who have seen them have advanced several theories about who were buried in the mounds. By the 1960s archaeologists were certain that the larger graves dated back to Dilmun and the smaller ones to Hellenistic times — specifically to the successors of Alexander the Great. In the last years of his life Alexander, after reaching India, intended to add Arabia to his conquests. Although he died before he could do so, his successors in Syria and Mesopotamia, the Seleucids, occupied territory at the head of the Gulf and built cities there in the Hellenistic pattern. The Danish expedition investigated two of them — one at Thaj in Saudi Arabia, the other on Failaka Island in Kuwait, where they unearthed typical Hellenistic temples, votive inscriptions to the Greek gods, and sherds of pottery from mainland Greece.

Archaeologists would also like to find the famous city called Gerrha, somewhere on the Arabian coast. Strabo, quoting earlier sources, reports that it was a Babylonian colony and that its merchants generally carried their merchandise, including incense, by land, a reference to the great caravans plodding north from what is now Yemen.

By this trade, Strabo adds, "both the Sabaeans and the Gerrhaei have become the richest of all tribes, and possess a great quantity of articles wrought in gold and silver, as couches, tripods, basins, drinking vessels, to which we must add the costly magnificence of their houses for the doors, walls and roofs are variegated with inlaid ivory, gold, silver and precious stones." Because of this wealth, the Seleucids under Antiochus III in 205 B.C. threatened to level Gerrha, and a leader of the city sent a

A tiny lapis lazuli figure and Dilmun pottery were found on Tarut Island.

large tribute of silver, pearls, and other precious gems along with a letter. This letter said, "Destroy not, O King, those two things which have been given of the gods, perpetual peace and freedom."

The exact location of Gerrha is still a mystery. Pliny described it as being opposite the island of Tylos, the classical name for Bahrain, "famous for the vast number of its pearls," and also refers to the region of Attene, 80 kilometers inland, possibly the oasis of al-Hasa. It is also possible that Gerrha is buried under the Jafurah sands where Seleucid coins have been found near the remains of an extensive irrigation system just inland from the Gulf. But these identifications are just conjecture; the location of Gerrha remains unknown.

There are other early sites. Jawan, for example, a tomb complex near the road from Dhahran to Ras Tanura, was excavated by an Aramco archaeologist with the agreement of the Government of Saudi Arabia. Objects found in the tomb, some of gold, belong to the middle of the first century A.D. Many inhabitants of eastern Arabia along the Gulf became Christians in the first few centuries of the Christian era, and Church sources say Christian bishoprics existed there from the third century on; one of them, a Nestorian bishopric, was located in Darin, the old port of Tarut Island. With the advent of Islam, however, all those Christian communities were absorbed into the Islamic state.

visited King Solomon. In the Quran she is described as ruling from "a magnificent throne" and in the Bible as coming to visit Solomon in a great caravan laden with "spices and very much gold and precious stones." This so-called Queen of Sheba was most probably the chieftain of the tribe of Saba, although she may have come from a Sabaean colony in the northern part of the peninsula rather than from the land of Saba in the south.

Other kingdoms in southern Arabia that later came to prominence around Saba were Ma'in, north of Saba; Awsan in the southwestern corner nearest Aden; Qataban, northeast of Awsan; and Hadhramaut, east of Qataban. This southern civilization in its later periods is often referred to as Himyaritic rather than Minaean or Sabaean, after the southwestern tribe of Himyar, which grew increasingly powerful.

The Sabaean civilization of southern Arabia, based on trade between India and the Mediterranean, produced works of art such as this relief of a ram.

This inscription is in Himyaritic, a language of the pre-Islamic southern Arabians.

To a great extent the prosperity of southern Arabia rested on trade — based largely on the sale of its own products (most notably aromatics) and on the transshipment of spices and other valuable cargoes from across the Arabian Sea. It is difficult today to grasp the importance of spices and aromatics in the ancient world. Spices were used to preserve food, or to flavor food preserved by smoking and drying and thus left tasteless. Aromatics were also important. They were used to adorn and perfume the body and were essential ingredients in the spiritual and political life of the times. They were used in religious rites, in the embalming and burial of the dead, in the consecration or "anointment" of the rulers, and — most importantly — as offerings in the great temples of Egypt and Canaan. In sum, spices and aromatics were valuable enough to justify the risks and costs of long trading voyages.

Some of the commodities that the people of southern Arabia traded in came from India and East Africa, but the most important — the rare and expensive aromatic gums called frankincense and myrrh — grew in Arabia Felix itself. In fact, although trade was of major importance, Sabaean prosperity rested to an even greater extent on agriculture, and to support this one of the engineering marvels of the age, the great dam of Marib, was built. Some 600 meters across, the dam spanned the Wadi Adhanah in the black hills near Marib, the capital of Saba, deflecting and

THE MIDDLE EAST IN THE FIFTH CENTURY B.C.

United under the Achaemenian kings, the Medes and Persians rapidly extend the boundaries of their empire. Reliefs in the fifth-century ceremonial capital of Persepolis graphically illustrate the cosmopolitan character of this empire, which extends from what is now Afghanistan and Pakistan through the Fertile Crescent to Egypt and northward through Asia Minor to the Danube. Only mainland Greece in the Age of Pericles successfully resists and obtains a peace agreement recognizing the independence of the city states and Greek colonies on the west coast of Asia Minor. The tolerant rule of the Persians encourages revival of the cities of the Levant, which have been successively occupied or destroyed by Assyrian and Babylonian forces.

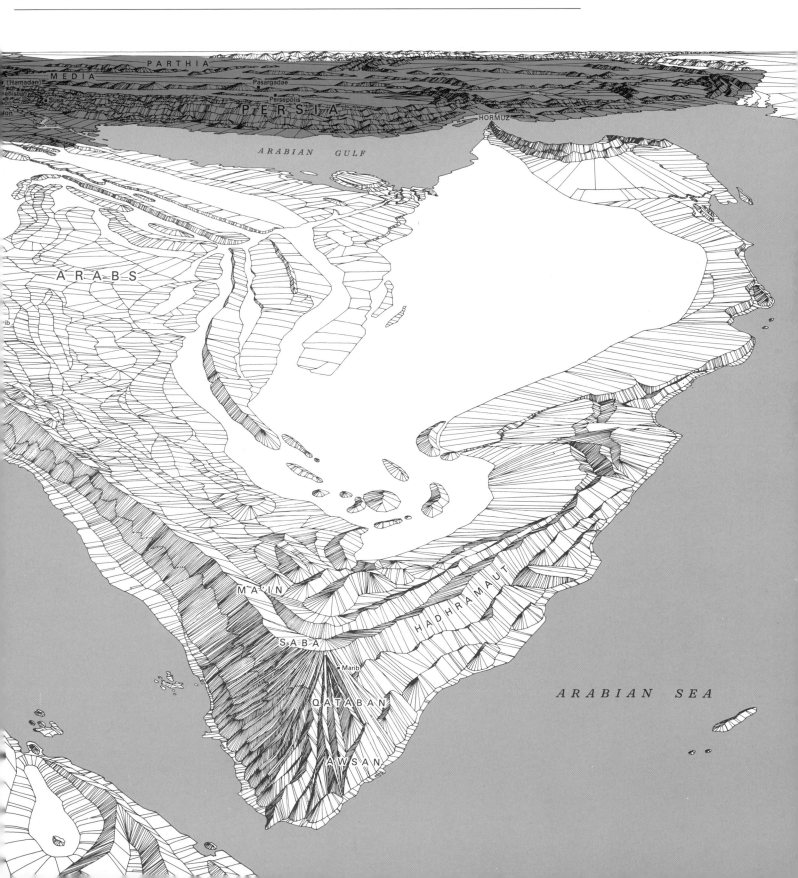

The Roman Empire proclaimed its might in the East as elsewhere by raising temples and other public buildings of imposing size and presence. Opposite page: The columns of the Temple of Jupiter at Baalbek in Lebanon are the largest of the ancient world. This page, clockwise from above: A ceremonial gate at Palmyra, an oasis and caravan station in the Syrian Desert. A carved head gazes down from a wall in Leptis Magna, built by the Romans in what is now Libya. The so-called Temple of Bacchus at Baalbek, which has withstood the onslaught of time better than most. A curved colonnade of classical pillars graces the forum at Jerash in today's Jordan.

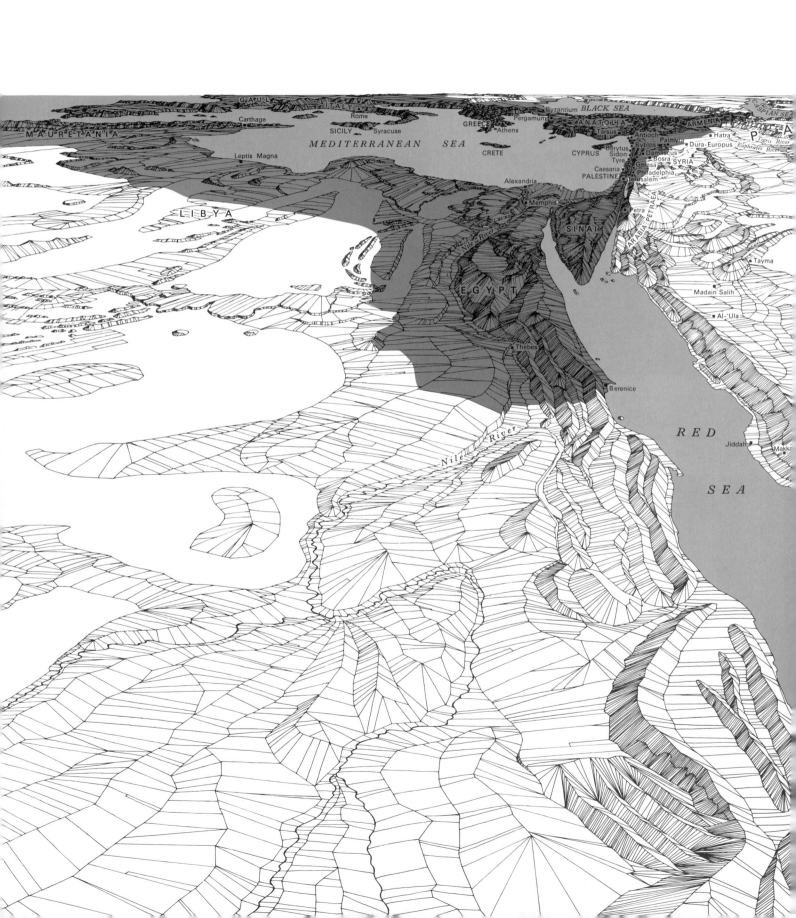

THE MIDDLE EAST AND THE ROMAN EMPIRE IN THE SECOND CENTURY A.D.

The Roman Empire achieves its maximum territorial extent, completely surrounding the Mediterranean. Expansion to the east is checked by the Parthians and the Syrian Desert. Conditions are chaotic in southern Arabia (known in classical times as Arabia Felix or "Happy Arabia"), since the overland trade routes on which it depended for its prosperity have been bypassed by the development of sea routes through the Red Sea to India. However, Gerrha on the Arabian Gulf and the caravan cities of the north continue to prosper. The caravan kingdom of the Nabataeans, which has extended its territory into north-central Arabia and established a city at Madain Salih, is annexed by the Roman Empire as Arabia Petraea or "Rocky Arabia."

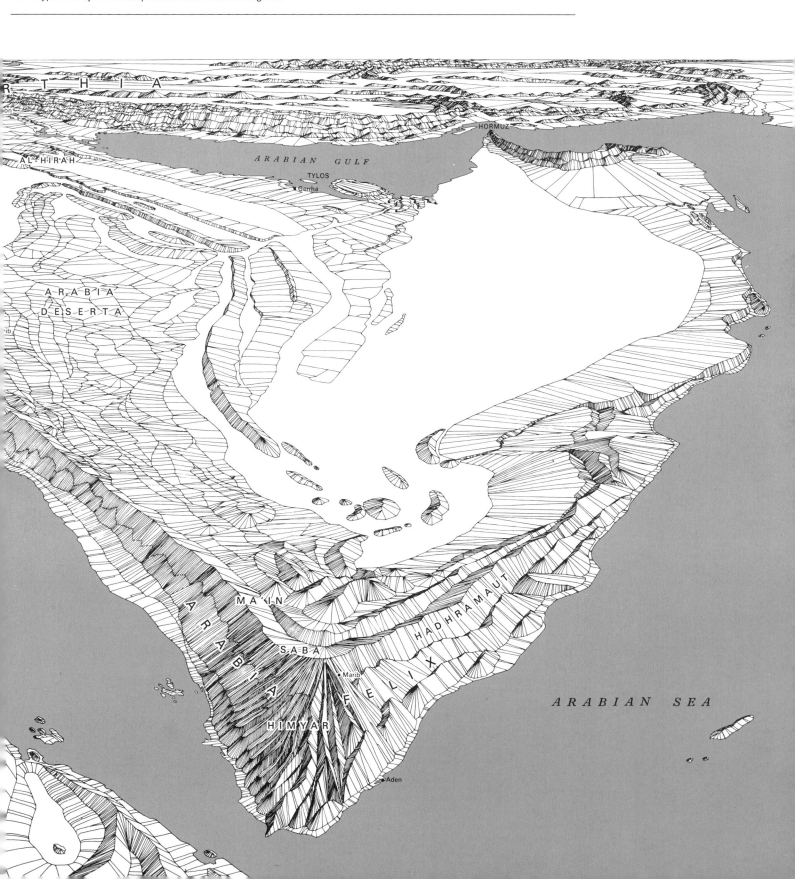

The Nabataeans, who controlled trade routes in northwestern Arabia from the fourth to the first century B.C., carved their tombs into sandstone cliffs at Petra, now in Jordan.

At Madain Salih, the Nabataeans established another important center, possibly at the frontier shared with the southern Arabians.

Palmyra, an ancient caravan port on the edge of the Syrian Desert, was important in biblical times as a center of east-west trade.

retaining the waters of seasonal flash floods, and, through an elaborate system of sluices and canals, distributing the water to the cultivated areas. The dam, in fact, was a cornerstone of the economy of the kingdom.

THE NORTH ARABIANS: To reach the Fertile Crescent and the Mediterranean the south Arabian merchants had to enlist the help of the Arabians of the north, who were masters of the desert. Only they knew the great wadi systems, which provided not only easier terrain for travel but also water near the surface in the ancient stream beds. Only they knew the location of desert wells and springs, the oasis way stations, the best grazing, and the alliances and hostilities of the tribes. And only they were familiar with the efficient use of the camel, which alone made desert travel possible.

The northern Arabians developed a great network of routes for their caravans. The main route from the south led from Yemen along the western side of the peninsula through Makkah and Yathrib (now called Medina) and thence via different branches into Mesopotamia, Syria, Palestine, and across Sinai into Egypt. Other routes branched off from this at the oasis of Najran, in the southwestern part of what is now Saudi Arabia, and at Makkah, to run north and east to Babylon and the rest of Mesopotamia and to the port of Gerrha on the Arabian Gulf. They also organized the caravans, often made up of hundreds of camels, into joint stock companies and contributed to the development of important commercial centers. These centers grew up in the oases where the camel trains halted, where customs duties were charged for the passage of merchandise, and where goods were often sold or redistributed — in places like Gerrha, Makkah, Yathrib, and al-'Ula, a south Arabian colony in northwestern Arabia. As a result the Arabians of the north gradually became equal partners with those of the south or even dominated them in the caravan trade.

Meanwhile, the isolation that had protected southern Arabia's commercial monopoly began to collapse. In the second century B.C. Hippalus, a Greek Egyptian, unlocked the secret of the monsoons and the ensuing increase in seaborne trade slowly but surely undermined and reduced the trade by caravan. The south also faced invasions from

Ethiopia, and once from Rome. About the same time, at the northern termini of the desert routes, other caravan cities had also been growing into centers of commercial importance to rival the established towns to the south. The most important of these were Petra and Palmyra.

Petra was the capital of a people called the Nabataeans, believed to be descended from the desert nomads. In the fourth century B.C. the Nabataeans had seized control of the northern portion of the rich trade routes running up the western coast of Arabia and, at their peak, extended their power from Madain Salih, north of Medina, to Damascus. But their main city was Petra in what is now Jordan. Both Petra and Madain Salih are remarkable for the beauty of their tombs. Rather than erecting buildings from separate structural elements, the Nabataeans carved their tombs out of sandstone cliffs, often imitating Egyptian, Greek, or Roman motifs. Petra — in the poet's phrase "a rose-red city half as old as time" — was hidden in an inaccessible valley in Wadi Musa and unknown to the West for centuries until rediscovered in 1812 by the Anglo-Swiss explorer John Burckhardt. Madain Salih, though close to the pilgrim route from Damascus to Makkah, was first seen and described by a Westerner, Charles Doughty, in 1876.

Palmyra — City of Palms — a desert port for caravans traveling to Mesopotamia, was known in the Bible as Tadmor, and that is its name in Arabic today. Inscriptions found there indicate that the population was mixed and that the inhabitants spoke Aramaic, Greek, and Latin. A large colony of merchants and caravaners from northern Arabia occupied a quarter of the city.

THE ROMANS: Because of their importance both Petra and Palmyra were absorbed into the orbit of the Romans, who from the first century B.C. had been moving eastward in a series of campaigns that would eventually bring much of the Middle East under their control. Rome incorporated Petra into the Roman Province called Arabia Petraea, and absorbed Palmyra in the first century A.D.

For Petra, Rome's victory was in effect the end of its history. But Palmyra proved more difficult to digest. In the third century, Odenathus, a prince of Palmyra, built up a strong autonomous state embracing all of Syria, northwestern Mesopotamia, and western Armenia. Odenathus' wife Zenobia, apparently of Arab stock to judge by her name (Zaynab), took the reins after his death and aggressively expanded Palmyrene power into Asia Minor, Mesopotamia, and Egypt and, declaring her son emperor, even had coins struck bearing his portrait and her own. This was too much for Rome — it meant the loss of the entire eastern empire — and in 272 Aurelian was sent to attack Palmyra. He destroyed Palmyrene power, captured Zenobia, and brought her to Rome decked in golden chains. She was then pensioned off and ended her days confined in a Roman villa.

If one Arab ruler went to Rome in chains, however, another became a Roman emperor. Called Philip the Arabian, he was a Syrian from Hawran who had served with the Roman legions against the Persians, was proclaimed emperor by his troops, and ruled the empire for years.

In the meantime Christianity had become a potent force spreading through the Roman empire and beyond. In the fourth century, for example, the kings of Ethiopia were converted and pockets of Christianity were established in southern Arabia. Najran became a Christian center and there were churches in Aden, San'a, Hadhramaut, and the island of Tarut near present-day Qatif. But there were also Judaic influences in the area and eventually they clashed. In the sixth century a Judaized king of Saba attacked Najran, in an attempt to impose Judaism on the area, and massacred the inhabitants in a place called al-Ukhdud ("The Ditch"), an event mentioned in the Quran: "destroyed were the inhabitants of The Ditch." In response Ethiopian forces crossed the Strait of Bab al-Mandab, crushed the army of Saba, and established a protectorate over southern Arabia.

At about the same time, from the third to sixth centuries, two other Arabian states had arisen to prominence — states whose wealth, culture, and martial exploits would be celebrated in Arabic verse for centuries. One was al-Manadhirah; the other was al-Ghasasinah, which rose from the ashes of Zenobia's ill-fated Palmyrene domains.

Al-Ghasasinah was a client of Byzantium and its ally in the endless wars between Byzantium and Persia. The Ghassanids were Arabs of Yemeni stock and Christians of the Monophysite persuasion. Al-Manadhirah, whose capital was al-Hira in what is now southern Iraq, bore the same relation to Sassanid Persia as al-Ghasasinah did to Byzantium. The people were Nestorian Christians but the rulers — called Lakhmids — remained pagan in deference to their Persian master, who identified Christianity with the Byzantine enemy.

A third Arab power on the peninsula was little-known Kindah, centered in al-Yamamah in southern Najd. Nominally a client of Himyar in the south, it at one time attacked and imposed some sort of hegemony over al-Hirah and was predominantly polytheistic.

THE RISE OF MAKKAH: None of those states, however, had any political influence in the area called the Hijaz, where a number of caravan cities had not only continued to develop as mercantile centers but had also managed to remain independent. Indeed, the Hijaz had brought the desert tribes into its own orbit. Yathrib was one of the caravan towns; Tabuk and Khaybar were others. But the greatest, and the one destined to play the most important role in the future of the Muslim world, was Makkah.

Makkah grew up around a shrine, the Ka'bah, whose origins go back to antiquity, but its wealth was founded on trade. It was the "commercial republic" of Arabia in the sixth century.

Although Makkah, as one of the northern Arabs' caravan cities, had established commercial contacts with Syria and Yemen, it was nevertheless relatively isolated. In the sixth century, however, there occurred a series of events that would catapult it into prominence. One of the events was the collapse of the great dam at Marib — which confirmed the decline of southern Arabia as a power and left a vacuum in the peninsula. At about the same time an Ethiopian governor of southern Arabia sent an expedition to conquer Makkah — thus effectively involving the city in the events of the day. But the most important event was the rise to power in Makkah itself of the tribe of Quraysh, into which, in about the year 570, a child called Muhammad would be born. Together, these events set the stage for the most important single development in the history of the Middle East: the advent of Islam.

Remnants of the great dam at Marib in Yemen recall the vast water management system which for more than a millennium diverted floodwaters to irrigate some 1,600 hectares (4,000 acres) of the prosperous city-state.

ISLAM AND ISLAMIC HISTORY

In or about the year 570 the child who would be named Muhammad and who would become the Prophet of one of the world's great religions, Islam, was born into a family belonging to Quraysh, the ruling tribe of Makkah, a city in the Hijaz region of northwestern Arabia.

Originally the site of the Ka'bah, a shrine of ancient origins, Makkah had with the decline of southern Arabia (see Chapter 1) become an important center of sixth-century trade with such powers as the Sassanians, Byzantines, and Ethiopians. As a result the city was dominated by powerful merchant families among whom the men of Quraysh were preeminent.

Muhammad's father, 'Abd Allah ibn 'Abd al-Muttalib, died before the boy was born; his mother, Aminah, died when he was six. The orphan was consigned to the care of his grandfather, the head of the clan of Hashim. After the death of his grandfather, Muhammad was raised by his uncle, Abu Talib. As was customary, Muhammad as a young child was sent to live for a year or two with a Bedouin family.

This custom, followed until recently by noble families of Makkah, Medina, Tayif, and other towns of the Hijaz, had important implications for Muhammad. In addition to enduring the hardships of desert life, he acquired a taste for the rich language so loved by the Arabs, whose speech was their proudest art, and learned the patience and forbearance of the herdsmen, whose life of solitude he first shared and then came to understand and appreciate.

About the year 590, Muhammad, who was then in his 20s, entered the service of a widow named Khadijah as a merchant actively engaged with trading caravans to the north. Sometime later Muhammad married Khadijah, by whom he had two sons — who did not survive — and four daughters.

During this period of his life Muhammad traveled widely. Then, in his late 30s, he began to retire to meditate in a cave on Mount Hira outside of Makkah, where the first of the great events of Islam took place. One day, as he sat in the cave, he heard a voice, later identified as that of the Angel Gabriel, which ordered him to:

Recite: In the name of thy Lord who created, Created man from a clot of blood.

In the still solitude of desert life, the Bedouins learn harsh lessons of patience, endurance, and survival.

Three times Muhammad pleaded his inability to do so, but each time the command was repeated. Finally, Muhammad recited the words of what are now the first five verses of the 96th surah or chapter of the Quran — words which proclaim God the Creator of man and the Source of all knowledge.

At first Muhammad divulged his experience only to his wife and his immediate circle. But as more revelations enjoined him to proclaim the oneness of God universally, his following grew, at first among the poor and the slaves, but later also among the most prominent men of Makkah. The revelations he received at this time and those he did so later are all incorporated in the Quran, the Scripture of Islam.

Not everyone accepted God's message transmitted through Muhammad. Even in his own clan there were those who rejected his teachings, and many merchants actively opposed the message. The opposition, however, merely served to sharpen Muhammad's sense of mission and his understanding of exactly how Islam differed from paganism. The belief in the unity of God was paramount in Islam; from this all else followed. The verses of the Quran stress God's uniqueness, warn those who deny it of impending punishment, and proclaim His unbounded compassion to those who submit to His will. They affirm the Last Judgment, when God, the Judge, will weigh in the balance the faith and works of each man, rewarding the faithful and punishing the transgressor. Because the Quran rejected polytheism and emphasized man's moral responsibility in powerful images, it presented a grave challenge to the worldly Makkans.

THE HIJRAH: After Muhammad had preached publicly for more than a decade, the opposition to him reached such a high pitch that, fearful for their safety, he sent some of his adherents to Ethiopia, where the Christian ruler extended protection to them, the memory of which has been cherished by Muslims ever since. But in Makkah the persecution worsened. Muhammad's followers were harassed, abused, and even tortured. At last, therefore, Muhammad told his followers they could go to the northern town of Yathrib, which was later to be renamed Medina ("The City"). Later, in the early fall of 622, he learned of a plot to murder him and, with his closest friend, Abu Bakr al-Siddiq, set off to join the emigrants.

In Makkah the plotters arrived at Muhammad's home to find that his cousin, 'Ali, had taken his place in bed. Enraged, the Makkans set a price on Muhammad's head and set off in pursuit. Muhammad and Abu Bakr, however, had taken refuge in a cave where, as they hid from their pursuers, a spider spun its web across the cave's mouth. When they saw that the web was unbroken, the Makkans passed by and Muhammad and Abu Bakr went on to Medina, where they were joyously welcomed by a throng of Medinans as well as the Makkans who had gone ahead to prepare the way.

This was the *Hijrah* — anglicized as Hegira — usually, but inaccurately, translated as "Flight" — from which the Muslim era is dated. In fact, the Hijrah was not a flight but a carefully planned migration which marks not only a break in history — the beginning of the Islamic era — but also, for Muhammad and the Muslims, a new way of life. Henceforth the organizational principle of the community was not to be mere blood kinship, but the greater brotherhood of all Muslims. The men who accompanied Muhammad on the Hijrah were called the *Muhajirun* — "those who made the Hijrah" or the "Emigrants" — while those in Medina who became Muslims were called the *Ansar* or "Helpers."

Muhammad was well acquainted with the situation in Medina. Earlier, before the Hijrah, he had met with envoys from the city. What the envoys saw and heard had impressed them and they had invited Muhammad to settle in Medina. After the Hijrah, Muhammad's exceptional qualities so impressed the Medinans that the rival tribes and their allies temporarily closed ranks as, on March 15, 624, Muhammad and his supporters moved against the pagans of Makkah.

The first battle, which took place near a well called Badr, now a small town southwest of Medina, had important effects. In the first place, the Muslim forces, outnumbered three to one, routed the Makkans. Secondly, the discipline displayed by the Muslims brought home to the Makkans, perhaps for the first time, the abilities of the man they had driven from their city.

A year later the Makkans struck back. Assembling an army of 3,000 men, they met the Muslims at Uhud, a ridge outside Medina. After an initial success the Muslims were driven back and the Prophet himself was wounded. As the Muslims were not completely defeated, the Makkans, with an army of 10,000, attacked Medina again two years later but with quite different results. At the Battle of the Trench, also known as the Battle of the Confederates, the Muslims scored a signal victory by introducing a new defense. On the side of Medina from which attack was expected they dug a trench too deep for the Makkan cavalry to clear without exposing itself to the archers posted behind earthworks on the Medina side. After an inconclusive siege, the Makkans were forced to retire. Thereafter Medina was entirely in the hands of the Muslims.

The Constitution of Medina — under which the clans accepting Muhammad as the Prophet of God formed an alliance, or federation — dates from this period. It showed that the political consciousness of the Muslim community had reached an important point; its members defined themselves as a community separate from all others. The Constitution of Medina also defined the role of non-Muslims in the community. Jews, for example, were part of the community; they were *dhimmis*, that is, protected people, as long as they conformed to its laws. This established an important precedent for the treatment of subject peoples during the later conquests. Christians and Jews, upon payment of a yearly tax, were allowed religious freedom and, while maintaining their status as non-Muslims, were associate members of the Muslim state. This status did not apply to polytheists, who could not be tolerated within a community that worshipped the One God.

Ibn Ishaq, one of the earliest biographers of the Prophet, says it was at about this time that Muhammad sent letters to the rulers of the earth — the King of Persia, the Emperor of Byzantium, the Negus of Abyssinia, and the Governor of Egypt among others — inviting them to submit to Islam. Nothing more fully illustrates the confidence of the small community, as its military power, despite the Battle of the Trench, was still negligible. But its confidence was not misplaced. Muhammad so effectively built up a series of alliances

A colonnade of lofty arches surrounds the courtyard at the Prophet's Mosque in Medina, after Makkah the second holiest city of Islam.

The Holy Quran

Islam appeared in the form of a book: the Quran. Muslims consider the Quran (sometimes spelled "Koran") to be the Word of God as transmitted by the Angel Gabriel, in the Arabic language, through the Prophet Muhammad. The Muslim view, moreover, is that the Quran supersedes earlier revelations; it is regarded as their summation and completion. It is the final revelation, as Muhammad is regarded as the final prophet — "the Seal of the Prophets."

In a very real sense the Quran is the mentor of millions of Muslims, Arab and non-Arab alike; it shapes their everyday life, anchors them to a unique system of law, and inspires them by its guiding principles. Written in noble language, this Holy Text has done more than move multitudes to tears and ecstasy; it has also, for almost 1,400 years, illuminated the lives of Muslims with its eloquent message of uncompromising monotheism, human dignity, righteous living, individual responsibility, and social justice. For countless millions, consequently, it has been the single most important force in guiding their religious, social, and cultural lives. Indeed, the Quran is the cornerstone on which the edifice of Islamic civilization has been built.

The text of the Quran was delivered orally by the Prophet Muhammad to his followers as it was revealed to him. The first verses were revealed to him in or about 610, and the last revelation dates from the last year of his life, 632. His followers at first committed the Quran to memory and then, as instructed by him, to writing. Although the entire contents of the Quran, the placement of its verses, and the arrangement of its chapters date back to the Prophet, as long as he lived he continued to receive revelations. Consequently, the Holy Text could only be collected as a single corpus — "between the two covers" — after the death of Muhammad.

This is exactly what happened. After the battle of al-Yamamah in 633, 'Umar ibn al-Khattab, later to become the second caliph, suggested to Abu Bakr, the first caliph, that because of the grievous loss of life in that battle, there was a very real danger of losing the Quran, enshrined as it was in the memories of the faithful and in uncollated fragments. Abu Bakr recognized the danger and entrusted the task of gathering the revelations to Zayd ibn Thabit, who as the chief scribe of the Prophet was the person to whom Muhammad frequently dictated the revelations in his lifetime. With great difficulty, the task was carried out and the first complete manuscript compiled from "bits of parchment, thin white stones — *ostracae* — leafless palm branches, and the memories of men." Later, during the time of 'Uthman, the third caliph, a final, authorized text was prepared and completed in 651, and this has remained the text in use ever since.

The contents of the Quran differ in substance and arrangement from the Old and New Testaments. Instead of presenting a straight historical narrative, as do the Gospels and the historical books of the Old Testament, the Quran treats, in allusive style, spiritual and practical as well as historical matters.

The Quran is divided into 114 surahs, or chapters, and the surahs are conventionally assigned to two broad categories: those revealed at Makkah and those revealed at Medina. The surahs revealed at Makkah — at the beginning of Muhammad's mission — tend to be short and to stress, in highly moving language, the eternal themes of the unity of God, the necessity of faith, the punishment of those who stray from the right path, and the Last Judgment, when all man's actions and beliefs will be judged. The surahs revealed at Medina are longer, often deal in detail with specific legal, social, or political situations, and sometimes can only be properly understood with a full knowledge of the circumstances in which they were revealed.

All the surahs are divided into *ayahs* or verses and, for

purposes of pedagogy and recitation, the Quran as a whole is divided into 30 parts, which in turn are divided into short divisions of nearly equal length, to facilitate study and memorization.

The surahs themselves are of varying length, ranging from the longest, Surah 2, with 282 verses, to the shortest, Surahs 103, 108, and 110, each of which has only three. With some exceptions the surahs are arranged in the Quran in descending order of length, with the longest at the beginning and the shortest at the end. The major exception to this arrangement is the opening surah, "al-Fatihah," which contains seven verses and which serves as an introduction to the entire revelation:

In the Name of God, the Merciful, the Compassionate.
Praise be to God, Lord of the Worlds;
The Merciful, the Compassionate;
Master of the Day of Judgment;
Thee only do we worship, and Thee alone we ask for help.
Guide us in the straight path,
The path of those whom Thou hast favored; not the path of those who earn Thine anger nor of those who go astray.

Non-Muslims are often struck by the range of styles found in the Quran. Passages of impassioned beauty are no less common than vigorous narratives. The sublime "Verse of the Throne" is perhaps one of the most famous:

God — There is no god but He,
The Living, the Everlasting;
Slumber seizes Him not, neither sleep;
To Him belongs all that is
In the heavens and the earth;
Who is there that can intercede with Him
Save by His leave?
He knows what lies before them
And what is after them,
Nor do they encompass anything of His knowledge
Except such as He wills;
His Throne extends over the heavens and earth;
The preserving of them wearies Him not;
He is the Most High, the All-Glorious.

Muslims regard the Quran as untranslatable; the language in which it was revealed — Arabic — is inseparable from its message and Muslims everywhere, no matter what their native tongue, must learn Arabic to read the Sacred Book and to perform their worship. The Quran of course is available in many languages, but these versions are regarded as interpretations rather than translations — partly because the Arabic language, extraordinarily concise and allusive, is impossible to translate in a mechanical, word-for-word way. The inimitability of the Quran has crystallized in the Muslim view of i'jaz or "impossibility," which holds that the style of the Quran, being divine, cannot be imitated: any attempt to do so is doomed to failure.

It must also be remembered that the Quran was originally transmitted orally to the faithful and that the Holy

Text is not meant to be read only in silence. From the earliest days it has always been recited aloud or, more accurately, chanted. As a result, several traditional means of chanting, or intoning, the Quran were found side by side. These methods carefully preserved the elaborate science of reciting the Quran — with all its intonations and its cadence and punctuation. As the exact pronunciation was important — and learning it took years — special schools were founded to be sure that no error would creep in as the traditional chanting methods were handed down. It is largely owing to the existence of these traditional methods of recitation that the text of the Quran was preserved without error. As the script in which the Quran was first written down indicated only the consonantal skeleton of the words, oral recitation was an essential element in the transmission of the text.

Because the circumstances of each revelation were thought necessary to correct interpretation, the community, early in the history of Islam, concluded that it was imperative to gather as many traditions as possible about the life and actions of the Prophet so that the Quran might be more fully understood. These traditions not only provided the historical context for many of the surahs — thus contributing to their more exact explication — but also contained a wide variety of subsidiary information on the practice, life, and legal rulings of the Prophet and his companions.

This material became the basis for what is called the sunnah, or "practice" of the Prophet — the deeds, utterances, and taqrir (unspoken approval) of Muhammad. Together with the Quran, the sunnah, as embodied in the canonical collections of traditions, the hadith, became the basis for the shari'ah, the sacred law of Islam.

Unlike Western legal systems, the shari'ah makes no distinction between religious and civil matters; it is the codification of God's Law, and it concerns itself with every aspect of social, political, economic, and religious life. Islamic law is thus different from any other legal system; it differs from canon law in that it is not administered by a church hierarchy; in Islam there is nothing that corresponds to a "church" in the Christian sense. Instead, there is the ummah — the community of the believers — whose cohesion is guaranteed by the sacred law. Every action of the pious Muslim, therefore, is determined by the Quran, by precedents set by the Prophet, and by the practice of the early community of Islam as enshrined in the shari'ah.

No description, however, can fully capture the overwhelming importance of the Quran to Muslims. Objectively, it is the central fact of the Islamic faith, the Word of God, the final and complete revelation, the foundation and framework of Islamic law, and the source of Islamic thought, language, and action. It is the essence of Islam. Yet it is, in the deeply personal terms of a Muslim, something more as well. In innumerable, almost indescribable ways, it is also the central fact of Muslim life. To a degree almost incomprehensible in the West it shapes and colors broadly, specifically, and totally the thoughts, emotions, and values of the devout Muslim's life from birth to death.

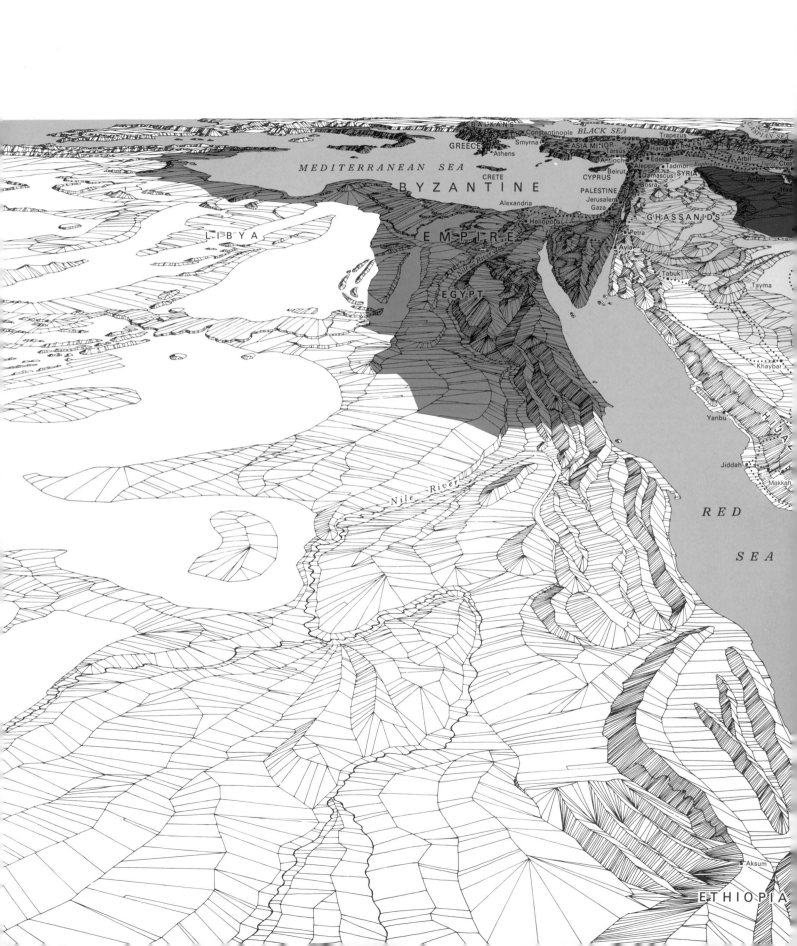

THE MIDDLE EAST AND THE MEDITERRANEAN IN THE SIXTH CENTURY A.D.

The Byzantine Empire, successor to the defunct Roman Empire to the west, reaches an accommodation with the Sassanian Empire, its archrival to the east, and recovers much of the territory of the Romans. The Ghassanids, a Christian Arab tribe in Syria, are clients of the Byzantine Empire while the pagan Lakhmid Dynasty, rulers of the Christian state of al-Hirah, are clients of Sassanid Persia. The Kingdom of Kindah encompasses central Arabia from the Nafud to the borders of Hadhramaut, while Makkah and other mercantile centers of the Hijaz in western Arabia remain independent. The Aksumite Empire of Ethiopia, which has acquired great prosperity from the maritime traffic through the Red Sea, dominates the languishing kingdoms of southern Arabia, which are further impoverished by the collapse of the great dam at Marib in 575. The strategic location of the Arabian Peninsula athwart the trade routes between eastern and western civilizations is the basis for its early development and continuing prominence.

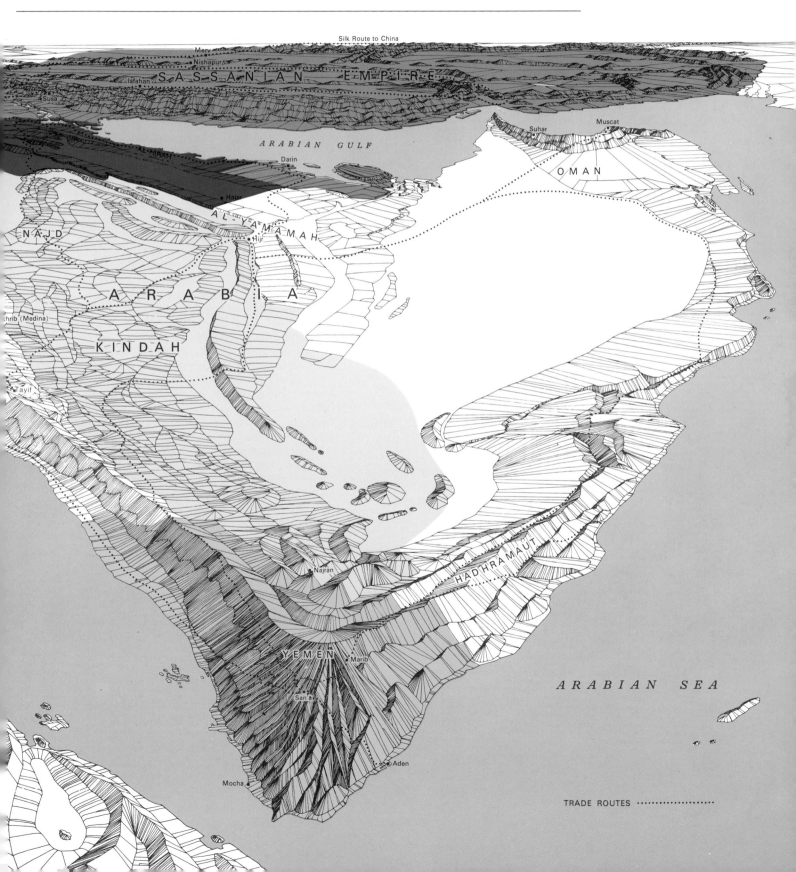

TRADE ROUTES ·····················

among the tribes — his early years with the Bedouins must have stood him in good stead here — that by 628 he and 1,500 followers were able to demand access to the Ka'bah during negotiations with the Makkans. This was a milestone in the history of the Muslims. Just a short time before, Muhammad had had to leave the city of his birth in fear of his life.

Now he was being treated by his former enemies as a leader in his own right. A year later, in 629, he reentered Makkah under terms agreed upon with the leaders of the city. However in 630 the Makkans breeched the agreement and Muhammad took the city without bloodshed and in a spirit of tolerance, establishing an ideal for future conquests. He also destroyed the idols in the Ka'bah, to put an end forever to pagan practices there. At the same time Muhammad won the allegiance of 'Amr ibn al-'As, the future conqueror of Egypt, and Khalid ibn al-Walid, the future "Sword of God," both of whom embraced Islam and joined Muhammad. Their conversion was especially noteworthy because these men had been among Muhammad's bitterest opponents only a short time before.

In one sense Muhammad's return to Makkah was the climax of his mission. In 632, just two years later, he was suddenly taken ill and on June 8 of that year, with his third wife 'Aishah in attendance, the Messenger of God "died with the heat of noon."

The death of Muhammad was a profound loss. To his followers this simple man from Makkah was far more than a beloved friend, far more than a gifted administrator, far more than the revered leader who had forged a new state from clusters of warring tribes. Muhammad was also the exemplar of the teachings he had brought them from God: the teachings of the Quran, which, for centuries, have guided the thought and action, the faith and conduct, of innumerable men and women, and which ushered in a distinctive era in the history of mankind. His death, nevertheless, had little effect on the dynamic society he had created in Arabia, and no effect at all on his central mission: to transmit the Quran to the world. As Abu Bakr put it: "Whoever worshipped Muhammad, let him know that Muhammad is dead, but whoever worshipped God, let him know that God lives and dies not."

The Ka'bah, spiritual axis of the Muslim world, stands in the courtyard of Makkah's Sacred Mosque.

Devout Muslims from all over the world gather for the pilgrimage to Makkah, for nearly 14 centuries one of the most impressive religious gatherings in the world.

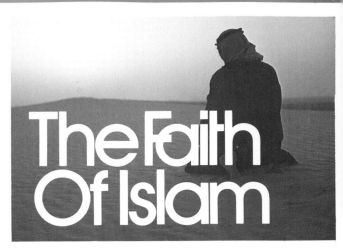

The Faith Of Islam

Islam, in Arabic, means "submission" — submission to the will of God. Faithful Muslims, therefore, submit unreservedly to God's will and obey His precepts as set forth in the Quran and transmitted to mankind by Muhammad, His Messenger.

Muslims believe that theirs is the only true faith. Islam, they say, was revealed through a long line of prophets inspired by God. Among them are Ibrahim (Abraham), patriarch of the Arabs through his first son Isma'il (Ishmael); Musa (Moses), who received the Torah (*Tawrah*); Dawud (David), who spoke through the Psalms (*Zabur*); and 'Isa (Jesus), who brought the Gospels (*Injil*). But the full and final revelation came through Muhammad, the last of all prophets, and was embodied in the Quran, which completes and supersedes all previous revelations.

As the chief source of Islamic doctrine and practice, the Quran is the main foundation of the *shari'ah*, the sacred law of Islam, which covers all aspects of the public and private, social and economic, and religious and political lives of all Muslims. In addition to the Quran the shari'ah has three sources: the *sunnah*, the practice of the Prophet; *ijma'*, the consensus of opinion; and *qiyas*, reasoning by analogy. The sunnah — which supplements and complements the Quran, the Word of God, and is next to it in importance — embodies the meticulously documented acts and sayings of the Prophet recorded in a body of writings called the *hadith*. Ijma' is the consensus of qualified jurists on matters not specifically referred to in the Quran or the sunnah. *Qiyas* is the application of human reasoning to extend the principles found in the two primary sources — the Quran and the sunnah — to cases involving matters unknown in the early years of Islam.

Systematized in the second and third centuries of the Muslim era (the eighth and ninth centuries A.D.), the shari'ah later developed into four major schools of jurisprudence: the Hanafi School, founded by Abu Hanifah; the Maliki School, founded by Malik ibn Anas; the Shafi'i School, founded by Muhammad al-Shaf'i; and the Hanbali School — the principal system of law applied in Saudi Arabia today — founded by Ahmad ibn Hanbal. Each of these men, all exceptional scholars, wrote or dictated long and learned commentaries upon which their schools of law were founded. Based on one or the other of these schools, learned officials called *qadis* administer the law in shari'ah courts. Despite the great body of tradition and law, however, the practice of Islam is essentially personal — a direct relationship between individuals and God. Although there are *imams,* who lead prayers and deliver sermons, there are no priests or ministers.

To practice their faith, Muslims must accept five primary obligations which Islam imposes. Called the Five Pillars of Islam, they are: the profession of faith (*shahadah*), devotional worship or prayer (*salah*), the religious tax (*zakah*), fasting (*sawm*), and the pilgrimage to Makkah (*hajj*).

The first pillar, the profession of faith, is the repetition of the statement, "There is no god but God; Muhammad is the Messenger of God" — in Arabic the euphonious "*La ilaha illa Allah; Muhammadun rasul Allah.*" It is a simple statement, yet also profound, for in it a Muslim expresses his complete acceptance of, and total commitment to, the message of Islam.

The second pillar, devotional worship or prayer, requires Muslims to pray five times a day — the dawn prayer, the noon prayer, the afternoon prayer, the sunset prayer, and the evening prayer — while facing toward the Ka'bah, the House of God, in Makkah. Like all Islamic ceremonies, prayer is simple and personal, yet also communal, and the wording of the prayers, the ablutions which are required before prayers, the number of bows, and other parts of the ritual are set out in detail.

The religious tax, the third pillar, is *zakah* in Arabic, a word that in the Prophet's lifetime came to suggest an obligatory religious tax. Like prayer, zakah is considered a form of worship. It enshrines the duty of social responsibility by which well-to-do Muslims must concern themselves about those less fortunate. The zakah prescribes payments of fixed proportions of a Muslim's possession for the welfare of the community in general and for its needy members in particular, whether Muslims or non-Muslims. This tax is often levied and disbursed by the state, but in the absence of a government collecting system it must be disbursed by the taxable Muslims themselves. In addition, all Muslims are encouraged to make voluntary contributions to the needy called *sadaqah.*

The fourth pillar is fasting during Ramadan, the ninth month of the Muslim year. Ordained in the Quran, the fast is an exacting act of deeply personal worship in which Muslims seek a richer perception of God and in which, as one writer puts it, Muslims assert that "man has larger needs than bread."

Ramadan begins with the sighting of the new moon, a moment usually signaled by cannon fire, after which abstention from eating and drinking, as well as physical continence, is obligatory every day between dawn and sunset. It is a rigorous fast, but its object is not mere abstinence and deprivation; it is, rather, the subjection of the passions and the purification of one's being so that the soul is brought nearer to God. Fasting is also an exercise in self-control and self-denial whereby one learns to appreciate the pangs of hunger that the poor often feel. The exercise of self-control extends far beyond refraining from food and drink; to make one's fast acceptable to God, one must also refrain from cursing, lying, cheating, and abusing or harming others.

Although rigorous, however, the fast, by Quranic injunction, also admits of a warm compassion. Those who are ill, or on an arduous journey, for example, may fast the

prescribed number of days at another time. Those for whom fasting is impossible may forego it; in compensation they should give stipulated alms to the needy.

The month of fasting is also joyous. In Muslim regions, in modern times, the faithful — at the sound of the sunset cannon or the call of the muezzin — break their fast, perform voluntary nocturnal worship (*tarawih*), and throng the streets in moods that are at once festive and, in the spirit of Ramadan, communal. For those who retire and rest after the day's fast there are, in some areas, men called *musahhirs* who, in the silent, predawn darkness beat muted drums and call the faithful to awake and eat before the long day's fast begins again.

The Ramadan cannon signals the time for sunset prayers and the breaking of the fast.

The last 10 days of Ramadan are particularly sacred because they include the anniversary of the night on which Muhammad received his first revelation from God — "the Night of Power" — and the appearance, on the final day, of the thin edge of the new moon announcing the end of Ramadan. At that moment the favor of God descends upon Muslims and, in a spirit of joyous achievement, they begin the three days of celebration called '*Id al-Fitr*, the Feast of the Breaking of the Fast. To cement social bonds further, Islam has instituted *zakat al-fitr*, an obligatory levy in the form of provisions or money for the poor, so that they can share in the joy of 'Id al-Fitr.

The fifth pillar of Islam is the pilgrimage to Makkah — the *hajj.* One of the most moving acts of faith in Islam, the hajj is, for those Muslims who can get to Makkah, the peak of their religious life, a moment when they satisfy a deep yearning to behold at least once the Ka'bah — the House of God and the physical focus of a lifetime of prayer. The hajj is at once a worldwide migration of the faithful and a remarkable spiritual happening that, according to Islamic tradition, dates back to Abraham, was affirmed by Muhammad, and then, by Muhammad's own pilgrimage, systematized into rites which are simple in execution but rich in meaning.

The hajj proper must be made between the eighth and thirteenth days of the twelfth month — Dhu al-Hijjah — of the Muslim year, but in one sense it begins when a Muslim

(Continued on page 52)

51

The Faith Of Islam

approaches Makkah, bathes, trims his nails and hair, discards jewelry and headgear, and puts on the *ihram* dress. This consists of two simple white seamless garments for men and a simple white dress and headcovering for women, symbolizing a state of purity; in donning it pilgrims make a declaration of pilgrimage and pronounce a devotional utterance called the *talbiyah:* "Here I am, O God, at Thy Service" — in Arabic the joyous cry *"Labbayk!"* After donning the ihram dress, the pilgrims may enter the haram, the sacred precinct surrounding Makkah, and then Makkah itself, where they perform the *tawaf* — the circling of the Ka'bah — and the *sa'y* — the running between two hills at al-Mas'a in Makkah. All this can be part of the *'umrah* or "lesser pilgrimage," often a prelude to the hajj but not an integral part of it. A distinction between the hajj and the 'umrah is that the 'umrah can be done at any time of the year, while the hajj must be performed on specified dates.

The major rites of the hajj begin on the eighth day of Dhu al-Hijjah when, with thunderous cries of *"Labbayk!"* the pilgrims pour out of Makkah to Mina, where, as the Prophet did, they meditate overnight. On the next day they proceed en masse to 'Arafat, even farther outside Makkah, and pray and meditate in what is the central rite of the pilgrimage: "the standing" — a few precious hours of profound self-examination, supplication, and penance in which, many say, a Muslim comes as close to God as he can on earth.

A pillar marks the Mount of Mercy, the rocky hill rising from the plain of 'Arafat.

Dressed in their simple *ihram* garments, all pilgrims are equal in the eyes of God.

At 'Arafat many actually do stand — from just after noon to sunset — but some also visit other pilgrims or the Mount of Mercy, where Muhammad delivered his farewell sermon. The standing is not the end of the hajj, but is the culmination of a Muslim's devotional life. As the Prophet said, "The best of prayers is the prayer of the Day of 'Arafat."

After sunset the pilgrims move to a place called Muzdalifah, where they gather stones for the "throwing of the pebbles" or "stoning of the pillars," and then pray and sleep. The third day of the pilgrimage, back at Mina, they enact a repudiation of evil by throwing the pebbles at a pillar held by many to represent Satan. According to one tradition it was in this area that Satan urged Abraham not to perform God's command to sacrifice his son Ishmael. At Mina too, begins *'Id al-Adha*, the great worldwide Feast of Sacrifice during which the pilgrims sacrifice animals — partly to commemorate Abraham's willingness to sacrifice his son and partly to symbolize a Muslim's willingness to sacrifice what is dearest to him . As Muslims throughout the world perform identical sacrifices on the same day, the Muslims at Mina in effect share their pilgrimage with Muslims everywhere.

As the pilgrims have now completed much of the hajj, Muslim men now clip their hair or shave their heads and women clip a symbolic lock to mark partial deconsecration. The pilgrims may also, at this point, remove the ihram dress.

In Makkah the rites are concluded by the tawaf of the return, the Circling of the Ka'bah seven times on foot, an act implying that all human activity must have God at the center. After the last circuit the pilgrims worship in the courtyard of the Mosque at the Place of Abraham, where the Patriarch himself offered prayer and, with Ishmael, stood while building the Ka'bah. The tawaf of the return is the last essential devotion of the pilgrimage; now the pilgrims have become *hajjis* — those who have completed the hajj. Most pilgrims also attempt to kiss, touch, or salute the *Hajar al-Aswad*, the Black Stone of the Ka'bah, a fragment of polished stone revered as a sign sent by God and a remnant of the original structure built by Abraham and Ishmael. Many also make the *sa'y* or running, a reenactment of a frantic search for water by Hagar when she and Abraham's son Ishmael were stranded in the valley of Makkah until the Angel Gabriel led them to water in the Well of Zamzam.

It is also customary for pilgrims to return to Mina between the eleventh and thirteenth days and cast their remaining pebbles at the three pillars there and then, in Makkah, make a farewell circling of the Ka'bah. Some may also visit the Mosque of the Prophet in Medina before returning to their homes throughout the world in the "sudden, glad stillness" of those who have stood at 'Arafat.

THE RIGHTLY GUIDED CALIPHS: With the death of Muhammad, the Muslim community was faced with the problem of succession. Who would be its leader? There were four persons obviously marked for leadership: Abu Bakr al-Siddiq, who had not only accompanied Muhammad to Medina 10 years before, but had been appointed to take the place of the Prophet as leader of public prayer during Muhammad's last illness; 'Umar ibn al-Khattab, an able and trusted Companion of the Prophet; 'Uthman ibn 'Affan, a respected early convert; and 'Ali ibn Abi Talib, Muhammad's cousin and son-in-law. To avoid contention among various groups, 'Umar suddenly grasped Abu Bakr's hand, the traditional sign of recognition of a new leader. Soon everyone concurred and before dusk Abu Bakr had been recognized as the *khalifah* of Muhammad. *Khalifah* — anglicized as caliph — is a word meaning "successor" but also suggesting what his historical role would be: to govern according to the Quran and the practice of the Prophet.

Abu Bakr's caliphate was short but important. An exemplary leader, he lived simply, assiduously fulfilled his religious obligations, and was accessible and sympathetic to his people. But he also stood firm when, in the wake of the Prophet's death, some tribes renounced Islam; in what was a major accomplishment, Abu Bakr swiftly disciplined them. Later, he consolidated the support of the tribes within the Arabian Peninsula and subsequently funneled their energies against the powerful empires of the East: the Sassanians in Persia and the Byzantines in Syria, Palestine, and Egypt. In short, he demonstrated the viability of the Muslim state.

The second caliph, 'Umar — appointed by Abu Bakr in a written testament — continued to demonstrate that viability. Adopting the title *Amir al-Muminin*, "Commander of the Believers," 'Umar extended Islam's temporal rule over Syria, Egypt, Iraq, and Persia in what from a purely military standpoint were astonishing victories. Within four years after the death of the Prophet the Muslim state had extended its sway over all of Syria and had, at a famous battle fought during a sandstorm near the River Yarmuk, blunted the power of the Byzantines — whose ruler Heraclius had shortly before disdainfully rejected the letter from the unknown Prophet of Arabia.

Even more astonishingly, the Muslim state administered the conquered territories with a tolerance almost unheard of in that age. At Damascus, for example, the Muslim leader Khalid ibn al-Walid signed a treaty which read as follows:

> This is what Khalid ibn al-Walid would grant to the inhabitants of Damascus if he enters therein: he promises to give them security for their lives, property and churches. Their city wall shall not be demolished, neither shall any Muslim be quartered in their houses. Thereunto we give them the pact of Allah and the protection of His Prophet, the caliphs and the believers. So long as they pay the poll tax, nothing but good shall befall them.

This tolerance was typical of Islam. A year after Yarmuk, 'Umar, in the military camp of al-Jabiyah on the Golan Heights, received word that the Byzantines were ready to surrender Jerusalem and rode there to accept the surrender in person. According to one account, he entered the city alone and clad in a simple cloak, astounding a populace accustomed to the sumptuous garb and court ceremonials of the Byzantines and Persians. He astounded them still further when he set their fears at rest by negotiating a generous treaty in which he told them:

> In the name of God ... you have complete security for your churches which shall not be occupied by the Muslims or destroyed.

This policy was to prove successful everywhere. In Syria, for example, many Christians who had been involved in bitter theological disputes with Byzantine authorities — and persecuted for it — welcomed the coming of Islam as an end to tyranny. And in Egypt, which 'Amr ibn al-'As took from the Byzantines after a daring march across the Sinai Peninsula, the Coptic Christians not only welcomed the Arabs, but enthusiastically assisted them.

This pattern was repeated throughout the Byzantine Empire. Conflict among Greek Orthodox, Syrian Monophysites, Copts, and Nestorian Christians contributed to the failure of the Byzantines — always regarded as intruders — to develop popular support, while the tolerance which Muslims showed toward

To many people in the early years of Islam, this Byzantine noble warrior was a symbol of tyranny.

Symbol of the oneness and centrality of God, the Ka'bah stands in the courtyard of Makkah's Sacred Mosque where at the season of the hajj the faithful gather for rituals that precede and end their pilgrimage. Facing page: A Pakistani youth in the pilgrim's simple robe reads his Quran; above him are pilgrims at the climax of their hajj, "standing" before God at 'Arafat near the spot where Muhammad delivered his farewell sermon. Below: Crowds at the small town of Mina cast pebbles at pillars that symbolize evil. This page, top: Hajjis spend one night camped at Muzdalifah between 'Arafat and Mina. Above: In the ceremony of sa'y they reenact the search for water by Hagar, wife of the patriarch Abraham.

Christians and Jews removed the primary cause for opposing them.

'Umar adopted this attitude in administrative matters as well. Although he assigned Muslim governors to the new provinces, existing Byzantine and Persian administrations were retained wherever possible. For 50 years, in fact, Greek remained the chancery language of Syria, Egypt, and Palestine, while Pahlavi, the chancery language of the Sassanians, continued to be used in Mesopotamia and Persia.

'Umar, who served as caliph for 10 years, ended his rule with a significant victory over the Persian Empire. The struggle with the Sassanid realm had opened in 687 at al-Qadisiyah, near Ctesiphon in Iraq, where Muslim cavalry had successfully coped with elephants used by the Persians as a kind of primitive tank. Now with the Battle of Nihavand, called the "Conquest of Conquests," 'Umar sealed the fate of Persia; henceforth it was to be one of the most important provinces in the Muslim Empire.

His caliphate was a high point in early Islamic history. He was noted for his justice, social ideals, administration, and statesmanship. His innovations left an enduring imprint on social welfare, taxation, and the financial and administrative fabric of the growing empire.

After the death of 'Umar an advisory council composed of Companions of the Prophet selected as the third caliph 'Uthman, during whose rule the first serious strains on Islamic unity would appear. 'Uthman achieved much during his reign. He pushed forward with the pacification of Persia, continued to defend the Muslim state against the Byzantines, added what is now Libya to the empire, and subjugated most of Armenia. 'Uthman also, through his cousin Mu'awiyah ibn Abi Sufyan, the governor of Syria, established an Arab navy which fought a series of important engagements with the Byzantines.

Of much greater importance to Islam, however, was 'Uthman's compilation of the text of the Quran as revealed to the Prophet. Realizing that the original message from God might be inadvertently distorted by textual variants, he appointed a committee to collect the canonical verses and destroy the variant recensions. The result was the text that is accepted to this day throughout the Muslim world.

These successes, however, were qualified by serious administrative weaknesses. 'Uthman was accused of favoritism to members of his family — the clan of Umayyah. Negotiations over such grievances were opened by representatives from Egypt but soon collapsed and 'Uthman was killed — an act that caused a rift in the community of Islam that has never entirely been closed.

This rift widened almost as soon as 'Ali, cousin and son-in-law of the Prophet, was chosen to be the fourth caliph. At issue, essentially, was the legitimacy of 'Ali's caliphate. 'Uthman's relatives — in particular Mu'awiyah, the powerful governor of Syria, where 'Ali's election had not been recognized — believed 'Ali's caliphate was invalid because his election had been supported by those responsible for 'Uthman's unavenged death. The conflict came to a climax in 657 at Siffin, near the Euphrates, and eventually resulted in a major division of Muslims between Sunnis or Sunnites and Shi'is (also called Shi'ites or Shi'ah), the "Partisans" of 'Ali — a division that was to color the subsequent history of Islam.

Actually the Sunnis and the Shi'is were agreed upon almost all the essentials of Islam. Both believed in the Quran and the Prophet, both followed the same principles of religion, and both observed the same rituals. However there was one prominent difference, which was essentially political rather than religious, and concerned the choice of the caliph or successor of Muhammad.

The majority of Muslims support the elective principle which led to the choice of Abu Bakr as the first caliph. This group is known as *ahl al-sunnah wa-l-jama'ah*, "the people of the Prophet's teachings and community," or Sunnis, who consider the caliph to be Muhammad's successor only in his capacity as ruler of the community. The main body of the Shi'is, on the other hand, believes that the caliphate — which they call the imamate or "leadership" — is nonelective. The caliphate, they say, must remain within the family of the Prophet — with 'Ali the first valid caliph. And while Sunnis consider the caliph a guardian of the *shari'ah*, the religious law, the Shi'is see the imam as a trustee inheriting and interpreting the Prophet's spiritual knowledge.

After the battle of Siffin, 'Ali — whose chief strength was in Iraq, with his capital at Kufa — began to lose the support of many of

A great horned headdress was the Sassanid symbol of royalty.

This eighth-century manuscript from Makkah or Medina is one of the two oldest known existing copies of the Quran.

his more uncompromising followers and in 661 he was murdered by a former supporter. His son Hasan was proclaimed caliph at Kufa but soon afterward deferred to Mu'awiyah, who had already been proclaimed caliph in Jerusalem in the previous year and who now was recognized and accepted as caliph in all the Muslim territories — thus inaugurating the Umayyad dynasty which would rule for the next 90 years.

The division between the Sunnis and the Shi'is continued to develop, however, and was widened in 680 when 'Ali's son Husayn tried to win the caliphate from the Umayyads and, with his followers, was killed at Karbala in Iraq. His death is still mourned each year by the Shi'is.

THE UMAYYADS:

The shift in power to Damascus, the Umayyad capital city, was to have profound effects on the development of Islamic history. For one thing, it was a tacit recognition of the end of an era. The first four caliphs had been without exception Companions of the Prophet — pious, sincere men who had lived no differently from their neighbors and who preserved the simple habits of their ancestors despite the massive influx of wealth from the conquered territories. Even 'Uthman, whose policies had such a divisive effect, was essentially dedicated more to the concerns of the next world than of this. With the shift to Damascus much was changed.

In the early days of Islam, the extension of Islamic rule had been based on an uncomplicated desire to spread the Word of God. Although the Muslims used force when they met resistance they did not compel their enemies to accept Islam. On the contrary, the Muslims permitted Christians and Jews to practice their own faith and numerous conversions to Islam were the result of exposure to a faith that was simple and inspiring.

With the advent of the Umayyads, however, secular concerns and the problems inherent in the administration of what, by then, was a large empire began to dominate the attention of the caliphs, often at the expense of religious concerns — a development that disturbed many devout Muslims. This is not to say that religious values were ignored; on the contrary, they grew in strength for centuries. But they were not always at the forefront and from the time of Mu'awiyah the caliph's role as "Defender of the Faith" increasingly required him to devote attention to the purely secular concerns which dominate so much of every nation's history.

Mu'awiyah was an able administrator, and even his critics concede that he possessed to a high degree the much-valued quality of *hilm* — a quality which may be defined as "civilized restraint" and which he himself once described in these words:

> I apply not my sword where my lash suffices, nor my lash where my tongue is enough. And even if there be one hair binding me to my fellowmen, I do not let it break: when they pull I loosen, and if they loosen I pull.

Nevertheless, Mu'awiyah was never able to reconcile the opposition to his rule nor resolve the conflict with the Shi'is. These problems were not unmanageable while Mu'awiyah was alive, but after he died in 680 the partisans of 'Ali resumed a complicated but persistent struggle that plagued the Umayyads at home for most of the next 70 years and in time spread into North Africa and Spain.

Medieval Muslims regarded the Great Mosque built by the Umayyads in Damascus as one of the wonders of the world.

The Umayyads, however, did manage to achieve a degree of stability, particularly after 'Abd al-Malik ibn Marwan succeeded to the caliphate in 685. Like the Umayyads who preceded him, 'Abd al-Malik was forced to devote a substantial part of his reign to political problems. But he also introduced much-needed reforms. He directed the cleaning and reopening of the canals that irrigated the Tigris-Euphrates Valley — a key to the prosperity of Mesopotamia since the time of the Sumerians — introduced the use of the Indian water buffalo in the riverine marshes, and minted a standard coinage which replaced the Byzantine and Sassanid coins, until then the sole currencies in circulation. 'Abd al-Malik's organization of government agencies was also important; it established a model for the later elaborate bureaucracies of the 'Abbasids and their successor states. There were specific agencies charged with keeping pay records; others concerned themselves with the collection of taxes. 'Abd al-Malik established a system of postal routes to expedite his communications throughout the far-flung empire. Most important of all, he

Facing al-Gharbiyah, the western minaret, a muezzin at the Umayyad Mosque calls believers to prayer.

KINGDOM OF THE FRANKS

ANDALUSIA
Tours
Toulouse
KINGDOM OF THE LOMBARDS
Constantinople BLACK SEA
Dabil
CASPIAN SEA

Malaga Cordoba Toledo
Granada
Tunis
Sinope
Tabriz
Mosul

Tangier
Kairouan
BYZANTINE EMPIRE
Antioch Aleppo Rakkah
Tigris River
Bat

MAGHRIB
Euphrates River

MEDITERRANEAN SEA
CYPRUS Tripoli Hama Homs SYRIA
Beirut Damascus
Karb

Tripoli
Barqa
PALESTINE
Jerusalem

L I B Y A
Alexandria
Gaza

Fustat
SINAI

E G Y P T
Tabuk

FEZZAN
al-Qusayr

Kha

Medir

Aswan
Yanbu

Nile River
RED

Jiddah
Makkah

SEA

Nile River

Aksum

E T H I O P I A

THE WORLD OF ISLAM IN THE EIGHTH CENTURY A.D.

The map has been completely transformed in the brief span of a hundred years by a new force in world affairs, the Arabs. Inspired by the newly revealed religion of Islam, they have crushed the Byzantines and other successors of the Roman Empire as they push west across North Africa and into Spain and have defeated the Sassanid Persians as they sweep eastward to the borders of China. Religious and administrative leadership is provided by the caliphate at Damascus under the Umayyads and at Baghdad under the 'Abbasids, who in 750 displace the Umayyad Dynasty (a branch of which continues to flourish for another 300 years in Spain). At a time when urban life and culture have virtually disappeared in Europe, society prospers and civilization flowers in the lands under the administration and protection of the Muslim caliphate.

introduced Arabic as the language of administration, replacing Greek and Pahlavi.

Under 'Abd al-Malik, the Umayyads expanded Islamic power still further. To the east they extended their influence into Transoxiana, an area north of the Oxus River in the former Soviet Union, and went on to reach the borders of China. To the west, they took North Africa in a continuation of the campaign led by 'Uqbah ibn Nafi' who founded the city of Kairouan — in what is now Tunisia — and from there rode all the way to the shores of the Atlantic Ocean.

These territorial acquisitions brought the Arabs into contact with previously unknown ethnic groups who embraced Islam and would later influence the course of Islamic history. The Berbers of North Africa, for example, who resisted Arab rule but willingly embraced Islam, later joined Musa ibn Nusayr and his general, Tariq ibn Ziyad, when they crossed the Strait of Gibraltar to Spain. The Berbers later also launched reform movements in North Africa which greatly influenced the Islamic civilization. In the East, Umayyad rule in Transoxiana brought the Arabs into contact with the Turks who, like the Berbers, embraced Islam and, in the course of time, became its staunch defenders. Umayyad expansion also reached the ancient civilization of India, whose literature and science greatly enriched Islamic culture.

In Europe, meanwhile, the Arabs had passed into Spain, defeated the Visigoths, and by 713 had reached Narbonne in France. In the next decades raiding parties continually made forays into France and in 732 reached as far as the Loire Valley, only 275 kilometers from Paris. There, at the Battle of Tours, or Poitiers, the Arabs were finally turned back by Charles Martel.

One of the Umayyad caliphs who attained greatness was 'Umar ibn 'Abd al-'Aziz, a man very different from his predecessors. Although a member of the Umayyad family, 'Umar had been born and raised in Medina, where his early contact with devout men had given him a concern for spiritual as well as political values. The criticisms that religious men in Medina and elsewhere had voiced of Umayyad policy — particularly the pursuit of worldly goals — were not lost on 'Umar who, reversing the policy of his predecessors, discontinued the levy of a poll tax on converts.

The minaret of the Great Mosque at Kairouan in Tunisia became the prototype for the majority of North African minarets.

Charles Martel checked Muslim expansion into Frankish territory.

Arabic Writing

The intricate pattern of the Quranic "Verse of the Throne" (see page 47), written in angular kufic characters in the form of a square, seems almost like a maze.

Another maze-like ornamental square of angular kufic is made up of the names of the 10 most revered Companions of Muhammad.

The Arabs gave to a large part of the world not only a religion — Islam — but also a language and an alphabet. Where the Muslim religion went, the Arabic language and Arabic writing also went. Arabic became and has remained the national language — the mother tongue — of North Africa and all the Arab countries of the Middle East.

Even where Arabic did not become the national language, it became the language of religion wherever Islam became established, since the Quran is written in Arabic, the Profession of Faith is to be spoken in Arabic, and five times daily the practicing Muslim must say his prayers in Arabic. Today, therefore, one can hear Arabic spoken — at least for religious purposes — from Mauritania on the Atlantic, across Africa and most of Asia, and as far east as Indonesia and the Philippines. Even in China (which has a Muslim population of over 40 million) and the Central Asian states of the former Soviet Union, Arabic can be heard in the *shahadah,* in prayer, and in the chanting of the Quran.

Of those people who embraced Islam but did not adopt Arabic as their everyday language, many millions have taken the Arabic alphabet for their own, so that today one sees the Arabic script used to write languages that have no basic etymological connection with Arabic. The languages of Iran, Afghanistan, and Pakistan are all written in the Arabic alphabet, as was the language of Turkey until some 50 years ago. It is also used in Kashmir and in some places in the Malay Peninsula and the East Indies, and in Africa it is used in Somalia and down the east coast as far south as Tanzania.

The *basmalah* ("In the name of God, the Merciful, the Compassionate" — the opening words of the Quran) is here done in an elaborate thuluth script with the letters joined so that the entire phrase is written without lifting the pen from the paper.

It is generally accepted that the Arabic alphabet developed from the script used for Nabataean, a dialect of Aramaic used in northern Arabia and what is now Jordan during roughly the thousand years before the start of the Islamic era. It seems apparent that Syriac also had some influence on its development. The earliest inscription that has been found that is identifiably Arabic is one in Sinai that dates from about A.D. 300. Another Semitic script which was in use at about the same time and which is found on inscriptions in southern Arabia is the origin of the alphabet now used for Amharic, the official language of Ethiopia.

The Arabic alphabet has 28 letters (additional letters have been added to serve the needs of non-Arabic languages that use the Arabic script, such as those of Iran and Pakistan), and each of the letters may have up to four different forms. All of the letters are strictly speaking consonants, and, unlike the Latin alphabet used for English and most European languages, Arabic writing goes from right to left.

Another significant difference is that the Arabic script has been used much more extensively for decoration and as a means of artistic expression. This is not to say that the Latin alphabet (and others such as the Chinese and Japanese, for instance) are not just as decorative and have not been used just as imaginatively. Since the invention of printing from type, however, calligraphy (which means literally "beautiful writing") has come to be used in English and the other European languages only for special documents and on special occasions and has declined to the status of a relatively minor art.

Another basmalah in ornamental thuluth script is written in the shape of an oval.

In the countries that use the Arabic alphabet, on the other hand, calligraphy has continued to be used not only on important documents but for a variety of other artistic purposes as well. One reason is that the cursive nature of the Arabic script and certain of its other peculiarities made its adaptation to printing difficult and delayed the introduction of the printing press, so that the Arab world continued for some centuries after the time of Gutenberg to rely on handwriting for the production of books (especially the Quran) and of legal and other documents. Metal typesetting and typewriters utilizing the Arabic alphabet were introduced in the eighteenth and nineteenth centuries, however, and the recent development of computers and photocomposers has allowed a return to the beauty and simplicity of Arabic writing.

Another reason for artistic calligraphy was a religious one. The Quran nowhere prohibits the representation of humans or animals in drawings or paintings, but as Islam expanded in its early years it inherited some of the prejudices against visual art of this kind that had already taken root in the Middle East. In addition, the early Muslims tended to oppose figural art (and in some cases all art) as distracting the community from the worship of God and hostile to the strictly unitarian religion preached by Muhammad, and all four of the schools of Islamic law banned the use of images and declared that the painter of animate figures would be damned on the Day of Judgment.

Wherever artistic ornamentation and decoration were required, therefore, Muslim artists, forbidden to depict human or animal forms, for the most part were forced to resort either to what has since come to be known as "arabesque" (designs based on strictly geometrical forms or patterns of leaves and flowers) or, very often, to calligraphy. Arabic calligraphy therefore came to be used not only in producing copies of the Quran (its first and for many centuries its most important use), but also for all kinds of other artistic purposes as well: on porcelain and metalware, for carpets and other textiles, on coins, and as architectural ornament (primarily on mosques and tombs but also, especially in later years, on other buildings as well).

(Continued on page 62)

Arabic Writing

At the start of the Islamic era two types of script seem to have been in use — both derived from different forms of the Nabataean alphabet. One was square and angular and was called *kufic* (after the town of Kufa in Iraq, though it was in use well before the town was founded). It was used for the first, handwritten copies of the Quran, and for architectural decoration in the earliest years of the Islamic Empire. The other, called *naskhi,* was more rounded and cursive and was used for letters, business documents, and wherever speed rather than elaborate formalism was needed. By the twelfth century kufic was obsolete as a working script except for special uses and in northwest Africa, where it developed into the *maghribi* style of writing still used in the area today. Naskhi, the rounded script, remained in use and from it most of the many later styles of Arabic calligraphy have been developed.

The basmalah is here written in ornamental "floriated" kufic (above) and in naskhi (below).

Calligraphy flourished during the Umayyad era in Damascus. During this period scribes began the modification of the original thick and heavy kufic script into the form employed today for decorative purposes, as well as developing a number of new scripts derived from the more cursive naskhi. It was under the 'Abbasids, however, that calligraphy first began to be systematized. In the first half of the tenth century the 'Abbasid vizier Ibn Muqlah completed the development of kufic, established some of the rules of shape and proportion that have been followed by calligraphers since his time, and was first to develop what became the traditional classification of Arabic writing into the "six styles" of cursive script: naskhi (from which most present day printing types are derived), *thuluth* (a more cursive outgrowth of naskhi), *rayhani* (a more ornate version of thuluth), *muhaqqaq* (a bold script with sweeping diagonal flourishes), *tawqi'* (a somewhat compressed variety of thuluth in which all the letters are sometimes joined to each other), and *ruq'ah* (the style commonly used today for ordinary handwriting in most of the Arab world). It was from these six, and from kufic, that later calligraphers, not only in the Arab world but in Iran, Turkey, and elsewhere as well, developed and elaborated other scripts.

The ruq'ah script is used for headlines and titles and is the everyday written script of most of the Arab world.

In Iran, for example, there came into use a particularly graceful and delicate script called *ta'liq*, in which the horizontal strokes of the letters are elongated and which is often written at an angle across the page. From ta'liq, in turn, another script called *nasta'liq* was derived which combines the Arabic naskhi and the Persian ta'liq into a beautifully light and legible script.

The graceful Persian ta'liq script is used in a sentence which starts "Beauty is a spell which casts its splendor upon the universe..."

It was in Ottoman Turkey, however, that calligraphy attained the highest development once the early creative flowering had faded elsewhere in the Middle East. So renowned were Ottoman calligraphers, in fact, that a popular saying was that "The Quran was revealed in Makkah, recited in Egypt, and written in Istanbul." The Ottomans were not content merely to improve and develop the types of script that they inherited from the Arabs and Persians but also added a number of new styles to the calligrapher's repertoire.

The diwani script (top) and the so-called "royal" diwani (below) were developed by Ottoman calligraphers for use on state documents.

One important addition by the Ottoman calligraphers was the script called *diwani,* so called from the word *diwan* (meaning state council or government office) since it was at first used primarily for documents issued by the Ottoman Council of State. It is an extremely graceful and very decorative script, with strong diagonal flourishes, though less easy to read than some other styles. After its development in Turkey, it spread to the Arab countries and is in use today for formal documents and also as architectural decoration.

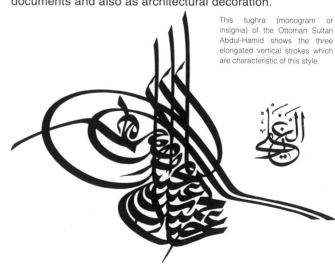

This tughra (monogram or insignia) of the Ottoman Sultan Abdul-Hamid shows the three elongated vertical strokes which are characteristic of this style.

Examples of more or less standard types of script such as these do not by any means exhaust the number of styles. Islamic calligraphers have experimented endlessly and have been extremely imaginative. Another distinctive Turkish contribution is the *tughra,* an elaborate and highly stylized rendering of the names of the Ottoman sultan, originally used to authenticate imperial decrees. The tughra later came to be used both in Turkey and by rulers of the Arab countries as a kind of royal insignia or emblem, on coins and stamps and wherever a coat of arms or royal monogram would be used by European governments.

In the muthanna or "doubled" style of calligraphy shown on the left each half of the design is a mirror image of the other. The basmalah in the thuluth script on the right has been written in the shape of an ostrich.

Another unusual variation of calligraphy, not often used nowadays, is the style called *muthanna* (Arabic for "doubled"). This is not really a type of script in itself but consists of a text in one of the standard scripts such as naskhi worked into a pattern in which one half is a mirror image of the other. Even more imaginative is what may be called pictorial calligraphy, in which the text (usually the profession of faith, a verse from the Quran, or some other phrase with religious significance) is written in the shape of a bird, animal, tree, boat, or other object. A Quranic verse in the kufic script, for example, may be written so that it forms the picture of a mosque and minarets.

The angular kufic script is here used to put a well-known religious expression into the form of a mosque with four minarets.

The art of calligraphy is still very much alive in the Arab world and wherever the Arabic alphabet is used. The list of everyday uses is almost endless: coins and paper money bear the work of expert calligraphers, wall posters and advertising signs in every town show the calligrapher's art, as do the cover and title page of every book, and the major headlines in every newspaper and magazine have been written by hand. Calligraphy — the art of "beautiful writing" — continues to be something that is not only highly prized as ornament and decoration but is immensely practical and useful as well.

This move reduced state income substantially, but as there was clear precedent in the practice of the great 'Umar ibn al-Khattab, the second caliph, and as 'Umar ibn 'Abd al-'Aziz was determined to bring government policy more in line with the practice of the Prophet, even enemies of his regime had nothing but praise for this pious man.

The last great Umayyad caliph was Hisham, the fourth son of 'Abd al-Malik to succeed to the caliphate. His reign was long — from 724 to 743 — and during it the Arab empire reached its greatest extent. But neither he nor the four caliphs who succeeded him were the statesmen the times demanded when, in 747, revolutionaries in Khorasan unfurled the black flag of rebellion that would bring the Umayyad Dynasty to an end.

Although the Umayyads favored their own region of Syria, their rule was not without accomplishments. Some of the most beautiful existing buildings in the Muslim world were constructed at their instigation — buildings such as the Umayyad Mosque in Damascus, the Dome of the Rock in Jerusalem, and the lovely country palaces in the deserts of Syria, Jordan, and Iraq. They also organized a bureaucracy able to cope with the complex problems of a vast and diverse empire, and made Arabic the language of government. The Umayyads, furthermore, encouraged such writers as 'Abd Allah ibn al-Muqaffa' and 'Abd al-Hamid ibn Yahya al-Katib, whose clear, expository Arabic prose has rarely been surpassed.

For all that, the Umayyads, during the 90 years of their leadership, rarely shook off their empire's reputation as a *mulk* — that is, a worldly kingdom — and in the last years of the dynasty their opponents formed a secret organization devoted to pressing the claims to the caliphate put forward by a descendant of al-'Abbas ibn 'Abd al-Muttalib, an uncle of the Prophet. By skillful preparation, this organization rallied to its cause many mutually hostile groups in Khorasan and Iraq and proclaimed Abu al-'Abbas caliph. Marwan ibn Muhammad, the last Umayyad caliph, was defeated and the Syrians, still loyal to the Umayyads, were put to rout. Only one man of importance escaped the disaster — 'Abd al-Rahman ibn Mu'awiyah al-Dakhil, a young prince who with a loyal servant fled to Spain and in 756 set up an Umayyad Dynasty there.

The shrine of the Dome of the Rock in Jerusalem, built in an area revered by Muslims, Christians, and Jews alike, covers the rock from which Muhammad is believed to have ascended to heaven with the Angel Gabriel.

Preeminent among artists of the Muslim world is the calligrapher, as it is his privilege to adorn the word of God. Opposite page: Ornamental kufic is used on a Quranic page that typifies the marriage of calligraphy and illumination, an art that reached its zenith in the fourteenth century. Cursive scripts on this page, clockwise from left: Fourteenth-century manuscript of a pharmaceutical text translated from Greek. Quran stand of wood, dated 1360, a fine example of Mongol art from western Turkestan. Section of the gold-embroidered *kiswah*, the black cloth covering of the Ka'bah, which is renewed each year at the time of the pilgrimage. Detail from a fourteenth-century Persian tile. Miniature from the *Shah-Nameh* (Book of Kings), illustrating the epic by the Persian poet Firdausi.

ISLAM IN SPAIN: By the time 'Abd al-Rahman reached Spain, the Arabs from North Africa were already entrenched on the Iberian Peninsula and had begun to write one of the most glorious chapters in Islamic history.

After their forays into France were blunted by Charles Martel, the Muslims in Spain had begun to focus their whole attention on what they called al-Andalus, southern Spain (Andalusia), and to build there a civilization far superior to anything Spain had ever known. Reigning with wisdom and justice, they treated Christians and Jews with tolerance, with the result that many embraced Islam. They also improved trade and agriculture, patronized the arts, made valuable contributions to science, and established Cordoba as the most sophisticated city in Europe.

A forest of 850 pillars connected by Moorish arches lines the great mosque of Cordoba.

By the tenth century, Cordoba could boast of a population of some 500,000, compared to about 38,000 in Paris. According to the chronicles of the day, the city had 700 mosques, some 60,000 palaces, and 70 libraries — one reportedly housing 500,000 manuscripts and employing a staff of researchers, illuminators, and bookbinders. Cordoba also had some 900 public baths, Europe's first streetlights and, eight kilometers outside the city, the caliphal residence, Madinat al-Zahra. A complex of marble, stucco, ivory, and onyx, Madinat al-Zahra took 40 years to build, cost close to one-third of Cordoba's revenue, and was, until destroyed in the eleventh century, one of the wonders of the age. Its restoration, begun in the early years of this century, is still under way.

By the eleventh century, however, a small pocket of Christian resistance had begun to grow, and under Alfonso VI Christian forces retook Toledo. It was the beginning of the period the Christians called the Reconquest, and it underlined a serious problem that marred this refined, graceful, and charming era: the inability of the numerous rulers of Islamic Spain to maintain their unity. This so weakened them that when the various Christian kingdoms began to pose a serious threat, the Muslim rulers in Spain had to ask the Almoravids, a North African Berber dynasty, to come to their aid. The Almoravids came and crushed the Christian uprising, but eventually seized control themselves. In 1147, the Almoravids were in turn defeated by another coalition of Berber tribes, the Almohads.

A pool in the Patio de los Arrayanes reflects the grandeur of the incomparable Alhambra.

Although such internal conflict was by no means uncommon — the Christian kingdoms also warred incessantly among themselves — it did divert Muslim strength at a time when the Christians were beginning to negotiate strong alliances, form powerful armies, and launch the campaigns that would later bring an end to Arab rule.

The Arabs did not surrender easily; al-Andalus was their land too. But, bit by bit, they had to retreat, first from northern Spain, then from central Spain. By the thirteenth century their once extensive domains were reduced to a few scattered kingdoms deep in the mountains of Andalusia — where, for some 200 years longer, they would not only survive but flourish.

It is both odd and poignant that it was then, in the last two centuries of their rule, that the Arabs created that extravagantly lovely kingdom for which they are most famous: Granada. It seems as if, in their slow retreat to the south, they suddenly realized that they were, as Washington Irving wrote, a people without a country, and set about building a memorial: the Alhambra, the citadel above Granada that one writer has called "the glory and the wonder of the civilized world."

The Alhambra was begun in 1238 by Muhammad ibn al-Ahmar who, to buy safety for his people when King Ferdinand of Aragon laid siege to Granada, once rode to Ferdinand's tent and humbly offered to become the king's vassal in return for peace.

It was a necessary move, but also difficult — particularly when Ferdinand called on him to implement the agreement by providing troops to help the Christians against Muslims in the siege of Seville in 1248. True to his pledge, Ibn al-Ahmar complied and Seville fell to the Christians. But returning to Granada, where cheering crowds hailed him as a victor, he disclosed his turmoil in that short, sad reply that he inscribed over and over on the walls of the Alhambra: "There is no victor but God."

Over the years, what started as a fortress slowly evolved under Ibn al-Ahmar's successors into a remarkable series of delicately lovely buildings, quiet courtyards, limpid pools, and hidden gardens. Later, after Ibn al-Ahmar's death, Granada itself was rebuilt and became, as one Arab visitor wrote, "as a silver vase filled with emeralds."

Science And Scholarship In Al-Andalus

For Europe and Western civilization the contributions of Islamic Spain were of inestimable value. When the Muslims entered southern Spain — which they called al-Andalus — barbarians from the north had overrun much of Europe and the classical civilization of Greece and Rome had gone into eclipse. Islamic Spain then became a bridge by which the scientific, technological, and philosophical legacy of the 'Abbasid period, along with the achievements of al-Andalus itself, passed into Europe.

In the first century of Islamic rule in Spain the culture was largely derived from that of the flourishing civilization being developed by the 'Abbasids in Baghdad. But then, during the reign of 'Abd al-Rahman III (912-961), Islamic Spain began to make its own contributions.

'Abd al-Rahman III was passionately interested in both the religious and the secular sciences. He was also determined to show the world that his court at Cordoba equaled in greatness that of the caliphs at Baghdad. Sparing neither time nor expense, he imported books from Baghdad and actively recruited scholars by offering handsome inducements. Soon, as a result, scholars, poets, philosophers, historians, and musicians began to migrate to al-Andalus. Soon, too, an infrastructure of libraries, hospitals, research institutions, and centers of Islamic studies grew up, establishing the intellectual tradition and educational system which made Spain outstanding for the next 400 years.

One of the earliest of the scholars drawn to al-Andalus was 'Abbas ibn Firnas, who came to Cordoba to teach music (then a branch of mathematical theory) and to acquaint the court of 'Abd al-Rahman with the recent developments in this field in Baghdad. Not a man to limit himself to a single field of study, however, Ibn Firnas soon began to investigate the mechanics of flight. He constructed a pair of wings out of feathers on a wooden frame and made the first attempt at flight, anticipating Leonardo da Vinci by some 600 years. Later, having survived the experiment with a back injury, he also constructed a famous planetarium. Not only was it mechanized — the planets actually revolved — but it simulated such celestial phenomena as thunder and lightning.

As in the 'Abbasid centers of learning, Islamic Spain's interest in mathematics, astronomy, and medicine was always lively — partly because of their obvious utility. In the tenth century Cordoban mathematicians began to make their own original contributions. The first original mathematician and astronomer of al-Andalus was Maslamah al-Majriti, who died in 1008. He had been preceded by competent scientists — men like Ibn Abi 'Ubaydah of Valencia, a leading astronomer in the ninth century. But al-Majriti was in a class by himself. He wrote a number of works on mathematics and astronomy, studied and elaborated the Arabic translation of Ptolemy's *Almagest*, and enlarged and corrected the astronomical tables of the famous al-Khwarazmi. He also compiled conversion tables in which the dates of the Persian calendar were related to Hijrah dates, so that for the first time the events of Persia's past could be dated with precision.

Al-Zarqali, known to the West as Arzachel, was another leading mathematician and astronomer who flourished in Cordoba in the eleventh century. Combining theoretical knowledge with technical skill, he excelled at the construction of precision instruments for astronomical use and built a water clock capable of determining the hours of the day and night and indicating the days of the lunar months. He also contributed to the famous Toledan Tables, a highly accurate compilation of astronomical data. Arzachel was famous as well for his *Book of Tables*. Many "books of tables" had been compiled before then, but his is an almanac containing tables which allow one to find the days on which Coptic, Roman, lunar, and Persian months begin, other tables which give the position of planets at any given time, and still others facilitating the prediction of solar and lunar eclipses. He also compiled valuable tables of latitude and longitude.

Another important scholar was al-Bitruji, who developed a new theory of stellar movement, based on Aristotle's thinking, in his *Book of Form*, a work that was later popular in the West. The names of many stars are still those given them by Muslim astronomers, such as Altair (from *al-tair*, "the flier"), Deneb (from *dhanab*, "tail"), and Betelgeuse (from *bayt al-jawza*, "the house of the twins" or "Gemini"). Other terms still in use today such as zenith, nadir, and azimuth are also derived from Arabic and so reflect the work of the Muslim astronomers of al-Andalus and their impact on the West.

Scientists of Islamic Spain also contributed to medicine, the Muslim science par excellence. Interest in medicine goes back to the very earliest times (the Prophet himself stated that there was a remedy for every illness),

(Continued on page 68)

Al-Andalus

and although the greatest Muslim physicians practiced in Baghdad, those in al-Andalus made important contributions too. Ibn al-Nafis, for example, discovered the pulmonary circulation of blood.

During the tenth century in particular, al-Andalus produced a large number of excellent physicians, some of whom studied Greek medical works translated at the famous House of Wisdom in Baghdad. Among them was Ibn Shuhayd, who in a fundamental work recommended drugs be used only if the patient did not respond to diet and urged that only simple drugs be employed in all cases but the most serious. Another important figure was Abu al-Qasim al-Zahrawi, the most famous surgeon of the Middle Ages. Known in the West as Abulcasis and Albucasis, he was the author of the *Tasrif*, a book that, translated into Latin, became the leading medical text in European universities during the later Middle Ages. Its section on surgery contains illustrations of surgical instruments of elegant, functional design and great precision.

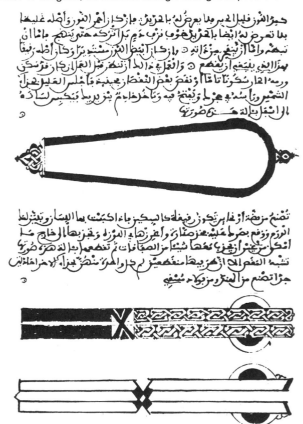

Tongue depressor and tooth extractors were illustrated in a surgical treatise by al-Zahrawi.

Other chapters describe amputations, opthalmic and dental surgery, and the treatment of wounds and fractures. Ibn Zuhr, known as Avenzoar, was the first to describe pericardial abcesses and to recommend tracheotomy when necessary as well as being a skilled practical physician, and Ibn Rushd wrote an important book on medical theories and precepts. The last of the great Andalusian physicians, Ibn al-Khatib, also a noted historian, poet, and statesman, wrote an important book on the theory of contagion in which

he said: "The fact infection becomes clear to the investigator, whereas he who is not in contact remains safe," and described how transmission is effected through garments, vessels, and earrings.

Islamic Spain made contributions to medical ethics and hygiene as well. One of the most eminent theologians and jurists, Ibn Hazm, insisted that moral qualities were mandatory in a physician. A doctor, he wrote, should be kind, understanding, friendly, and able to endure insults and adverse criticism. Furthermore, he went on, a doctor should keep his hair and fingernails short, wear clean clothes, and behave with dignity.

As an outgrowth of medicine, Andalusian scientists also interested themselves in botany. Ibn al-Baytar, for example, the most famous Andalusian botanist, wrote a book called *Simple Drugs and Food,* an alphabetically arranged compendium of medicinal plants, most of which were native to Spain and North Africa, and which he had spent a lifetime gathering. In another treatise Ibn al-'Awwam lists hundreds of species of plants and gives precise instructions regarding their cultivation and use. He writes, for example, of how to graft trees, produce hybrids, stop blights and insect pests, and make perfume.

Another important field of study in al-Andalus was the study of geography. Partly out of economic and political considerations, but mostly out of an all-consuming curiosity about the world and its inhabitants, the scholars of Islamic Spain started with works from Baghdad and went on to add such contributions as a basic geography of al-Andalus by Ahmad ibn Muhammad al-Razi and a description of the topography of North Africa by Muhammad ibn Yusuf al-Warraq. Another contributor to geography was al-Bakri, an important minister at the court of Seville but also an accomplished linguist and litterateur. One of his two important geographical works is devoted to the geography of the Arabian Peninsula, with particular attention to the elucidation of its place-names. It is arranged alphabetically, and lists the names of villages, towns, wadis, and monuments which he culled from the hadith and from histories. The other was an encyclopedia of the entire world, arranged by country, with each entry preceded by a short historical introduction. It included descriptions of the people, customs, and climate of each country, the principal features, the major cities, and even anecdotes.

In the study of geography such figures as Ibn Jubayr, an Andalusian traveler, and the most famous traveler of all, Ibn Battutah, also made important contributions. Born in North Africa, then in the cultural orbit of Islamic Spain, Ibn Battutah traveled extensively for 28 years and produced a travel book that proved to be a rich source for both historians and geographers. It included invaluable information on people, places, navigation, caravan routes, roads, and inns. But the most famous geographer of the period was al-Idrisi, who studied in Cordoba. After traveling widely, al-Idrisi settled in Sicily and wrote a systematic geography of the world, usually known as the *Book of Roger* after his patron Roger II, the Norman King of Sicily. The information contained in the Book of Roger was also engraved on a silver planisphere, a disc-shaped map that was one of the wonders of the age.

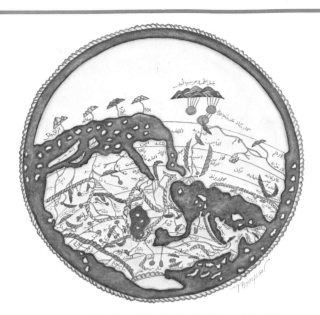

Al-Idrisi's planisphere is considered the first scientific map of the world.

A Moorish-built tower soars above the Guadalquivir River in Seville.

Innumerable scholars in al-Andalus also devoted themselves to the study of history and linguistic sciences, the prime "social sciences" cultivated by the Arabs, and brought them to a high level. Ibn al-Khatib, for example, who distinguished himself in almost all branches of learning, produced more than 50 works on travel, medicine, poetry, music, politics, and theology, as well as writing the finest history of Granada that has survived. The most original mind of the period, however, was undoubtedly Ibn Khaldun, the first historian to develop and explicate general laws governing the rise and decline of civilizations. In the *Prolegomena*, an introduction to a huge, seven-volume universal history — an introduction longer than some of the volumes — Ibn Khaldun approached history as a science and challenged the logic of many accepted historical accounts. In a sense, he was the first modern philosopher of history.

Another great area of Andalusian intellectual activity was philosophy, where an attempt was made to deal with intellectual problems posed by the introduction of Greek philosophy into the context of Islam. One of the first to deal with this was Ibn Hazm, who as the author of more than 400 books has been described as "one of the giants of the intellectual history of Islam." There were other philosophers too, such as Ibn Bajjah, known to the West as Avempace, who was also a physician, and Ibn Tufayl, the author of *Hayy ibn Yaqzan*, the story of a child growing up in complete solitude on a desert island who, entirely by his own intellectual efforts, discovers for himself the highest physical and metaphysical realities. It was however, Averroes — Ibn Rushd — who earned the greatest reputation. He was an ardent Aristotelian and his works had a lasting effect, in their Latin translation, on the development of Western philosophy.

The list of Islamic Spain's contributions to the West, in fact, is almost endless. In addition to Islamic Spain's contributions in mathematics, economy, medicine, botany, geography, history, and philosophy, al-Andalus also developed and applied important technological innovations: the windmill and new techniques in the crafts of metalworking, weaving, and building.

Meanwhile, outside Granada, the Christian kings waited. In relentless succession they had retaken Toledo, Cordoba, and Seville. Only Granada survived. Then, in 1482, in a trivial quarrel, the Muslim kingdom split into two hostile factions and, simultaneously, two strong Christian sovereigns, Ferdinand and Isabella, married and merged their kingdoms. As a result, Granada fell 10 years later. On January 2, 1492 — the year they sent Columbus to America — Ferdinand and Isabella hoisted the banner of Christian Spain above the Alhambra and Boabdil, the last Muslim king, rode weeping into exile with the bitter *envoi* from his aged mother, "Weep like a woman for the city you would not defend like a man!"

In describing the fate of Islam in Spain, Irving suggested that the Muslims were then swiftly and thoroughly wiped out. Never, he wrote, was the annihilation of a people more complete. In fact, by emigration to North Africa and elsewhere, many Muslims carried remnants of the Spanish era with them and were thus able to make important contributions to the material and cultural life of their adopted lands.

Much of the emigration, however, came later. At first, most Muslims simply stayed in Spain; cut off from their original roots by time and distance they quite simply had no other place to go. Until the Inquisition, furthermore, conditions in Spain were not intolerable. The Christians permitted Muslims to work, serve in the army, own land, and even practice their religion — all concessions to the importance of Muslims in Spain's still prosperous economy. But then, in the period of the Inquisition, all the rights of the Muslims were withdrawn, their lives became difficult, and more began to emigrate. Finally, in the early seventeenth century, most of the survivors were forcibly expelled.

THE 'ABBASIDS: In the Middle East during these centuries, the 'Abbasids, after their victory over the Umayyads, had transformed the Umayyads' Arab empire into a multinational Muslim empire. They moved the capital of the empire from Syria to Iraq, where they built a new capital, Baghdad, from which, during the next five centuries, they would influence many of the main events of Islamic history.

Astride the Tigris, present-day Baghdad stands in the vicinity of the 'Abbasid capital, a fabulous city of mosques, mansions, and libraries.

THE 'ABBASID CALIPHATE AND RIVAL STATES AT END OF THE TENTH CENTURY A.D.

Although most of Islam still acknowledges the religious authority of the 'Abbasid caliphate, political power is dispersed among a number of independent states. The largest of these is controlled by the Fatimids who, starting out from Tunisia early in the century, have advanced across North Africa to Egypt and Palestine and down the west coast of Arabia to Hadhramaut. Persian culture has revived under the Samanids. The Hamdanids dominate Syria and northern Mesopotamia. But the center of culture and learning has shifted from Baghdad to the caliphate of Cordoba, where an Umayyad Dynasty still flourishes.

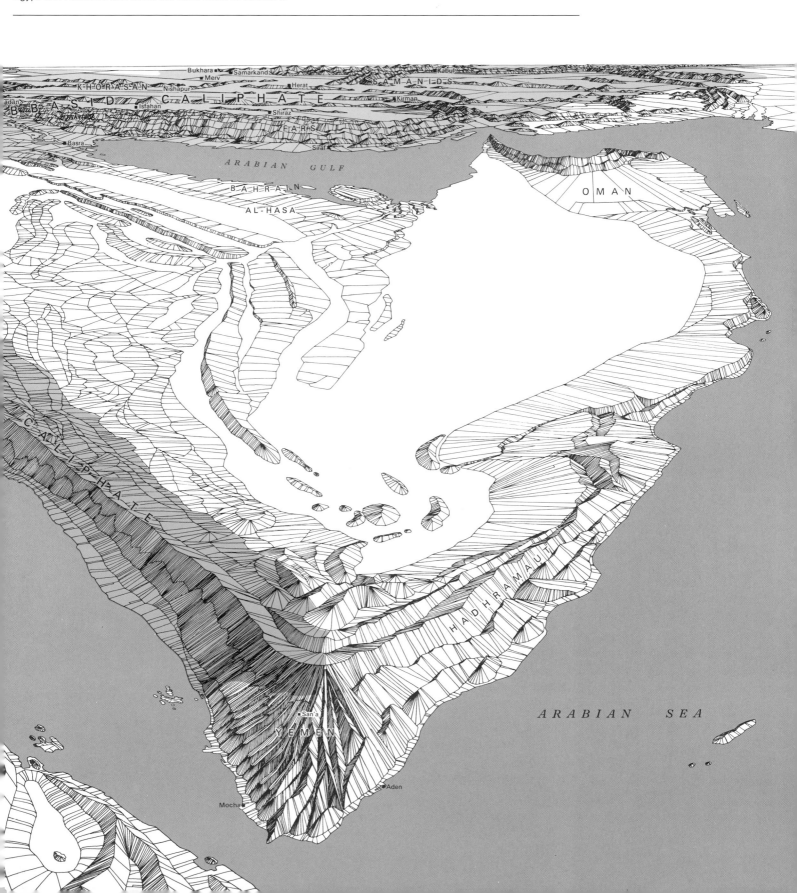

The Great Mosque of the Umayyads in Damascus dates from the early eighth century, and numerous works of rebuilding have not changed its fundamental character.

Golden domes and gold-topped minarets highlight the mosque of al-Kazimayn in Baghdad, built in the early sixteenth century.

The mosque of Bibi Khanum, named for Tamerlane's favorite wife, was once the most imposing building of Samarkand.

In the early period of 'Abbasid rule, al-Mansur, the second caliph of the dynasty, continued the reorganization of the administration of the empire along the lines that had been laid down by his Umayyad predecessor, 'Abd al-Malik. Much of the 'Abbasid administration, for example, was left in the hands of well-educated Persian civil servants, many of whom came from families that had traditionally served the Sassanid kings. The important office of *wazir* or vizier, chief counselor, may well have developed from Sassanid models. The vizier was much more than an advisor; indeed, when the caliph was weak, a capable vizier became the most powerful man in the empire.

The creation of the office of the vizier was only one of the innovations the 'Abbasids brought to statecraft. Another was the development of the Umayyad postal system into an efficient intelligence service; postmasters in outlying provinces were the eyes and ears of the government, and regular reports were filed with the central government on everything from the state of the harvest to the doings of dissident sects. Under the 'Abbasids, too, a whole literature was created for the use and training of the clerical classes that had come into being. Since all government business was by now transacted in Arabic, manuals of correct usage were written for the instruction of non-Arabic speakers who had found government employment. There was also a vast literature on the correct deportment of princes, as well as anthologies of witty sayings and anecdotes with which to enliven one's epistolary style.

In some ways the 'Abbasids were more fortunate than the Umayyads. When, for example, al-Mansur died in 775 after a reign of 20 years, his son, al-Mahdi, inherited a full treasury and an empire that was more devoted to trade than war.

The developments in trade, indeed, are among the achievements of the 'Abbasids that are too often overlooked. Because Islamic rule unified much of the Eastern world, thus abolishing many boundaries, trade was freer, safer, and more extensive than it had been since the time of Alexander the Great. Muslim traders, consequently, established trading posts as far away as India, the Philippines, Malaya, the East Indies, and China.

From the eighth to the eleventh centuries this trade was largely concerned with finding and importing basic necessities—grain, metals, and wood. To obtain them, of course, the Muslims had to export too, often using the imports from one region as exports to another: pearls from the Gulf, livestock from the Arabian Peninsula (particularly Arabian horses and camels), and — one of the chief products — cloth. The Muslims also traded medicines, an offshoot of 'Abbasid advances in medical science, as well as paper and sugar.

This expansion of commercial activity led to other developments too. One was a system of banking and exchange so sophisticated that a letter of credit issued in Baghdad could be honored in Samarkand in Central Asia or Kairouan in North Africa. The demands of trade also generated development of crafts. From Baghdad's large urban population, for example, came craftsmen of every conceivable sort: metalworkers, leatherworkers, bookbinders, papermakers, jewelers, weavers, druggists, bakers, and many more. As they grew in importance to the economy, these craftsmen eventually organized themselves into mutual-benefit societies which in some ways were similar to later Western guilds and which offered many social services: lodging travelers, engaging in pious works such as caring for orphans, and endowing schools. Because of this growth in commerce the 'Abbasids also developed a system by which a *muhtasib*, an inspector, made sure that proper weights and measures were given and that dishonest practices of all sorts were avoided.

THE GOLDEN AGE: The early 'Abbasids were also fortunate in the caliber of their caliphs, especially after Harun al-Rashid came to the caliphate in 786. His reign is now the most famous in the annals of the 'Abbasids — partly because of the fictional role given him in *The Thousand and One Nights* (portions of which probably date from his reign), but also because his reign and those of his immediate successors marked the high point of the 'Abbasid period. As the Arab chronicles put it, Harun al-Rashid ruled when the world was young, a felicitous description of what in later times has come to be called the Golden Age of Islam.

The Golden Age was a period of unrivaled intellectual activity in all fields: science, technology, and (as a result of intensive study of the Islamic faith) literature — particularly biography, history, and linguistics. Scholars, for example, in collecting and reexamining the *hadith* — the sayings and actions of the Prophet — compiled immense biographical detail about the Prophet and other information, historic and linguistic, about the Prophet's era. This led to such memorable works as *Sirat Rasul Allah*, the "Life of the Messenger of God," by Ibn Ishaq, later revised by Ibn Hisham; one of the earliest Arabic historical works, it was a key source of information about the Prophet's life and also a model for other important works of history such as al-Tabari's *Annals of the Apostles and the Kings* and his massive commentary on the Quran.

'Abbasid writers also developed new genres of literature such as *adab*, the embodiment of sensible counsel, sometimes in the form of animal fables; a typical example is *Kalilah wa-Dimnah*, translated by Ibn al-Muqaffa' from a Pahlavi version of an Indian work. Writers of this period also studied tribal traditions and wrote the first systematic Arabic grammars.

During the Golden Age Muslim scholars also made important and original contributions to mathematics, astronomy, medicine, and chemistry. They collected and corrected previous astronomical data, built the world's first observatory, and developed the astrolabe, an instrument that was once called "a mathematical jewel." In medicine they experimented with diet, drugs, surgery, and anatomy, and in chemistry, an outgrowth of alchemy, isolated and studied a wide variety of minerals and compounds.

Important advances in agriculture were also made in the Golden Age. The 'Abbasids preserved and improved the ancient network of wells, underground canals, and water-wheels, introduced new breeds of livestock, hastened the spread of cotton, and, from the Chinese, learned the art of making paper, a key to the revival of learning in Europe in the Middle Ages.

The Golden Age also, little by little, transformed the diet of medieval Europe by introducing such plants as plums, artichokes, apricots, cauliflower, celery, fennel, squash, pumpkins, and eggplant, as well as rice, sorghum, new strains of wheat, the date palm, and sugarcane.

Many of the advances in science, literature, and trade which took place during the Golden Age of the 'Abbasids and which would provide the impetus for the European Renaissance reached their flowering during the caliphate of al-Mamun, son of Harun al-Rashid and perhaps the greatest of all the 'Abbasids. But politically the signs of decay were already becoming evident. The province of Ifriqiyah — North Africa west of Libya and east of Morocco — had fallen away from 'Abbasid control during the reign of Harun al-Rashid, and under al-Mamun other provinces soon broke loose also. When, for example, al-Mamun marched from Khorasan to Baghdad, he left a trusted general named Tahir ibn al-Husayn in charge of the eastern province. Tahir asserted his independence of the central government by omitting mention of the caliph's name in the mosque on Friday and by striking his own coins — acts which became the standard ways of expressing political independence. From 821 onward Tahir and his descendants ruled Khorasan as an independent state, with the tacit consent of the 'Abbasids.

Al-Mamun died in 833, in the town of Tarsus, and was succeeded by his brother, al-Mu'tasim, under whose rule the symptoms of decline that had manifested themselves earlier grew steadily worse. As he could no longer rely on the loyalty of his army, al-Mu'tasim recruited an army of Turks from Transoxiana and Turkestan. It was a necessary step, but its outcome was dominance of the caliphate by its own praetorian guard. In the years following 861, the Turks made and unmade rulers at will, a trend that accelerated the decline of the central authority. Although the religious authority of the 'Abbasid caliphate remained unchallenged, the next four centuries saw political power dispersed among a large number of independent states: Tahirids, Saffarids, Samanids, Buwayhids, Ziyarids, and Ghaznavids in the east; Hamdanids in Syria and northern Mesopotamia; and Tulunids, Ikhshidids, and Fatimids in Egypt.

Some of these states made important contributions to Islamic culture. Under the Samanids, the Persian language, written in the Arabic alphabet, first reached the level of a literary language and poets like Rudaki, Daqiqi, and Firdausi flourished. The Ghaznavids

This Persian miniature depicts students with a teacher of astronomy — one of the sciences to which scholars of the Golden Age made great contributions.

Muslim scientists developed the astrolabe, an instrument used long before the invention of the sextant to observe the position of celestial bodies.

Books of fables, often illustrated, served a dual purpose — to instruct and to entertain.

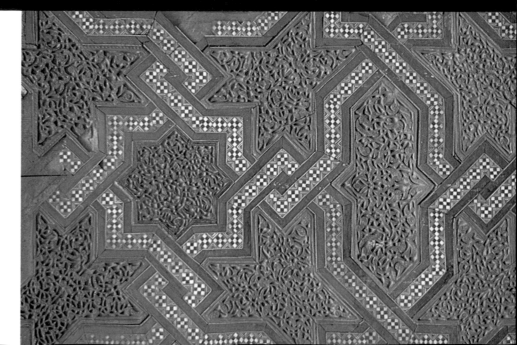

Artisans in far reaches of the Muslim world, born to differing traditions but united in service of a common faith, wove the rich fabric of Islamic art with a common repertory of motifs. Above: Calligraphy, itself an art, decorates a tenth-century bowl from Samarkand. Right: Geometric interlacings in a detail from a wooden pulpit in a twelfth-century Moroccan mosque. Facing page, beginning below: Ivory horn made in Sicily, late eleventh century, when Fatimid influence reached Italy. Gold pendant with cloisonné inlay from Egypt, twelfth century. The seventeenth-century bottle from Kirman in Persia combines Chinese-inspired elements with Islamic arabesque. Sixteenth-century crimson brocade reflects the opulence of the Ottoman court.

patronized al-Biruni, one of the greatest and most original scholars of medieval Islam, and the Hamdanids, a purely Arab dynasty, patronized such poets as al-Mutanabbi and philosophers like the great al-Farabi, whose work kept the flame of Arab culture alive in a difficult period. But in historical terms, only the Fatimids rivaled the preceding dynasties.

THE FATIMIDS: The most stable of the successor dynasties founded in the ninth and tenth centuries was that of the Fatimids, a branch of the Shi'is. The Fatimids won their first success in North Africa, where they established a rival caliphate at Raqqadah near Kairouan and, in 952, embarked on a period of expansion that within a few years took them to Egypt.

For a time the Fatimids aspired to be rulers of the whole Islamic world, and their achievements were impressive. At their peak they ruled North Africa, the Red Sea coast, Yemen, Palestine, and parts of Syria. The Fatimids built the Mosque of al-Azhar in Cairo — from which developed al-Azhar University, now the oldest university in the world and perhaps the most influential Islamic school of higher learning. Fatimid merchants traded with Afghanistan and China and tried to divert some of Baghdad's Arabian Gulf shipping to the Red Sea.

But the Fatimid's dreams of gaining control of the Islamic heartland came to nothing, partly because many of the other independent states refused to support them and partly because they, like the 'Abbasids in Baghdad, lost effective control of their own mercenaries. Such developments weakened the Fatimids, but thanks to a family of viziers of Armenian origin they were able to endure until the Ayyubid succession in the second half of the twelfth century — even in the face of the eleventh-century invasion by the Seljuk Turks.

THE SELJUK TURKS: Although individual Turkish generals had already gained considerable, and at times decisive, power in Mesopotamia and Egypt during the tenth and eleventh centuries, the coming of the Seljuks signaled the first large-scale penetration of Turkish elements into the Middle East. Descended from a tribal chief named Seljuk,

Founded in 970, the mosque of al-Azhar in Cairo is one ot the earliest and finest examples of the Egyptian style in Islamic architecture.

The Quran, the primary document of the Islamic faith, is the first Arabic book. Its style, at once vigorous, allusive, and concise, deeply influenced later compositions in Arabic, as it continues to color the mode of expression of native speakers of Arabic, Christian as well as Muslim, both in writing and in conversation.

The Quran also largely determined the course of Arabic literature. The earliest Arabic prose came into being not from literary motives, but to serve religious and practical needs, above all the need to fully understand the Islamic revelation and the circumstances of the first Muslim community in the Hijaz. The sayings and actions of the Prophet and his Companions were collected and preserved, at first by memory and then by writing, to be finally collected and arranged by such men as al-Bukhari and Muslim in the ninth century. This material, the *hadith*, not only provided the basic texts from which Islamic law was elaborated, but also formed the raw material for historians of the early Muslim community. Since each hadith, or "saying," is a first-person narrative, usually by an eyewitness of the event described, it has an immediacy and freshness that has come down unimpaired through the centuries. The personalities of the narrators — Abu Bakr, 'Umar, 'Aishah, and a host of others — are just as vivid as the events described, for the style of each hadith is very personal.

The hadith also determined the characteristic form of such works as Ibn Ishaq's *Life of the Messenger of God,* originally written in the middle of the eighth century. In this book, hadith dealing with the life of the Prophet are arranged in chronological order, and the comments of the author are kept to a minimum. Events are seen through the eyes of the people who witnessed them; three or four versions of the same event are often given, and in each case the "chain of transmission" of the hadith is given, so that the reader may judge its authenticity.

During Umayyad times, a number of historians wrote monographs on specific historical, legal, and religious questions, and in each case these authors seem to have adhered to the hadith method of composition. Although few of the works of these writers have survived in their entirety, enough has been preserved by later incorporation in such vast works as the *Annals* of al-Tabari to give us an idea not only of their method of composition, but also of their wide-ranging interests.

The practice of prefacing a chain of authorities to each hadith led to the compilation of vast biographical dictionaries, like the *Book of Classes* of the early ninth-century author Ibn Sa'd, which includes a biography of the Prophet and a great deal of information on notable personalities in Makkah and Medina during his lifetime. Works such as this allowed readers to identify and judge the veracity of transmitters of hadith; later, the content of biographical dictionaries was broadened to include poets, writers, eminent reciters of the Quran, scientists, and the like. These biographical dictionaries are often lively reading, and are a mine of information about social and political circumstances in the Islamic world.

The spread of Islam naturally found chroniclers, such as al-Waqidi, who wrote in the late eighth and early ninth centuries, and al-Baladhuri, who composed his well-known *Book of the Conquests* in the ninth century. These books, like the hadith, were written for practical motives.

Arabic Literature

Al-Waqidi was interested in establishing the exact chronology of the spread of Islam in the Arabian Peninsula and adjoining areas, while al-Baladhuri was interested in legal and tax problems connected with the settlement of new lands. Their books nevertheless are classics of their kind and, aside from containing much interesting information, they have passages of great descriptive power.

By the ninth century, the method of compiling history from hadith and carefully citing the authorities for each tradition — a process which had resulted in books of unwieldy length — was abandoned by some authors, like al-Dinawari and al-Ya'qubi, who omitted the chains of transmitters and combined hadith to produce a narrative. The result was greater readability and smaller compass, at the sacrifice of richness and complexity. The works of al-Dinawari and al-Ya'qubi, unlike those of their predecessors, aimed to entertain as well as instruct; they are "literary" productions. This form of light history reached its apogee in the tenth century in al-Mas'udi's brilliant and entertaining *Meadows of Gold and Mines of Gems*, a comprehensive encyclopedia of history, geography, and literature. The literary productions of these men would not, however, have been possible without the careful collections of historical hadith made by their predecessors.

Just as the writing of history began from practical rather than literary motives, so the collection and preservation of Arabic poetry was undertaken by scholars with, at first, little interest in its artistic merit. The linguists and exegetes of Kufa and Basra began collecting this poetry in the eighth century because of the light it threw on unusual expressions and grammatical structures in the Quran and the hadith. Editions and commentaries were prepared of the poems of 'Antarah, Imru al-Qays, and many others, and thus the works of the early poets were preserved for later generations.

The Quran apart, poetry has always been considered the highest expression of literary art among the Arabs. Long before the coming of Islam, Bedouin poets had perfected the forms of panegyric, satire, and elegy. Their poetry obeys strict conventions, both in form and content, which indicates that it must have had a long period of development before it was finally committed to writing by scholars.

The principal form used by the desert poets was the qasidah or ode, a poem of variable length rhyming in the last syllable of each line. The qasidah begins with a description of the abandoned encampment of the poet's beloved and goes on to an account of his anguish at her absence and his consuming love for her. The poet then describes an arduous journey across the desert and ends the qasidah with an appeal to the generosity of his host. Although the subject matter is almost invariable, the language is very complex and of great precision.

In the Hijaz during the first century of Islam, contemporary with the first hadith scholars, a group of poets broke with the past and introduced new forms and subjects. Men like 'Umar ibn Abi Rabi'ah wrote realistic and urbane verse, and a school of poetry which expressed the themes of Platonic love grew up around the poet Jamil ibn Mu'ammar, better known as Jamil al-'Udhri. The lives and works of these poets of the Umayyad period are preserved in the entertaining tenth-century anthology by Abu al-Faraj al-Isfahani, the *Book of Songs*.

The Umayyad court in Damascus patronized poets and musicians. It was also the scene of the development of the type of Arabic literature called *adab*. Adab is usually translated as "belles-lettres," which is slightly misleading. This literature, at least in its inception, was created to serve the practical end of educating the growing class of government ministers in the Arabic language, manners and deportment, history, and statecraft. Works in Sanskrit, Pahlavi, Greek, and Syriac began to find their way into Arabic at this time. 'Abd al-Hamid ibn Yahya al-Katib, an Umayyad official, and the creator of this genre, defined its aims as follows: "Cultivate the Arabic language so that you may speak correctly; develop a handsome script which will add luster to your writings; learn the poetry of the Arabs by heart; familiarize yourself with unusual ideas and expressions; read the history of the Arabs and the Persians, and remember their great deeds." 'Abd Allah ibn al-Muqaffa', a contemporary of 'Abd al-Hamid ibn Yahya, translated the history of the ancient kings of Persia into Arabic, as well as *Kalilah wa-Dimnah*, an Indian book of advice for princes cast in the form of animal fables. His works are the earliest surviving examples of Arabic art prose and are still used as models in schools throughout the Middle East.

By the ninth century, Arabic literature had entered its
(Continued on page 78)

Arabic Literature

classical age. The various genres had been defined — adab, history, Quranic exegesis, geography, biography, poetry, satire, and many more. Al-Jahiz was perhaps the greatest stylist of the age, and one of the most original personalities. He wrote more than 200 books, on every conceivable subject; he was critical, rational, and always amusing. The most gifted of al-Jahiz's contemporaries was probably Ibn Qutaybah, also a writer of encyclopedic learning. His *Book of Knowledge*, a history of the world beginning with the creation, is the earliest work of its kind and later had many imitators.

The tenth century witnessed the creation of a new form in Arabic literature, the *maqamat*. This was the title of a work by al-Hamadhani, called Badi' al-Zaman, "The Wonder of the Age." His *Maqamat* ("Sessions") is a series of episodes written in rhymed prose concerning the life of Abu al-Fath al-Iskandari, a sort of confidence trickster. These stories are packed with action, and were immediately popular. A hundred years later, al-Hamadhani was imitated by al-Hariri, whose *Maqamat* is filled with obscure words, alliteration, puns, and wild metaphors. This purely Arab form can be compared with the Spanish picaresque novels, which it may have influenced.

Rhymed prose, which had come to be used even in government documents, was employed by Abu al-'Ala al-Ma'arri in his *Message of Forgiveness*. Al-Ma'arri lived in the eleventh century, leading an ascetic life in his native Syrian village. Blind from the age of four, he possessed a prodigious memory and great intellectual curiosity. *The Message of Forgiveness* is cast in the form of a journey to paradise; the narrator there interrogates the scholars of the past regarding their lives and works, receiving surprising and often ironic responses. The book represents a high point of classical Arabic literature.

One of the other great figures of late classical literature was the poet al-Mutanabbi, whose verbal brilliance has always been admired by Arab critics, although it is difficult for those whose native tongue is not Arabic to appreciate it fully.

The period between the fall of Baghdad to the Mongols in 1258 and the nineteenth century is generally held to be a period of literary as well as political decline for the Arabs. It is true that during these 500 years Arabic writers were more preoccupied with the preservation of their literary heritage than with the development of new forms and ideas. Faced with the massive destruction of books by the invasions of Genghis Khan and Hulagu and later of Tamerlane, scholars compiled digests and abridgments of works that had survived in order to ensure their continued existence.

There were also some original works, however. Ibn Battutah, the greatest traveler of the Middle Ages, lived in the fourteenth century, and his *Travels* provide a fascinating picture of the Muslim world, from the islands of the Indian Ocean to Timbuktu. Ibn Khaldun, like Ibn Battutah a native of North Africa, lived in the late fourteenth and early fifteenth centuries. His *Prolegomena* is a work of brilliance and originality; the author analyzes human society

in terms of general sociological laws and gives a lucid account of the factors that contribute to the rise and decline of civilizations.

This postclassical period also saw the composition of popular romances, such as the *Romance of 'Antar*, based on the life of the famous pre-Islamic poet; the *Romance of the Bani Hilal*, a cycle of stories and poems based on the migration of an Arabian tribe to North Africa in the eleventh century; and many more. These romances could be heard recited in coffee shops from Aleppo to Marrakesh until very recently. The most famous popular work of all, *The Thousand and One Nights*, assumed its present form during the fifteenth century.

A revival of Arabic literature began in the nineteenth century, and coincided with the first efforts of Arabic-speaking nations to assert their independence of Ottoman rule. Napoleon, during his occupation of Egypt in the late eighteenth century, introduced a printing press with fonts of Arabic type, and Muhammad 'Ali, ruler of Egypt from 1805 to 1848, initiated a series of projects to modernize Egypt. He encouraged the use of Arabic in schools and government institutions, and established a printing press. Selected Egyptian students were sent to study in France, and on their return assigned to undertake translations of Western technical manuals on agriculture, engineering, mathematics, and military tactics. These works, together with many classics of Arabic literature, were printed at the government press at Bulaq and had a profound impact on intellectuals in the Arab East.

Another factor in the literary revival was the swift growth of journalism in Lebanon and Egypt. Starting in the late 1850s, newspapers were soon available through the Middle East. By 1900 well over 150 newspapers and journals were being published. These journals had a great influence on the development and modernization of the written Arabic language; their stress on substance rather than style did much to simplify Arabic prose and bring it within the comprehension of everyone.

One of the first leaders of the Arabic literary renaissance was the Lebanese writer and scholar Butrus al-Bustani, whose dictionary and encyclopedia awakened great interest in the problems of expressing modern Western ideas in the Arabic language. His nephew Sulayman translated Homer's *Iliad* into Arabic, thus making one of the first expressions of Western literature accessible to the Arabic-reading public. Other writers, such as the Egyptian Mustafa al-Manfaluti, adapted French romantic novels to the tastes of the Arab public.

The historical novel, in the hands of Jurji Zaydan, proved immensely popular, perhaps because of the intense interest Arabs have always had in their past. But the first Arabic novel that can rank with European productions is Muhammad Husayn Haykal's *Zaynab,* set in Egypt and dealing with local problems.

One of the greatest figures in modern Arabic literature is Taha Husayn (1889-1973). Blind from an early age, he wrote movingly of his life and beloved Egypt in his autobiography, *al-Ayyam,* "The Days." Taha Husayn was a graduate of both al-Azhar and the Sorbonne, and his voluminous writings on Arabic literature contributed a new critique of this vast subject. It was his near-contemporary compatriot Naguib Mahfouz, however, who placed modern Arabic works firmly on the world literary stage by winning the 1988 Nobel Prize for Literature. By the time Mahfouz was awarded this honor, he had authored 30 novels and 10 volumes of short stories ranging in style from stark realism to spiritual allegory.

The novel and the short story were not the only new forms introduced to the Arabic-reading public. The drama, first in the form of translations of Western work, then of original compositions, was pioneered by Ahmad Shawqi and came to maturity in the hands of Tawfiq al-Hakim. Al-Hakim's long career and devotion to the theater did much to make this one of the liveliest arts of the Middle East.

The history of modern Arabic poetry, with its many schools and contending styles, is almost impossible to summarize. Traditional forms and subjects were challenged by 'Abbas Mahmud al-'Aqqad, Mahmud Shukri, and Ibrahim al-Mazini, who strove to introduce nineteenth century European themes and techniques into Arabic, not always with success. Lebanese poets were in the forefront of modernist verse, and one of them, Gibran Kahlil Gibran, proved very popular in the West. Poets are now experimenting with both old and new techniques, although discussions of form have given way to concern for content. The exodus of Palestinians from their native land has become a favorite theme, often movingly handled.

In Saudi Arabia, it was not until well into the twentieth century that literary movements in neighboring lands made themselves felt. Poetry, of course, has been cultivated in Arabia since the pre-Islamic period. Some modern writers, such as 'Abd Allah ibn Khamis, hold strictly to classical forms and serious subject matter. Others have adopted new techniques and content. Hasan al-Qurashi, Tahir Zamakhshari, Muhammad Hasan Faqi and Mahrum (the pen name of Amir 'Abd Allah al-Faysal), to name a few, have won renown for their poetry throughout the Arab world. Muhammad Hasan Faqi's poetry is introspective and philosophical, while the verse of the three others is lyrical and romantic. Ghazi al-Gosaibi is distinguished by a fresh imagination that expresses itself in both Arabic and English verse. The long tradition of Bedouin folk poetry still thrives today and has been adopted in new compositions such as those by the Amirs Khalid al-Faysal and Badr ibn 'Abd al-Muhsin.

Two novels by the late Hamid al-Damanhuri have been well received. They are *Thaman al-Tadhiyah,* "The Price of Sacrifice," and *Wa-Marrat al-Ayyam,* "And the Days Went By." The short story has also become a flourishing genre, exemplified by, among others, the works of Khalil al-Fuzay' and 'Abd al-'Aziz al-Mishri. Other Saudi writers have gained high reputations in journalistic and scholarly pursuits. Such offerings include the lively and sometimes controversial literary criticism of 'Abd Allah al-Ghudhami, the wide-ranging writings of 'Aziz Diya, 'Abd Allah Jifri, and Abu 'Abd al-Rahman al-Zahiri, as well as the numerous volumes of Saudi ethnographic and geographic lore by national prizewinner Hamad al-Jasir.

The last 40 years i have also seen the active participation of Saudi women in the country's literary and journalistic activities with the appearance of such talents as Khayriyah al-Saqqaf, Qumashah al-'Ulayyan, and Sultanah al-Sudayri. Overall, the rising number of high quality works by Saudi writers bodes well for the future of literary developments in the kingdom.

whose homeland lay beyond the Oxus River near the Aral Sea, the Seljuks not only developed a highly effective fighting force but also, through their close contacts with Persian court life in Khorasan and Transoxiana, attracted a body of able administrators. Extending from Central Asia to the Byzantine marches in Asia Minor, the Seljuk state under its first three sultans — Tughril Beg, Alp-Arslan, and Malikshah — established a highly cohesive, well-administered Sunni state under the nominal authority of the 'Abbasid caliphs at Baghdad.

One of the administrators, the Persian Nizam-al-Mulk, became one of the greatest statesmen of medieval Islam. For 20 years, especially during the rule of Sultan Malikshah, he was the true custodian of the Seljuk state. In addition to having administrative abilities, he was an accomplished stylist whose book on statecraft, *Siyasat-Namah,* is a valuable source for the political thought of the time. In it he stresses the responsibilities of the ruler: for example, if a man is killed because a bridge is in disrepair, it is the fault of the ruler, for he should make it his business to apprise himself of the smallest negligences of his underlings. Nizam-al-Mulk, furthermore, was a devout and orthodox Muslim who established a system of *madrasahs* or theological seminaries (called *nizamiyah* after the first element of his name) to provide students with free education in the religious sciences of Islam, as well as in the most advanced scientific and philosophical thought of the time. The famous theologian al-Ghazali whose greatest work, the *Revival of the Sciences of Religion,* was a triumph of Sunni theology taught for a time at the nizamiyah schools at Baghdad and at Nishapur. Nizam-al-Mulk was the patron of the poet and astronomer 'Umar al-Khayyam (Omar Khayyam), whose verses, as translated by Edward FitzGerald in the nineteenth century, have become as familiar to English readers as the sonnets of Shakespeare.

After the death of Malikshah in 1092, internal conflict among the young heirs led to the fragmentation of the Seljuks' central authority into smaller Seljuk states led by various members of the family, and still smaller units led by regional chieftains, no one of whom was able to unite the Muslim world as still another force appeared in the Middle East: the Crusaders.

The *Rubaiyat* of Omar Khayyam owes its popularity in the West to the translation by FitzGerald.

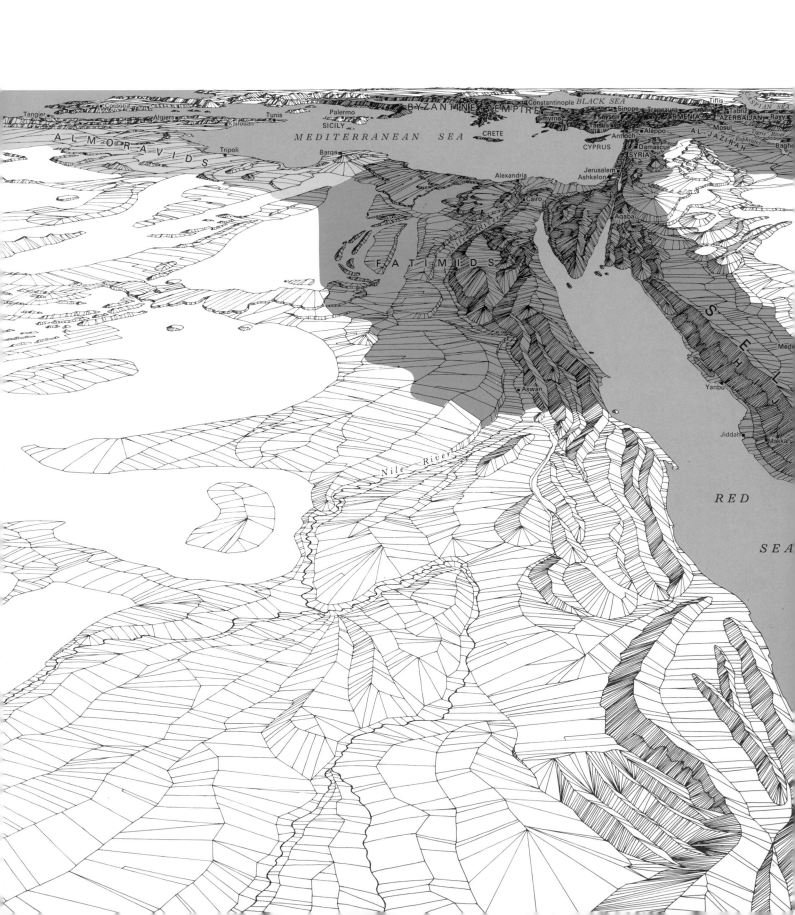

Tangier
Cordoba
Algiers
Kairouan
Tunis
A L M O R A V I D S
Tripoli
Barqa
MEDITERRANEAN SEA
SICILY
Palermo
CRETE
BYZANTINE EMPIRE
Constantinople
Smyrna
Tarsus
CYPRUS
Antioch
Aleppo
Damascus
SYRIA
BLACK SEA
Sinope
Trapezunt
Kayseri
ARMENIA
Tiflis
CASPIAN SEA
Tabriz
AZERBAIJAN
Rayy
AL-JAZIRAH
Mosul
Tigris River
Euphrates
Bagh
Alexandria
Cairo
Jerusalem
Ashkelon
Aqaba
Nile River
FATIMIDS
Aswan
Yanbu
Jiddah
Makkah
Med
H I J
RED
SEA
Nile River

THE MIDDLE EAST AND THE MEDITERRANEAN ON THE EVE OF THE CRUSADES

The Seljuks from beyond the Oxus, the first Turkic people to penetrate the Middle East in force, have adopted Islam and quickly overrun most of the 'Abbasid lands east of the Mediterranean, including Oman and the Hijaz, and have wrested most of Anatolia from the Byzantine Empire. The Fatimids have successfully resisted the Seljuks and retain control in Egypt and Sinai. In the west, Berber nomads, the Almoravids, have established themselves in Morocco and succeed the Umayyads in a reduced area in Spain. To the east, Transoxiana under the Qarakhanids is a Seljuk vassal state and the Ghaznavids of Khorasan and Afghanistan are expanding Muslim rule in India. At the end of the eleventh century, as the Seljuk central authority is being eroded by internal conflict, the initially successful invading forces of the Crusaders arrive on the scene.

THE CRUSADERS: To Arab historians, the Crusaders were a minor irritant, their invasion one more barbarian incursion, not nearly as serious a threat as the Mongols were to prove in the thirteenth and fourteenth centuries.

The First Crusade began in 1095 after the Byzantines — threatened by Seljuk power — appealed to Pope Urban II for military aid. Pope Urban, hoping to divert the Christian kings and princes from their struggles with each other, and perhaps also seeing an opportunity to reunite the Eastern and Western churches, called for a "Truce of God" among the rulers of Europe and urged them to take the Holy Land from the Muslims.

Considered dispassionately, the venture was impossible. The volunteers — a mixed assemblage of kings, nobles, mercenaries, and adventurers — had to cross thousands of kilometers of unfamiliar and hostile country and conquer lands of whose strength they had no conception. Yet so great was their fervor that in 1099 they took Jerusalem, establishing along the way principalities in Antioch, Edessa, and Tripoli. Although unable to fend off the Crusaders at first — even offering the Crusaders access to Jerusalem if they would come as pilgrims rather than invaders — the Muslims eventually began to mount effective counterattacks. They recaptured Aleppo and besieged Edessa, thus bringing on the unsuccessful Second Crusade.

In the meantime the Crusaders — or Franks, as the Arabs called them — had extended their reach to the borders of Egypt, where the Fatimids had fallen after 200 years. There they faced a young man called Salah al-Din (Saladin) who had founded still another new dynasty, the Ayyubids, and who was destined to blunt the thrust of the Crusaders' attack. In 1187 Saladin counterattacked, eventually recapturing Jerusalem. The Europeans mounted a series of further crusading expeditions against the Muslims over the next hundred years or so, but the Crusaders never again recovered the initiative. Confined to the coast, they ruled small areas until their final defeat at the hands of the Egyptian Mamluks at the end of the thirteenth century.

Although the Crusades achieved no lasting results in terms of military conquest, they were important in the development of trade, and their long-range effects on Western society — on everything from feudalism to fashion — are inestimable. Ironically, they also put an end to the centuries-old rivalry between the Arabs and Byzantines. By occupying Constantinople, the capital of their Christian allies, in the Fourth Crusade, the Crusaders achieved what the Arabs had been trying to do from the early days of Islam. Although the Byzantine Empire continued until 1453, when Constantinople fell to the Ottoman Turks, it never recovered its former power after the Fourth Crusade, and subsisted only in the half-light of history during its remaining years.

For the West, however, the Crusaders' greatest achievement was the opening of the eastern Mediterranean to European shipping. The Venetians and Genoese established trading colonies in Egypt, and luxury goods of the East found their way to European markets. In the history of the Middle Ages, this was far more important than ephemeral conquests. Control of the Eastern trade became a constantly recurring theme in later relations between the European countries and the East, and in the nineteenth century was to lead to widespread Western intervention.

THE MONGOLS AND THE MAMLUKS: In the thirteenth century still another threat to the Muslim world appeared in the land beyond the Oxus: the Mongols. Led by Genghis Khan, a confederation of nomadic tribes which had already conquered China now attacked the Muslims. In 1220 they took Samarkand and Bukhara. By mid-century they had taken Russia, Central Europe, northern Iran, and the Caucasus, and in 1258, under Hulagu Khan, they invaded Baghdad and put an end to the remnants of the once-glorious 'Abbasid Empire. The ancient systems of irrigation were destroyed and the devastation was so extensive that agricultural recovery, even in the twentieth century, is still incomplete. Because a minor scion of the dynasty took refuge with the Mamluks in Egypt, the 'Abbasid caliphate continued in name into the sixteenth century. In effect, however, it expired with the Mongols and the capture of Baghdad. From Iraq the Mongols pressed forward into Syria and then toward Egypt where, for the first time, they faced adversaries who refused to quail before their vaunted power. These were the Mamluks, soldier-slaves from the Turkish steppe area north of the Black and Caspian Seas with a

The most imposing of the many fortresses built by the Crusaders, the elegant Krak des Chevaliers in Syria (top) held out against the Muslims for over a century and a half. The Crusader castle at Sidon in Lebanon (bottom) was abandoned after the final defeat of the Crusader Kingdom of Jerusalem.

later infusion of Circassians from the region of the Caucasus Mountains.

The Mamluks had been recruited by the Ayyubids and then, like the Turkish mercenaries of the 'Abbasid caliphs, had usurped power from their enfeebled masters. Unlike their predecessors, however, they were able to maintain their power, and they retained control of Egypt until the Ottoman conquest in 1517. Militarily formidable, they were also the first power to defeat the Mongols in open combat when, in 1260, the Mongols moved against Palestine and Egypt. Alerted by a chain of signal fires stretching from Iraq to Egypt, the Mamluks were able to marshal their forces in time to meet, and crush, the Mongols at 'Ayn Jalut near Nazareth in Palestine.

In the meantime the Mongols, like so many of the peoples who had come into contact with Islam, had begun to embrace it. At the dawn of the fourteenth century, Ghazan Khan Mahmud officially adopted Islam as the religion of the state, and for a time peace descended on the eastern portion of the Mongol empire. During this period the Mongols built mosques and schools and patronized scholarship of all sorts. But then, in 1380, a new Turko-Mongol confederation was hammered together by another world conqueror: Tamerlane, who claimed descent from Genghis Khan. Under Tamerlane, the Mongol forces swept down on Central Asia, India, Iran, Iraq, and Syria, occupying Aleppo and Damascus and threatening — but not defeating — the Mamluks. Once again, however, the Muslims survived their invaders. Tamerlane died on his way to conquer China, and his empire melted away.

Politically and economically, the Mongol invasions were disastrous. Some regions never fully recovered and the Muslim empire, already weakened by internal pressures, never fully regained its previous power. The Mongol invasions, in fact, were a major cause of the subsequent decline that set in throughout the heartland of the Arab East. In their sweep through the Islamic world the Mongols killed or deported numerous scholars and scientists and destroyed libraries with their irreplaceable works. The result was to wipe out much of the priceless cultural, scientific, and technological legacy that Muslim scholars had been preserving and enlarging for some 500 years.

THE LEGACY: The foundation of this legacy was the astonishing achievements of Muslim scholars, scientists, craftsmen, and traders during the 200 years or so that are called the Golden Age. During this period, from 750 to 950, the territory of the Muslim Empire encompassed present-day Iran, Syria, Iraq, Egypt, Palestine, North Africa, Spain, and parts of Turkey and drew to Baghdad peoples of all those lands in an unparalleled cross-fertilization of once isolated intellectual traditions.

Geographical unity, however, was but one factor. Another was the development of Arabic, by the ninth century, into the language of international scholarship as well as the language of the Divine Truth. This was one of the most significant events in the history of ideas.

A third important factor was the establishment in Baghdad of a paper mill. The introduction of paper, replacing parchment and papyrus, was a pivotal advance which had effects on education and scholarship as far-reaching as the invention of printing in the fifteenth century. It made it possible to put books within the reach of everyone.

The Mamluks, originally a class of soldier-slaves, seized power in Egypt in the thirteenth century and stood fast against the Mongols.

Unlike the Byzantines, with their suspicion of classical science and philosophy, the Muslims were enjoined by the Prophet to "seek learning as far as China" — as, eventually, they did. In the eighth century, however, they had a more convenient source: the works of Greek scientists stored in libraries in Constantinople and other centers of the Byzantine empire. In the ninth century the Caliph al-Mamun, son of the famous Harun al-Rashid, began to tap that invaluable source. With the approval of the Byzantine emperor, he dispatched scholars to select and bring back to Baghdad Greek scientific manuscripts for translation into Arabic at *Bayt al-Hikmah*, "the House of Wisdom."

Bayt al-Hikmah was a remarkable assemblage of scholar-translators who undertook a Herculean task: to translate into Arabic all of what had survived of the philosophical and scientific tradition of the ancient world and incorporate it into the conceptual framework of Islam.

As the early scholars in the Islamic world agreed with Aristotle that mathematics was the basis of all science, the scholars of the House of Wisdom first focused on mathematics. Ishaq

The fierce Tamerlane, one of history's greatest conquerors, is also remembered as a patron of architecture and the arts.

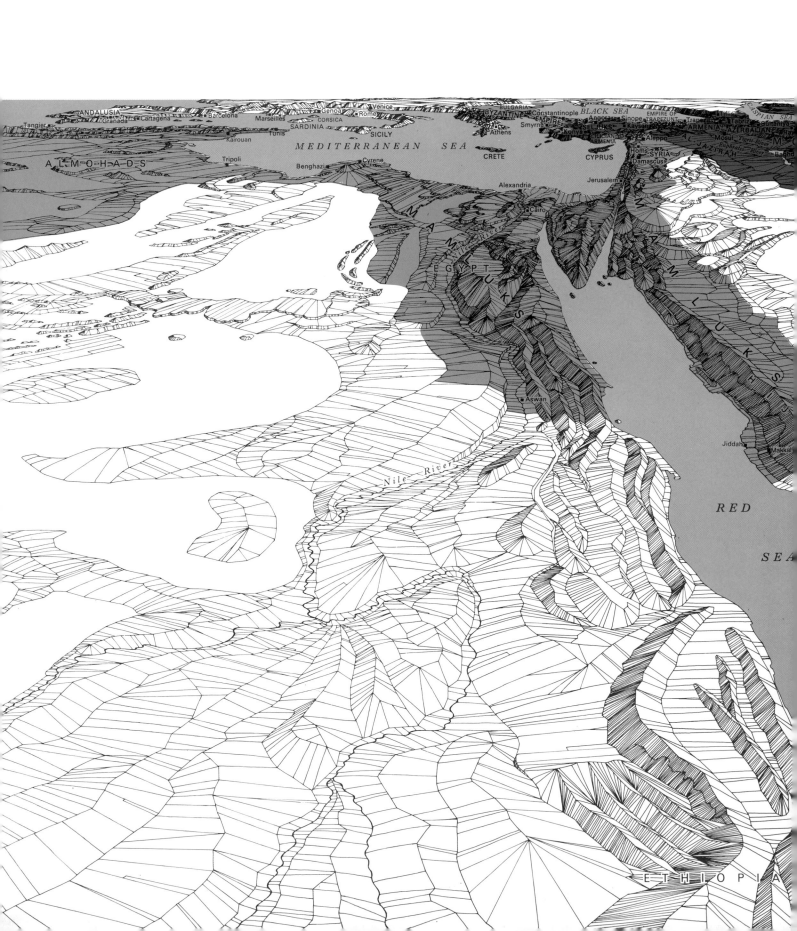

ANDALUSIA
Tangier Granada Cartagena Barcelona Marseilles Genoa Venice Rome BULGARIA Constantinople BLACK SEA EMPIRE OF Tiflis CASPIAN SEA
Tunis SARDINIA CORSICA Rome BYZANTINE Angora Sinope TRABEZUNT Trebizond Tabriz Sultaniya
ALMOHADS Kairouan SICILY Athens Smyrna EMPIRE SELJUKS Kaysari ARMENIA AZERBAIJAN Mosul M
Tripoli MEDITERRANEAN SEA CRETE Konya CILICIAN Aleppo AL JAZIRAH Tigris River Bag
Benghazi Cyrene CYPRUS Homs SYRIA Euphrates River
Alexandria Damascus
Jerusalem
Cairo
M
A
Nile River EGYPT M M
L A
U M
K L
S U
K
Aswan S
Nile River Jiddah
Makkah
RED
SEA
ETHIOPIA

THE MUSLIM WORLD IN THE MONGOL PERIOD

At the end of the thirteenth century the Mongols have swept across Asia, leaving devastation in their wake and creating for a brief time the greatest empire in history, stretching from Korea to Poland. Their advance has finally been checked by the Mamluks in Egypt, and the Mongol Empire starts to break up almost immediately, with Mesopotamia and Persia being ruled by the Il-Khanid Dynasty of Mongols. In the west a Berber dynasty, the Almohads, rules in Morocco, and the Nasrids rule in the Kingdom of Granada, all that remains of the once exten-sive Umayyad domains in Spain. The Byzantine Empire never recovers after being occupied by the Crusaders in 1202, but the tiny Empire of Trapezunt, an offshoot of the Byzantine Empire, prospers from trade, having been bypassed by both the Seljuks and the Mongols. At the end of the century the Mamluks take Acre, ending Christian rule in the Levant and the era of the Crusades. Although Europeans no longer occupy territory in the eastern Mediterranean, its waters are open to their shipping and provide access to eastern trade.

A sophisticated water clock featured trumpeters, half life-size, who sounded the hour.

Mechanical figures atop a bloodletting device recorded the amount drawn from a patient.

ibn Hunayn and Thabit ibn Qurrah, for example, prepared a critical edition of Euclid's Elements, while other scholars translated a commentary on Euclid originally written by a mathematician and inventor from Egypt, and still others translated at least 11 major works by Archimedes, including a treatise on the construction of a water clock. Other translations included a book on mathematical theory by Nichomachus of Gerasa, and works by mathematicians like Theodosius of Tripoli, Apollonius Pergacus, Theon, and Menelaus, all basic to the great age of Islamic mathematical speculation that followed.

The first great advance on the inherited mathematical tradition was the introduction of "Arabic" numerals, which actually originated in India and which simplified calculation of all sorts and made possible the development of algebra. Muhammad ibn Musa al-Khwarazmi seems to have been the first to explore their use systematically, and wrote the famous *Kitab al-Jabr wa-l-Muqabalah*, the first book on algebra, a name derived from the second word in his title. One of the basic meanings of *jabr* in Arabic is "bone-setting," and al-Khwarazmi used it as a graphic description of one of the two operations he uses for the solution of quadratic equations.

The scholars at *Bayt al-Hikmah* also contributed to geometry, a study recommended by Ibn Khaldun, the great North African historian, because "it enlightens the intelligence of the man who cultivates it and gives him the habit of thinking exactly." The men most responsible for encouraging the study of geometry were the sons of Musa ibn Shakir, al-Mamun's court astronomer. Called Banu Musa — "the sons of Musa" — these three men, Muhammad, Ahmad, and al-Hasan, devoted their lives and fortunes to the quest for knowledge. They not only sponsored translations of Greek works, but wrote a series of important original studies of their own, one bearing the impressive title *The Measurement of the Sphere, Trisection of the Angle, and Determination of Two Mean Proportionals to Form a Single Division between Two Given Quantities*.

The Banu Musa also contributed works on celestial mechanics and the atom, helped with such practical projects as canal construction, and in addition recruited one of the greatest of the ninth-century scholars, Thabit ibn Qurrah.

During a trip to Byzantium in search of manuscripts, Muhammad ibn Musa happened to meet Thabit ibn Qurrah, then a money changer but also a scholar in Syriac, Greek, and Arabic. Impressed by Thabit's learning, Muhammad personally presented him to the caliph, who was in turn so impressed that he appointed Thabit court astrologer. As Thabit's knowledge of Greek and Syriac was unrivaled, he contributed enormously to the translation of Greek scientific writing and also produced some 70 original works — in mathematics, astronomy, astrology, ethics, mechanics, music, medicine, physics, philosophy, and the construction of scientific instruments.

Although the House of Wisdom originally concentrated on mathematics, it did not exclude other subjects. One of its most famous scholars was Hunayn ibn Ishaq, Ishaq's father — known to the West as Joanitius — who eventually translated the entire canon of Greek medical works into Arabic, including the Hippocratic oath. Later a director of the House of Wisdom, Hunayn also wrote at least 29 original treatises on medical topics and a collection of 10 essays on ophthalmology which covered, in systematic fashion, the anatomy and physiology of the eye and the treatment of various diseases which afflict vision. The first known medical work to include anatomical drawings, the book was translated into Latin and for centuries was the authoritative treatment of the subject in both Western and Eastern universities.

Others prominent in Islamic medicine were Yuhanna ibn Masawayh, a specialist in gynecology, and the famous Abu Bakr Muhammad ibn Zakariya al-Razi — known to the West as Rhazes. According to a bibliography of his writings, al-Razi wrote 184 works, including a huge compendium of his experiments, observations, and diagnoses with the title *al-Hawi*, "The All-Encompassing."

A fountainhead of medical wisdom during the Islamic era, al-Razi, according to one contemporary account, was also a fine teacher and a compassionate physician, who brought rations to the poor and provided nursing for them. He was also a man devoted to common sense, as the titles of two of his works suggest: *The Reason Why Some Persons and the Common People Leave a Physician Even If He Is Clever,*

Modern Arabic (western)	1	2	3	4	5	6	7	8	9	0
Early Arabic (western)										
Arabic letters used as numerals	ا	ب	ج	د	ه	و	ز	ح	ط	ي
Modern Arabic (eastern)	١	٢	٣	٤	٥	٦	٧	٨	٩	٠
Early Arabic (eastern)										
Early Devanagari (Indian)										
Later Devanagari (tenth-century Sanskrit)	१	२	३	४	५	६	७	८	९	०

Arabic Numerals

The system of numeration employed throughout the greater part of the world today was probably developed in India, but because it was the Arabs who transmitted this system to the West the numerals it uses have come to be called Arabic.

After extending Islam throughout the Middle East, the Arabs began to assimilate the cultures of the peoples they had subdued. One of the great centers of learning was Baghdad, where Arab, Greek, Persian, Jewish, and other scholars pooled their cultural heritages and where in 771 an Indian scholar appeared, bringing with him a treatise on astronomy using the Indian numerical system.

Until that time the Egyptian, Greek, and other cultures used their own numerals in a manner similar to that of the Romans. Thus the number 323 was expressed like this:

Egyptian	999	∩∩	III
Greek	HHH	△△	III
Roman	CCC	XX	III

The Egyptians actually wrote them from right to left, but they are set down above from left to right to call attention to the similarities of the systems.

The Indian contribution was to substitute a single sign (in this case meaning "3" and meaning "2") indicating the number of signs in each cluster of similar signs. In this manner the Indians would render Roman CCC XX III as:

This new way of writing numbers was economical but not flawless. The Roman numeral CCC II, for instance, presented a problem. If a 3 and a 2 respectively were substituted for the Roman clusters CCC and II, the written result was 32. Clearly, the number intended was not 32 but 302. The Arab scholars perceived that a sign representing "nothing" or "nought" was required because the place of a sign gave as much information as its unitary value did. The place had to be shown even if the sign which showed it indicated a unitary value of "nothing." It is uncertain whether the Arabs or the Indians filled this need by inventing the zero, but in any case the problem was solved: now the new system could show neatly the difference between XXX II(32) and CCC II (302).

If the origin of this new method was Indian, it is not at all certain that the original shapes of the Arabic numerals also were Indian. In fact, it seems quite possible that the Arab scholars used their own numerals but manipulated them in the Indian way. The Indian way had the advantage of using much smaller clusters of symbols and greatly simplifying written computations. The modern forms of the individual numbers in both eastern Arabic and western Arabic, or European, appear to have evolved from letters of the Arabic alphabet.

The Semites and Greeks traditionally assigned numerical values to their letters and used them as numerals. This alphabetical system is still used by the Arabs, much as Roman numerals are used in the West for outlines and in enumerating kings, emperors, and popes.

The new mathematical principle on which the Arabic numerals were based greatly simplified arithmetic. Their adoption in Europe began in the tenth century after an Arabic mathematical treatise was translated by a scholar in Spain and spread throughout the West.

Architectural monuments spanning a thousand years bear witness to the spread of Islam. Seen at left is a unique fusion of Roman and Asian elements, Jerusalem's Dome of the Rock, built in 691-692. Continuing clockwise, facing page: Purity of line characterizes the late twelfth-century Kutubiyah Mosque of the Berbers in Marrakesh. Sixteenth-century Sultan Selim Mosque at Edirne is the apogee of Ottoman Turkish architecture, soaring space enclosed with a massive dome. Persia's greatest contribution to ornament, gloriously colored enameled tile, faces the dome and stalactite portal of Shaykh Lutf Allah Mosque, built in the early 1600s on Isfahan's vast royal plaza. The peak of Mogul architecture and possibly the most famous work of all times and cultures is the dazzling Taj Mahal mausoleum built at Agra in 1629. Above: Watercourses and fountains make an oasis of the Alhambra palace built at Granada in the fourteenth century. Here incredibly light and elegant elements of Islamic decoration find their highest realization.

and *A Clever Physician Does Not Have the Power to Heal All Diseases, for That is Not within the Realm of Possibility.*

The scholars at the House of Wisdom, unlike their modern counterparts, did not "specialize." Al-Razi, for example, was a philosopher and a mathematician as well as a physician and al-Kindi, the first Muslim philosopher to use Aristotelian logic to support Islamic dogma, also wrote on logic, philosophy, geometry, calculation, arithmetic, music, and astronomy. Among his works were such titles as *An Introduction to the Art of Music, The Reason Why Rain Rarely Falls in Certain Places, The Cause of Vertigo,* and *Crossbreeding the Dove.*

Another major figure in the Islamic Golden Age was al-Farabi, who wrestled with many of the same philosophical problems as al-Kindi and wrote *The Perfect City,* which illustrates to what degree Islam had assimilated Greek ideas and then impressed them with its own indelible stamp. This work proposed that the ideal city be founded on moral and religious principles from which would flow the physical infrastructure. The Muslim legacy included advances in technology too. Ibn al-Haytham, for example, wrote *The Book of Optics,* in which he gives a detailed treatment of the anatomy of the eye, correctly deducing that the eye receives light from the object perceived and laying the foundation for modern photography. In the tenth century he proposed a plan to dam the Nile. It was by no means theoretical speculation; many of the dams, reservoirs, and aqueducts constructed at this time throughout the Islamic world still survive.

Muslim engineers also perfected the waterwheel and constructed elaborate underground water channels called *qanats.* Requiring a high degree of engineering skill, *qanats* were built some 15 meters underground with a very slight inclination over long distances to tap underground water and were provided with manholes so that they could be cleaned and repaired.

Agricultural advances are also part of the Muslim legacy. Important books were written on soil analysis, water, and what kinds of crops were suited to what soil. Because there was considerable interest in new varieties — for nutritive and medicinal purposes — many new plants were introduced: sorghum, for

At Hama in Syria, antique wooden wheels still lift the waters of the Orontes to gardens, baths, and cooling fountains.

example, which had recently been discovered in Africa.

The introduction of numerous varieties of fruits and vegetables and other plants to the West via the Islamic empire was, however, largely the result of the vast expansion of trade during the Golden Age. This trade was vital; in the central lands of the 'Abbasid empire natural resources such as metals and wood were scarce, and increases in urban populations had outstripped the capacity of the agricultural system to support them. The 'Abbasids, therefore, were forced to develop extensive and complicated patterns of trade. To obtain food, for example, Baghdad had to import wheat from Syria and Egypt, rice from the Fayyum in Egypt, southern Morocco, and Spain, and olive oil from Tunisia. Called "a forest of olive trees," Tunisia exported so much olive oil that its port of Sfax was called "the port of oil."

To obtain scarce metals the 'Abbasids had to turn elsewhere. They imported the technologically advanced "ondanique" steel from India, for example, and then processed it at such famous centers of weapons manufacture as Damascus and Toledo, both of which cities won fame for their blades. The 'Abbasids also imported iron from Europe, tin from the British Isles and Malaya, and silver from northern Iran, Afghanistan, and the Caucasus. For gold, once the vast quantities in the treasuries of the conquered countries were exhausted, they turned to several sources. One was the gold mines of the Hijaz which were reopened around 750, reworked for about 400 years, and then, in 1931, explored again by Karl Twitchell, who was searching for minerals in that area on behalf of King 'Abd al-'Aziz of Saudi Arabia.

For these necessities the 'Abbasid traders exchanged a wide variety of products: pearls, livestock, paper, sugar, and (a specialty of the Islamic world) luxurious cloth. The traditional cloths were wool and linen — the latter an Egyptian specialty since ancient times — but cotton, which was introduced into upper Iraq about the time of the Prophet, later spread with Islam around the Mediterranean, to Syria, North Africa, Spain, Sicily, Cyprus, and Crete.

The cloth trade produced a number of auxiliary exports: gold and silver thread for embroidery, gum from the Sudan for glazing, and needles, looms, and dyestuffs. Closely connected with the trade in dyestuffs was the trade in medicines, an offshoot of 'Abbasid advances in medicine and the spread of hospitals in all major Islamic cities. As scientific research and translation of medical texts from India and possibly even China expanded the earlier pharmacopoeia, ingredients for medicines were brought from all over the known world and also reexported.

Because the religious, political, and military achievements of the Islamic period loom so large in the history of the world, the extraordinary cultural, scientific, technological, and commercial achievements are frequently obscured or overlooked. Yet these advances were, in fact, of enduring significance to mankind as a whole. The destruction by the Mongols of many of these achievements and of much of what the Muslims had accomplished by the end of the Golden Age was a tragic loss for the West and for the world as a whole.

THE OTTOMANS: During the second Mongol invasion, Tamerlane had met and very nearly annihilated another rising power: the Ottomans. Under a minor chieftain named Othman, groups of Turkish-speaking peoples in Anatolia were united in the Ottoman confederation which, by the second half of the fourteenth century, had conquered much of present-day Greece and Turkey and was threatening Constantinople.

The Ottoman state was born on the frontier between Islam and the Byzantine Empire. Turkish tribes, driven from their homeland in the steppes of Central Asia by the Mongols, had embraced Islam and settled in Anatolia on the battle lines of the Islamic world, where they formed the Ottoman confederation. They were called *ghazis*, warriors for the faith, and their highest ambition was to die in battle for their adopted religion.

In addition to their military abilities the Turks seem to have been endowed with a special talent for organization. Towards the end of the Ottoman Empire, this talent fossilized into bureaucracy—and a moribund bureaucracy at that. But at the beginning, when its institutions were responsive to the needs of the people and the state, the Ottoman Empire was a model of administrative

efficiency. This, together with a series of brilliant sultans — culminating in the redoubtable Suleiman the Magnificent — established the foundations of an empire that at its height was comparable to that of the Romans.

The first important step in the establishment of this empire was taken in 1326 when the Ottoman leader Orhan captured the town of Bursa, south of the Sea of Marmara, and made it his capital.

It was probably during the reign of Orhan that the famous institution of the Janissaries, a word derived from the Turkish *yeni cheri* ("new troops"), was formed. An elite corps of slave soldiers conscripted from the subject population of the empire, they were carefully selected on the basis of physique and intelligence, educated, trained, introduced to Islam, and formed into one of the most formidable military corps ever known. At a later period the Janissaries became so powerful that they made and unmade sultans at their will, and membership in the corps was a sure road to advancement.

Orhan's successor, Murad I, who launched naval attacks upon the Aegean coasts of Europe, established himself on the European shores of the Bosporus, and crushed a Balkan coalition. The next Ottoman leader was Bayazid I, who besieged Constantinople and routed the armies dispatched by an alarmed Europe to raise the siege.

It was at this point in history that Tamerlane and his Mongols advanced into Anatolia and very nearly crushed the Ottomans forever. They recovered, however, and later, under the leadership of a new sultan, Murad II, besieged Constantinople for the second time. They were repulsed, but by 1444 they had advanced into Greece and Albania, leaving Constantinople isolated though unconquered. Murad II was succeeded by Mehmed (Muhammad) II, called "The Conqueror" because on May 29, 1453, after his artillery finally breached Constantinople's massive walls, the city fell.

After the fall of Constantinople, and during the sixteenth century, the Ottoman system evolved the centralized administrative framework by which the sultans maintained effective control over the extraordinarily diverse peoples in the vast empire.

An important part of this framework was the *millet* system — essentially a division of the

The Janissaries, soldiers of an elite force organized by the Turks in the fourteenth century, were famed as the world's best-disciplined infantry.

The mosque at Kyustendil in Bulgaria was founded during Ottoman rule.

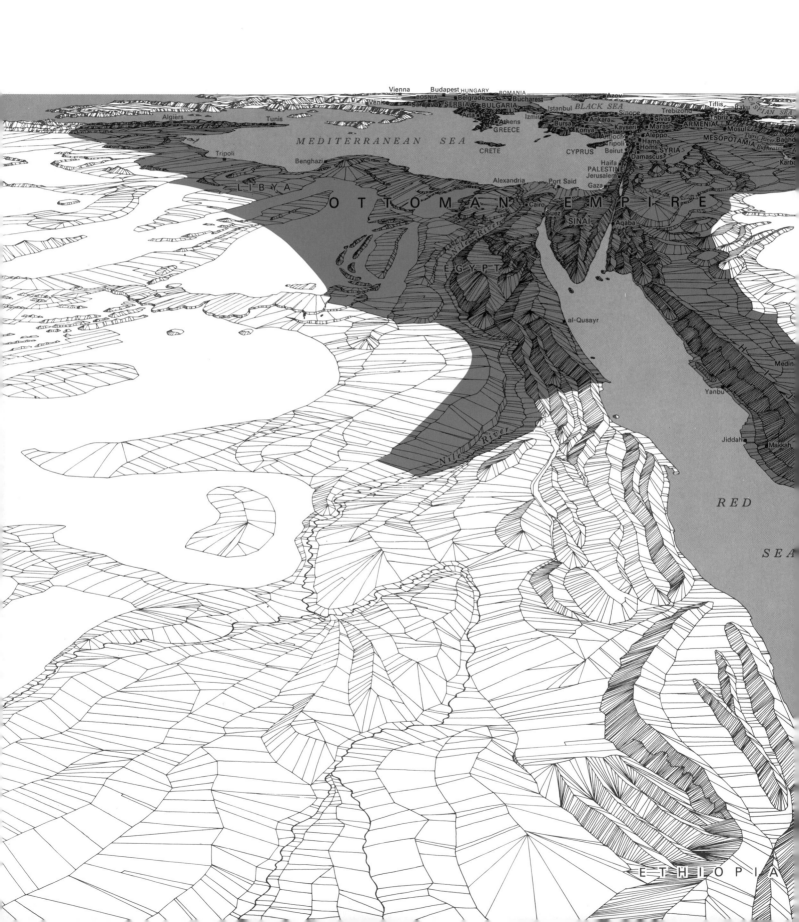

Vienna   Budapest HUNGARY   ROMANIA
                                              CRIMEA   Azov
Venice   BOSNIA Belgrade   Bucharest   CASPIAN SEA
                Sarajevo SERBIA BULGARIA   Istanbul BLACK SEA   Sinope   Trebizond   Tiflis Baku
Algiers   Arta   Sivas   ARMENIA
          Tunis   Athens Izmir   Bursa Ankara   Kayseri Marash   Mosul
                              GREECE   Konya   Antioch Aleppo   MESOPOTAMIA Tigris River Baghd
                        MEDITERRANEAN   SEA   Tripoli Hama Homs SYRIA   Euphrates
                                   CRETE   CYPRUS   Beirut Damascus   Karb
          Tripoli   Haifa
                    Benghazi   PALESTINE
              LIBYA   Alexandria   Port Said Jerusalem
                                                              Gaza
                    OTTOMAN   EMPIRE   Cairo
                                              Suez   SINAI
                                              Blue River   Aqaba
                                        White River
                        EGYPT

                                                                                      Medin
                                   al-Qusayr
                                                                          Yanbu

                                   Nile   River   RED
                                                                    Jiddah   Makkah

                                                                                      SEA

                                                                                ETHIOPIA

THE OTTOMAN EMPIRE AT ITS GREATEST EXTENT

In the sixteenth century three Muslim empires are at or close to the pinnacle of their power and brilliance: the Ottomans under Suleiman the Magnificent, Safavid Persia under Shah Abbas the Great, and Mogul India under Akbar the Great. The Ottoman Turks have conquered and maintain effective control over diverse peoples in a vast empire stretching from Persia almost to the gates of Vienna and along the north coast of Africa to Algiers. In the Arabian Peninsula the Ottomans penetrate to al-Hasa on the Arabian Gulf and to Mocha on the Red Sea. However, the sharifs of Makkah and Medina are virtually independent. Throughout this period the Ottomans contest control of the Arabian Gulf and the Red Sea with the Portuguese, who establish themselves in Bahrain, Muscat, and Hormuz and assist Ethiopia in repulsing the Turks from the coast of East Africa.

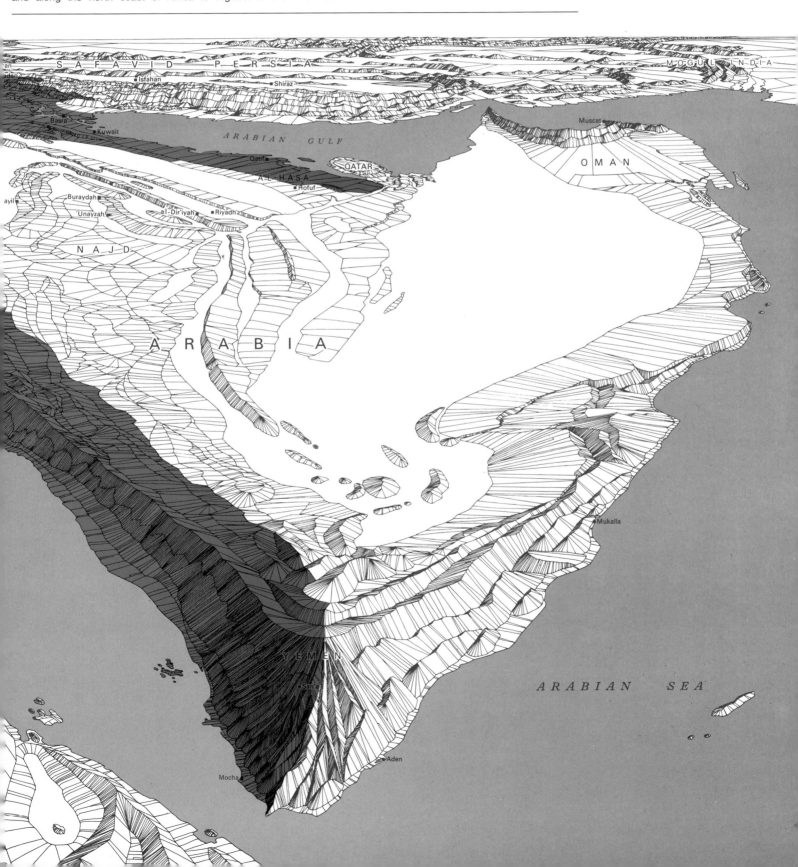

Suleiman the Magnificent, his sovereignty stretching from Budapest to Baghdad and from Mesopotamia to Morocco, was the leading sovereign of the sixteenth century.

The Hijaz Railway, completed by the Turks in 1908, linked Damascus with Medina, nearly 1,300 kilometers to the south.

Mustafa Kemal, also known as "Ataturk," was leader of the nationalist movement that established the secular Turkish Republic.

empire into a communal system based upon religious affiliation. Each *millet* was relatively autonomous, was ruled by its own religious leader, and retained its own laws and customs. The religious leader, in turn, was responsible to the sultan or his representatives for such details as the payment of taxes. There were also, however, organizations which united the diverse peoples. Particularly important were the guilds of artisans which often cut across the divisions of religion and location.

There was also a territorial organization of the empire, at the upper levels of which was a unit called the *muqata'ah* under the control of a noble or administrator who could keep some portion of the state revenues derived from it. The amount varied with the importance of the individual noble or administrator, and he could use it as he saw fit. Such rights were also given to some administrators or governors in place of, or in addition to, salaries, thus insuring a regular collection of revenues and reducing record keeping.

The Ottoman Empire reached its peak in size and splendor under the sultan called Suleiman the Magnificent, who ruled from 1520 to 1566 and was known to the Turks as Suleiman the Law-Giver. But from the middle of the sixteenth century on the empire began to decline. This process got under way as the office of the Grand Vizier gradually assumed more power and indifferent sultans began to neglect administration. Another factor was that the Janissaries became too strong for the sultans to control. The sultans were further weakened when it became customary to bring them up and educate them in isolation and without the skills necessary to rule effectively.

Some sultans later regained power through political maneuverings and by playing off factions against one another, but as a result administration was paralyzed. When Europe found a new route to India — thus eliminating the traditional transshipment of goods through the Arab regions of the empire, revenues began to fall, triggering inflation, corruption, administrative inefficiency, and fragmentation of authority.

Temporary reforms under various sultans, and the still formidable, if weakened, military prowess of the Ottomans helped maintain their empire. As late as 1683, for example, they besieged Vienna. Nevertheless,

the decline continued. Because of the increasingly disruptive part played by the Janissaries the empire, in a series of eighteenth-century wars, slowly lost territory. Because of administrative paralysis, local governors became increasingly independent and, eventually, revolts broke out. Even the various reform movements were balked, and with the invasion of Egypt in 1798 by France it became obvious that the once powerful empire was weakening.

In 1824 Mahmud II finally broke the power of the Janissaries, brought in German advisers to restructure the army, and launched a modernization program. He also brought the semiautonomous rulers in various provinces under control, with the exception of the defiant and able Muhammad 'Ali in Egypt. On the death of Mahmud, his sons continued his efforts with a series of reforms called the *tanzimat*. Some of these were no more than efforts to placate European powers — which by then had great influence on the empire's policies — but others, in education and law, were important. Again, however, the effects were temporary and the empire continued to lose territory through rebellion or foreign intervention.

By the early years of the twentieth century the Ottoman Empire was clearly in decline and was referred to as the "Sick Man of Europe." There were, however, some positive accomplishments in this period, such as the Hijaz Railway. Building the railway was undertaken in 1900 by Sultan Abdul-Hamid, as a pan-Islamic project. Completed in 1908, it permitted thousands of Muslims to make the pilgrimage in relative comfort and safety. It also helped to give the Ottoman government more effective control over its territories in western Arabia.

During the early twentieth century too, a group called the Young Turks forced the restoration of the constitution (which had been suspended by Abdul-Hamid), eventually deposed the sultan, and again attempted to modernize the Ottoman state. The Turkish defeat in the First World War (in which the Ottoman Empire sided with Germany and the Central Powers) finally discredited the Young Turks, however, and paved the way for the success of a new nationalist movement under the leadership of an army officer named Mustafa Kemal, later known as Ataturk or "Father of the Turks." The nationalist government under Ataturk, dedicated to leading

Turkey in the direction of secularism and Westernization, abolished the sultanate, declared a republic, and eventually (in 1924) abolished the caliphate as well.

THE COMING OF THE WEST: The Western world had for centuries been gradually penetrating most of the areas that had once been part of the Muslim empire, and in the latter part of the nineteenth century, in the vacuum left by the long decay and decline of the Ottoman Empire, European powers came to dominate the Middle East.

Among the first Europeans to gain a foothold in the Middle East were the Venetians who, as early as the thirteenth century, had established trading posts in what are now Lebanon, Syria, and Egypt, and who controlled much of the shipping between Arab and European ports. Then, in 1497, five years after Ferdinand and Isabella ended Islamic rule in Spain, Vasco da Gama led a fleet of four Portuguese ships around Africa and in 1498 found a new sea route to India from Europe. Dutch, British, and French frigates and merchantmen followed and began establishing trading outposts along the shores of the Indian Ocean, eventually undercutting both Venetian shipping and the Mediterranean trade on which the Middle East had thrived for millennia.

The process of European penetration was gradual and complex; but there were, nevertheless, clearly identifiable turning points. In the sixteenth century, for example, the Ottoman Empire voluntarily granted a series of concessions called the "Capitulations" to European powers — concessions which gave the Europeans decided advantages in foreign trade in the empire. Another turning point was the invasion of Egypt in 1798 by Napoleon Bonaparte. Hoping to cut Britain's lines to India and cripple its maritime and economic power, Napoleon crushed the Mamluks (who governed Egypt under Ottoman suzerainty) and briefly occupied the country. By defeating Egypt, then still part of the Ottoman Empire, Napoleon exposed the inner weaknesses, both military and administrative, of the sultans, shattered the myth of Ottoman power, and inaugurated more than 150 years of direct political intervention by the West.

Europe's worldwide nineteenth-century search for raw materials, markets, military bases, and colonies eventually touched most of what had been the Arab empire. In 1820 Great Britain imposed a pact on Arab tribes on the coast of the Arabian Gulf; in the 1830s France occupied Algeria; in 1839 Britain occupied Aden, at the strategic entrance to the Red Sea; and in 1869 Ferdinand de Lesseps, with the backing of the French emperor, completed what would become, and still is, one of the key shipping arteries of the world, the Suez Canal.

Western culture spread with Western economic and political control. In Lebanon missionaries from several countries founded a network of schools and universities. By introducing modern Western ideas these fostered the growth of Arab nationalism, contributed to the revival of Arabic literature, and provided a powerful impulse toward modernization. In addition to education, contact with the West led to improvements in medical care and the introduction of Western techniques in agriculture, commerce, and industry. For the most part, however, Western domination tended to benefit the nations of Europe at the expense of the Arab world. Although the Suez Canal, for example, has been of immense value to Egypt, the profits for nearly a century went to European shareholders in the company that managed the canal. Western and Western-stimulated efforts to modernize parts of the Middle East, moreover, often led Middle Eastern rulers to incur debts which led to European financial control and then to European political domination. It was such a series of steps that ended with France occupying Tunisia in 1881 and Britain taking control of Egypt in 1882. Later, in emulation, Italy in 1911 seized Libya.

Resistance to European penetration took several forms. In the cities, Arab intellectuals debated whether modernization or a return to their roots would be the more effective path to the removal of foreign dominance and, consequently, to independence. Elsewhere, Muslim leaders such as the Mahdi in the Sudan and 'Abd al-Qadir al-Jazairi in Algeria took direct action. These struggles were later romanticized and distorted in a wave of books and films on, for example, Gordon of Khartoum and the French Foreign Legion. Still other intellectuals, such as the Egyptian Muhammad 'Abduh and his Syrian disciple Rashid Rida, undertook to reform the Muslim educational

When Vasco da Gama sailed a new sea route from Europe to India, he was guided by an illustrious Arab navigator.

Charles George Gordon, earlier nicknamed "Chinese Gordon," died while leading British forces in Sudan.

Islam numbers many millions of adherents outside the Middle Eastern countries. At top, this page, is the Mopti Mosque in the West African republic of Mali. The religious teacher directly above is Taiwanese. At top right is the mosque in Washington, D.C., a landmark for the 3 million Muslims of North America. Right: Sunset silhouettes a minaret in Sarajevo, an embattled city where some 80 mosques have borne witness to the long Islamic heritage of Bosnia and Herzegovina. Facing page: One of the largest mosques of the Far East is in Bandar Seri Bagawan, capital of the Sultanate of Brunei in Southeast Asia.

system and to restate Islamic values in terms of modern concepts — needs deeply felt by most Muslim thinkers of the nineteenth and twentieth centuries.

Western penetration also drew the Middle East into the First World War, when the Ottoman Empire sided with Germany, and Great Britain, in response, encouraged and supported the Arab Revolt against the Turks. By promising aid — and ultimate independence from the Ottomans — Great Britain encouraged the Arabs to launch a daring guerrilla campaign against Turkish forces, a campaign widely publicized in the writings of T. E. Lawrence ("Lawrence of Arabia").

Arab attacks against the Turks, in which the British soldier-scholar known as "Lawrence of Arabia" played a well-publicized role, were a valuable contribution to the victory of the Allied forces in the Middle East in World War I.

By diverting Turkish strength and blocking the Turkish-German route to the Red Sea and India, the Arab Revolt contributed substantially to the Allied victory, but it did not result in full independence for the Arab lands. Instead, France and Great Britain secretly agreed to partition most of the Arab provinces of the Ottoman Empire between them and eventually obtained mandates from the League of Nations: Britain over Iraq, Palestine, and Transjordan; France over Syria and Lebanon. The mandates were inconsistent with British promises to the Arabs and contrary to the recommendations of President Wilson's King-Crane Commission, a group sent to the Middle East in 1919 to ascertain the wishes of the Arab peoples.

The mandates, however, were granted, thus extending Western control of the Middle East and also setting the stage for one of the most tragic and intractable conflicts of modern times: the conflict over Palestine which has, since 1948, ignited four wars, sent masses of Palestinian Arabs into exile, contributed to the energy crisis of 1973, and, in 1975, fueled the civil war in Lebanon.

The conflict over Palestine actually goes back to 1896, when Theodor Herzl published a pamphlet called *Der Judenstaat* ("The Jewish State"), in which he advocated British-backed Jewish colonization in Argentina or Palestine — with the hope of eventually creating a sovereign Jewish state. Herzl's writings and personal advocacy led to the formal development of Zionism, a political movement dedicated to the creation of such a state, and eventually focusing on Palestine. The Zionist claim to Palestine was mainly based on the fact that there had been periods of Hebrew rule in

Lord Balfour, British Foreign Secretary in 1917, issued the historic declaration which promised British support for a Jewish "national home."

Canaan and the land west of the Jordan River between 1300 B.C. and A.D. 70.

The Arabs considered this claim to be without substance. Palestine, they pointed out, had been part of the Islamic world almost continually for 12 centuries; from 636 to the First World War. In 1917, however, Lord Balfour, the British Foreign Secretary, issued the Balfour Declaration, which promised British support for the establishment of a "national home for the Jewish people" in Palestine providing that "nothing shall be done which may prejudice the civil and religious rights of existing non-Jewish communities" — a reference to the Arabs, who then were 92 percent of the population. The declaration was interpreted by key Zionist leaders as support for a sovereign Jewish state, but both Winston Churchill and Lord Balfour himself later said publicly that "a national home" meant a cultural or religious center, a view that America's King-Crane Commission independently presented. Establishment of a national home did not imply a Jewish state, the commission said.

In the wake of the Balfour Declaration, and during the British mandate, Jewish immigration increased. So, in proportion, did sporadic strife between Arabs and Jews. Immigration nevertheless continued and in the 1930s — with the rise of Adolf Hitler — increased still further. As British efforts to control it generated widespread disapproval in the West and stimulated underground warfare by militant Zionist units against British forces, Britain eventually placed the problem in the hands of the United Nations, which in 1947 voted to partition Palestine into Jewish and Arab states.

Fighting then flared up in Palestine. Six months later, when Britain withdrew and the State of Israel was proclaimed, the Arabs went to war against the newly declared nation. As Jewish forces were victorious — and as stories spread that some 250 Arab civilians had been massacred in a village called Deir Yassin — thousands of Palestinians fled, among the first of today's more than 4 million refugees and exiles. The United Nations negotiated a truce, but fighting became endemic and war broke out again in 1956, 1967, and 1973. The 1967 war triggered underground warfare by Palestinian militants, whose attacks were primarily aimed at Israel, but also included strikes in Europe. A number of United

Nations Resolutions were passed calling for peace, the return of the refugees to their homes, Israeli withdrawal from occupied territories, and the establishment of permanent boundaries.

No tangible progress was made, however, until 1977, when President Anwar Sadat of Egypt traveled to Jerusalem and, in an unprecedented initiative, appeared before the Israeli parliament. In March 1979 — after extensive mediation by President Jimmy Carter — Egypt and Israel signed a treaty to which the United States was also a signatory. This ended the long-standing state of war between Egypt and Israel but did nothing to resolve the status of the Palestinians. Efforts toward an overall solution, including a peace plan put forward by Saudi Arabia, continued, but the situation was further complicated by Israel's invasion of Lebanon in 1982. In late 1987 Palestinians living in the West bank and Gaza began the *intifadah*, or civil uprising, against Israeli occupation forces.

The *intifadah* and other developments in the Middle East succeeded in drawing world attention to the Arab-Israeli problem and in October 1991 discussions on holding comprehensive Middle East peace talks got under way in Madrid under the sponsorship of the United States and the then-Soviet Union. In the summer of 1993 an unexpected breakthrough occurred. Palestinians and Israelis, after six months of secret negotiations, achieved not only mutual recognition for the first time, but also an interim plan for Palestinian self-rule in the Gaza Strip and Jericho, as well as agreement on further, staged negotiations. This accord was signed in Washington in September 1993 under the auspices of President Bill Clinton. In July 1994 PLO Chairman Yasser Arafat moved his headquarters to the newly autonomous areas. Another step in Arab-Israeli relations was taken later that year when King Hussein of Jordan and Prime Minister Yitzhak Rabin of Israel met on their common border to sign a treaty of peace between their nations. In the mid-1990s the basic questions underlying the decades-old conflict — the nature and extent of the future Palestinian state, the status of Jerusalem, and Israel's continued occupation of other Arab territories — were still unresolved. Hope had risen, however, that the long-elusive settlement of the Palestine problem might, in time, be reached.

REVIVAL IN THE ARAB EAST: Elsewhere in the Arab world, meanwhile, the last vestiges of European political dominance were being eliminated. Egypt, for example, after ousting in 1952 a royal dynasty going back to the 1800s, forced the British to relinquish control of the Suez Canal and withdraw from the country. Algeria, 10 years later, won its independence from France after six years of bitter warfare. Even earlier, Iraq, Syria, and Lebanon had broken their ties with Britain and France.

This tumultuous period also saw an increase in the influence of the United States and the Soviet Union in the Middle East. Neither power had played a major role in the early phases of penetration, but this changed as they developed conflicting interests in the region. Their rivalry was a major polarizing force in Middle East politics until the dissolution of the Soviet Union in 1991.

In the same period, the Arab countries themselves, voluntarily and pragmatically, continued to adopt Western techniques, forms, and to some extent concepts. Most Arab countries, for example, have embraced the concept of the sovereign nation state and Western patterns of political administration: parliaments, political parties, and constitutions. Many, too, have adopted Western legal codes, have accepted international and regional organizations and international courts as means of dealing with other nations, and have organized and equipped their armed forces along Western lines. In recent years, most Arab countries have also adopted the modern industrial economy as a national goal and introduced modern techniques of agriculture and modern methods of transport and mass communications, and invested vast sums in education.

If Western influences are important in the Middle East, however, they are by no means paramount. Western forms have been adapted as much as they have been adopted, and healthy hybrid forms and concepts abound. More importantly, traditional values are still deeply cherished and promoted. In sum, modernization has not been entirely synonymous with Westernization. By the final decades of the 20th century the Arabs had emerged as full and independent participants in the affairs of the world. In the forefront was Saudi Arabia, the heartland of Islam, and the site of the momentous events which initiated Islamic history 14 centuries ago.

Inaugurated in 1971 amid hopes it would revolutionize the life of Egypt, the Aswan High Dam is, in its own way, a monument as great as any left by the pharaohs.

Chronological Char[t]

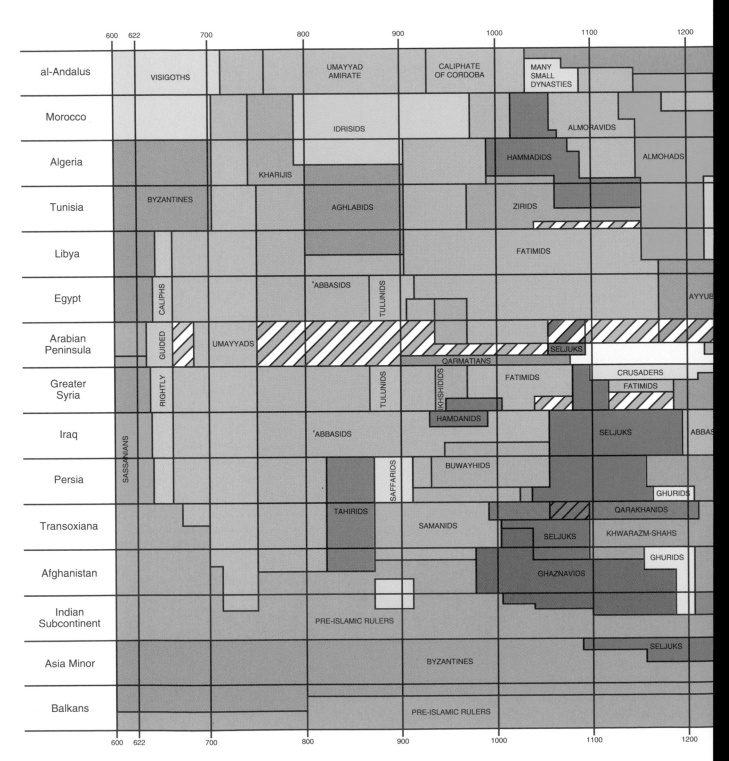

	600	622	700	800	900	1000	1100	1200
al-Andalus		VISIGOTHS		UMAYYAD AMIRATE		CALIPHATE OF CORDOBA	MANY SMALL DYNASTIES	
Morocco				IDRISIDS			ALMORAVIDS	
Algeria			KHARIJIS			HAMMADIDS		ALMOHADS
Tunisia		BYZANTINES		AGHLABIDS		ZIRIDS		
Libya						FATIMIDS		
Egypt		CALIPHS		'ABBASIDS	TULUNIDS			AYYUB
Arabian Peninsula		RIGHTLY GUIDED	UMAYYADS		QARMATIANS	SELJUKS		
Greater Syria		RIGHTLY		TULUNIDS	IKHSHIDIDS	FATIMIDS	CRUSADERS FATIMIDS	
					HAMDANIDS			
Iraq	SASSANIANS			'ABBASIDS			SELJUKS	ABBAS
Persia				SAFFARIDS	BUWAYHIDS			GHURIDS
Transoxiana				TAHIRIDS	SAMANIDS		QARAKHANIDS KHWARAZM-SHAHS SELJUKS	
Afghanistan						GHAZNAVIDS		GHURIDS
Indian Subcontinent				PRE-ISLAMIC RULERS				
Asia Minor					BYZANTINES			SELJUKS
Balkans					PRE-ISLAMIC RULERS			

622, BEGINNING OF THE ISLAMIC ERA

100

Of Islamic History

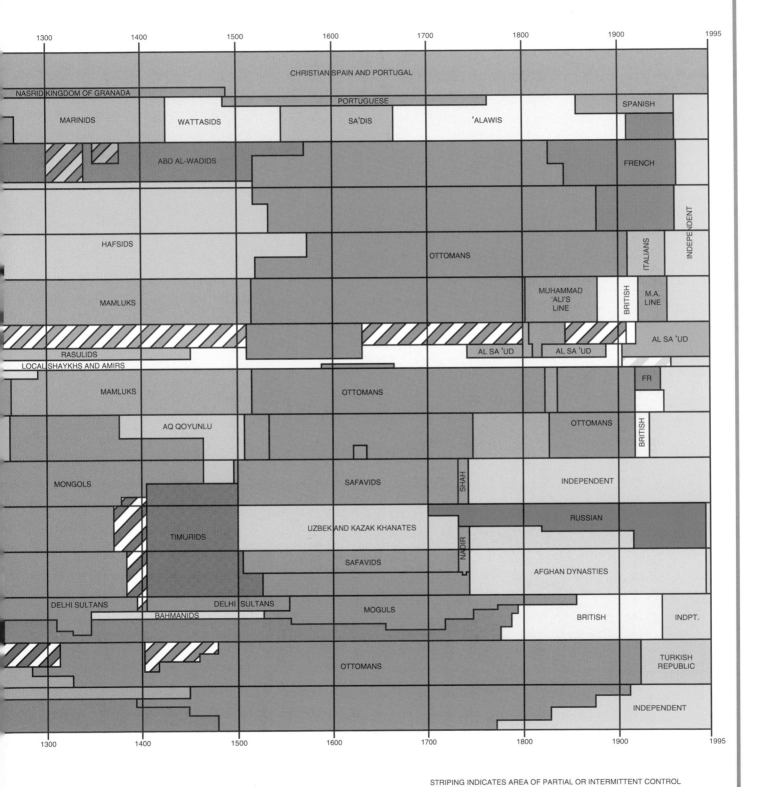

1300 | 1400 | 1500 | 1600 | 1700 | 1800 | 1900 | 1995

CHRISTIAN SPAIN AND PORTUGAL

NASRID KINGDOM OF GRANADA

PORTUGUESE

SPANISH

MARINIDS | WATTASIDS | SA'DIS | 'ALAWIS

ABD AL-WADIDS

FRENCH

INDEPENDENT

HAFSIDS

OTTOMANS

ITALIANS

MAMLUKS

MUHAMMAD 'ALI'S LINE

BRITISH

M.A. LINE

AL SA 'UD

RASULIDS

AL SA 'UD

AL SA 'UD

LOCAL SHAYKHS AND AMIRS

FR

MAMLUKS

OTTOMANS

OTTOMANS

BRITISH

AQ QOYUNLU

MONGOLS

SAFAVIDS

SHAH

INDEPENDENT

UZBEK AND KAZAK KHANATES

RUSSIAN

TIMURIDS

SAFAVIDS

NADIR

AFGHAN DYNASTIES

DELHI SULTANS | DELHI SULTANS | MOGULS | BRITISH | INDPT.

BAHMANIDS

OTTOMANS

TURKISH REPUBLIC

INDEPENDENT

1300 | 1400 | 1500 | 1600 | 1700 | 1800 | 1900 | 1995

STRIPING INDICATES AREA OF PARTIAL OR INTERMITTENT CONTROL

101

SAUDI ARABIA:
THE HEARTLAND

Salt flats called *sabkhahs* — sometimes partly covered by encroaching sand — are a common sight in low-lying areas.

Hummocks of sand held by desert shrubs characterize *dikakah* terrain.

Massive sand dunes of the Rub' al-Khali.

In the history of Saudi Arabia, location, terrain, and climate have all been significant factors. Because the peoples of the Arabian Peninsula were cut off from Africa and Iran by narrow seas and separated from the northern civilizations by difficult terrain, and because they lived, to a large extent, in or by forbidding deserts, they developed cultures in relative isolation from those of the Tigris, Euphrates, and Nile river valleys. Coupled with a harsh climate and the peninsula's particularly demanding terrain, this isolation substantially affected what is now Saudi Arabia. Nevertheless Arabia belongs to and is part of a huge and relatively uniform geographical unit stretching from the west coast of Africa across 10,000 kilometers to the Indus Valley.

The Arabian Peninsula is shaped like the head of an axe, the butt attached to the Asian continent in the north and the edge facing the Arabian Sea, an arm of the Indian Ocean. In the west the Red Sea divides the peninsula from Africa and in the east the Arabian Gulf divides it from Iran.

Close to the Red Sea coast, a range of mountains drops sharply toward the sea on the west, while eastward the land slopes gradually across the peninsula to the Arabian Gulf. Near the center of Arabia the slope is interrupted by a series of parallel escarpments, like folds in a piece of cloth, with steep western faces; but on the eastern side the land again assumes a gentle slope until, on the shores of the Arabian Gulf, it merges almost imperceptibly with the sea. There are local mountains, gravel plains, buttes and mesas, great stretches of sand, lava fields, oases, and a network of great and small wadis — riverbeds that are usually dry. There are mountains too along the southern coast, in southern Yemen, and in Oman farther to the east.

Geographically the Arabian Peninsula may be divided into a number of regions, each characterized by distinctive terrain: the coasts, the sand deserts, the plateaus, the escarpments, and the mountains.

ARABIAN GULF COASTAL REGIONS: The strip of land bordering the western and southern parts of the Arabian Gulf is low and sandy and the configuration of the coast is extremely irregular. For some distance from the coast the sea is shallow and full of shoals and in many places the waterfront shifts back and forth over several kilometers at the whim of tide and wind. Shallow fords connect some islands with the mainland at low tide, while under some conditions coastal flats are covered by the sea for several kilometers inland. Sand spits often change their form as a result of wind and waves.

One feature commonly found in this area is the salt flat — called *sabkhah* in Arabic. Salt flats occur in depressions in which water, a meter or so below the surface, evaporates and leaves a deposit of salt. This salt, mixed with silt, forms a thin crust which dries into a flat, sometimes cracked or puffy, surface. In dry weather packed tracks on this surface provide excellent driving for automobiles. But when rain falls the sabkhahs turn into impassable morasses; after a short period of drying they form a thin, deceptively dry crust hiding the mud below.

The northeastern coast of Arabia, from Jubail to Kuwait Bay, consists primarily of low, rolling plains covered with a thin mantle of sand which supports bushes and grass — sparse in some places, dense in others. The roots hold the sand in hummocks forming terrain called in Arabic *'afjah* or *dikakah*. Southwest of Kuwait this bumpy terrain merges with the gravel plains of the Dibdibah through which the broad valley called al-Batin runs northeast to the channels of the Tigris and Euphrates. In early days al-Batin was a main route between Iraq and the heart of Arabia.

From Jubail southward lies a fairly wide belt of drifting sand piled up into large dunes. Scrub palms, and green oases such as Qatif and al-Hasa, break the monotony of the barren land and show the presence of water under the sand. Widening to the south, this sand belt merges with the sand area known as the Jafurah, which in turn runs into the great sand desert of the Rub' al-Khali, "the Empty Quarter" — the region of the highest and most impressive dunes in the peninsula and one of the great sand areas of the world.

The sands of the Jafurah lie upon flat gravel plains called *hadabahs*. To the west of the sands the plains spread out in broad, gravelly sheets from the oasis of al-Hasa in the north to an area some distance south of Yabrin.

West of the coastal strip sands is a barren plateau called the Summan, 80 to 240 kilometers wide, where old stream channels and other forms of erosion have cut the rock plain into irregular terrain. From this plateau isolated buttes and mesas as well as extensive tablelands project into the coastal lowlands.

THE GREAT NAFUD AND THE DAHNA:

In the north of the peninsula lies an expanse of reddish sand, known as the Nafud, covering approximately 57,000 square kilometers and reaching almost to the Jordan border. Nafud is the term used for a sand region in northern Saudi Arabia, and what is known to Westerners as the Great Nafud is called simply *al-Nafud* by the Bedouins. It contains a few watering places and affords good grazing for camels, sheep, and goats in winter and spring.

From the Great Nafud in the north a great arc of sand, one of the distinctive geographical features of Saudi Arabia, sweeps down to the Rub' al-Khali in the south. The middle of this arc, a long narrow belt of sand called the Dahna, is approximately 1,300 kilometers long. The sands here are from medium to fine in grain, and the color, because of iron oxide (hematite) stain, approaches a deep orange, especially in the morning and late afternoon light. Trailing off to the south of the Great Nafud and generally parallel to the Dahna are other large stretches of sand lying west of the Tuwayq Escarpment.

THE EMPTY QUARTER:

In the southern part of Saudi Arabia is the immense body of sand known as the Rub' al-Khali or Empty Quarter. Occupying a basin bordered by the mountains of Oman on the east, the plateau behind the coastal escarpments on the south, and the foothills of the mountains of Yemen and 'Asir on the southwest, the Rub' al-Khali is approximately 1,200 kilometers long and has a maximum width of nearly 650 kilometers. It covers an area of about 650,000 square kilometers (bigger than France, Belgium, and Holland together) and is the largest continuous body of sand in the world.

The sands of the Empty Quarter are not all of one type. Some areas are covered by comparatively stabilized sand sheets. Across other areas mobile dunes move with the prevailing winds and are shaped by the winds into varying forms. Dunes may be star-shaped, dome-shaped, or crescent-shaped. Masses of sand sometimes form long single or parallel veins or ridges — *'uruq* in Arabic — and sand mountains of complex arrangement attain heights of 150 to 250 meters.

Most of the Rub' al-Khali is uninhabited except after the infrequent rains, when Bedouins move in to take advantage of the pasturage. A few parts support scattered nomads, especially along the regular north-south crossing which follows a string of water wells between longitudes 50° and 52° east. Bedouins also frequent the northeastern section, which contains many wells of salty water satisfactory for the watering of camels. The Bedouins call the area *al-Ramlah* or *al-Rimal*, the words for sandy areas. The Rub' al-Khali, formerly crossed only by hardy Bedouins, was one of the last regions on earth to be explored by Westerners. Bertram Thomas, the first such explorer, made a trip in 1931 from Salalah on the Arabian Sea to the peninsula of Qatar on the Arabian Gulf. In 1932 H. St. John Philby explored the central and northwestern portions, which he described in his book *The Empty Quarter*. Yet another Englishman, Wilfred Thesiger, made several trips through unexplored parts of the Rub' al-Khali from 1945 onward.

NORTHERN SAUDI ARABIA:

The northern region of Saudi Arabia, lying between the Great Nafud and the Jordanian and Iraqi borders, is geographically part of the great Syrian Desert. It is made up of rock, gravel plains, and lava beds.

The plateau between the northern extremity of the Great Nafud and the junction of the borders of Saudi Arabia, Jordan, and Iraq attains an elevation of 1,000 meters and forms the divide between the drainage east to the Euphrates and west to Wadi al-Sirhan, a great depression 320 kilometers long, 30 to 50 kilometers wide, and 300 meters below the level of the plateau. For thousands of years one of the most important trade routes connecting the Mediterranean and central Arabia passed through Wadi al-Sirhan and the oasis of al-Jawf.

ESCARPMENT REGION:

West of the sands of the Dahna lies Najd, the heartland of Saudi

The rocky Summan plateau parallels the coastal lowlands.

The narrow Dahna sand belt stretches in a long arc from the Great Nafud southward to the Rub' al Khali.

Saudi Arabia, while arid for the most part, exhibits a varied geography that in places belies its sand desert image. Rugged on the western side, with stark peaks rising boldly from a narrow coastal plain, the land tilts roughly eastward until, sandy and low, its soft edge merges almost imperceptibly with the waters of the Arabian Gulf. Most spectacular are the southwestern coastal highlands of 'Asir, seen at left above, where bare mountains of jagged igneous rock some 3,000 meters high plunge to terraced green fields and cultivated lowland valleys. Below, far left: Sinuous dunes of the Rub' al-Khali often interconnect in a network of overlapping sand ridges to form a nearly impenetrable barrier to the south; not until this century was the region crossed by travelers other than hardy Bedouins. Center: Eroded limestone formations break the monotony of the central plateau region as several parallel escarpments curve through central Arabia in a gener- ally north-south direction; most prominent is the Tuwayq Escarpment, whose cliffs seen here rise an average of 860 meters above sea level. Left: Lines of deep-rooted tamarisk trees keep dunes from engulfing farmland at the oasis town of Buraydah in northern Najd, where fingers of sand project south from the Great Nafud desert. Directly above: Sea, sand, and shallow shoals blend into graceful patterns along the eastern shore of the peninsula where, from the air, it is sometimes difficult to tell where the land ends and the water begins.

Arabia. About 300 kilometers wide, this region is dominated by several escarpments which are steep on the western side with gentle eastern slopes; among them are the 800-kilometer-long Tuwayq Escarpment and the 'Aramah Escarpment. The Tuwayq Escarpment has an average elevation of 860 meters above sea level with a maximum of 1,100 meters. The top of the escarpment is about 250 meters above the level of the plains to the west. Sedimentary formations that outcrop along the eastern watershed of the Tuwayq Escarpment include the water-bearing strata that support irrigation of important agricultural areas to the east, such as al-Kharj. The 'Aramah Escarpment, about 550 meters above sea level, is less imposing; it is only about 125 meters higher than the plain to the west of it.

The remaining escarpments, which are roughly parallel to the two dominant ones, are lower and not so long. Some, however, appear to be major elevations when contrasted with the adjacent featureless plains.

CENTRAL PLATEAU REGION: Within the arc formed by the Great Nafud and the Tuwayq Escarpment, and reaching as far as the western coastal mountains, lies a broad central plateau about 500 kilometers wide. Most of this area is part of the ancient land mass of Arabia and was not covered by the sea which deposited layers of sediment that make up the surface of eastern Saudi Arabia. The rocks of the central plateau are largely of igneous and metamorphic origin, and over its western half there are extensive lava beds of comparatively recent date.

The elevation of the central plateau ranges from 1,100 to 1,400 meters, although there are depressions as low as 900 meters and ridges higher than 1,800 meters. The plateau has a gentle, gradual slope eastward from the coastal mountains. Short streams of high gradient have carved deep gorges in the steep-walled western flanks of the coastal range. Longer patterns of drainage have developed on the gentler eastern slopes. Of the latter, the greatest is Wadi al-Rumah, which through its extension, al-Batin, runs northeast to the vicinity of Basra in Iraq. Three large valleys, Ranyah, Bishah, and Tathlith, join to flow in rare times of flood into Wadi al-Dawasir, which trends eastward to the Rub' al-Khali.

Water from deep springs supports rich agricultural development in the al-Kharj area.

Wildlife

The Arabian Peninsula is the habitat of a surprising variety of mammals, insects, reptiles, and birds; and the Arabian Gulf, as a five-year marine biology survey in the 1970s showed, plays host to possibly 4,000 species of marine life.

MAMMALS: Some of the peninsula's more beautiful mammals, unfortunately, now exist only in the most isolated areas and others survive only in zoos. Once, for example, three species of gazelles — called in Arabic, the *rim,* the *'ifri,* and the *idm* — roamed the plains and sands of Arabia in large herds and the oryx *(Oryx leucoryx)* — the antelope which may have given rise to the unicorn legends — ranged widely through the Rub' al-Khali and the Great Nafud deserts. By the mid-1950s, however, hunters using trucks and cars had almost exterminated the gazelle, and few, if any, oryx were left outside zoos or private herds.

This disturbing trend was fortunately reversed with stricter hunting controls and, in 1986, with the establishment of Saudi Arabia's National Commission for Wildlife Conservation and Development. By the 1990s, the commission had clearly demonstrated success in its projects to set aside protected wildlife preserves, reintroduce locally endangered species, and establish breeding programs for selected threatened animals.

These measures should ensure the survival of most of the larger Arabian mammals, which exist in surprising variety. For example jackals and two species of desert fox still survive, as do the ibex, or mountain goat, the wolf, the striped hyena, the baboon, the leopard, and possibly the

The desert hedgehog is well protected with a coat of sharp spines.

cheetah, though all but the jackal and the fox are rare. The Arabian tahr, a unique wild goat, still inhabits cliff faces in the mountains of Oman where a national park has been established to preserve it from extinction. There are also occasional sightings of leopards in the highlands of southwestern Arabia and Oman.

Arabia's smaller mammals include the strikingly patterned black-and-white ratel or honey badger; the hyrax —

a plump, furry, rodent-like vegetarian the size of a beaver; the mongoose; the porcupine; and the desert hedgehog. Hares, smaller than American jackrabbits, are found throughout the country and are still often hunted in the traditional way — with falcons and salukis.

One of the most interesting desert mammals is the jerboa, which resembles the American kangaroo rat and is its ecological equivalent. Described by Charles Doughty in his classic *Arabia Deserta* as "a small white aery creature... of a pitiful beauty," it is also one of the most highly adapted desert mammals; at least two species of jerboa can survive in true sand desert, hopping on their hind legs. Sand cats, two species of spiny mouse, gerbils, *jirds* (sand rats), and bats also inhabit the deserts.

BIRDS: Bird life is also surprisingly extensive. Vultures, partridges, and ravens are common in Arabia, as is the owl — which some Bedouins say was originally a wailing woman, seeking her lost child in the wilderness. Along the coasts there are flamingos, egrets, pelicans, cormorants, and many smaller shore birds. In the oases the most common bird is the white-cheeked bulbul, a songbird with a sulfur-yellow bottom. There is also an occasional ring-necked parakeet, with its raucous screech and distinctive long tail. The hoopoe — cinnamon-colored, with a hand-

Until he moves, this winter visitor is nearly invisible against the brush.

some black-and-white crest — is a harbinger of the rainy winter season, and the plaintive voice of *Umm Salim*, the two-striped lark, is familiar to desert travelers, as are sightings of the crested lark and desert lark.

Altogether, more than 250 non-native bird species from all over the Palearctic region winter in Arabia or pass across the peninsula on their way to wintering grounds in Africa, and their seasonal presence brightens the watered localities of the kingdom. Cuckoos, thrushes, warblers, swallows, wagtails, and wheatears, along with hordes of shore birds and waterfowl, are among the seasonal migrants.

Eagles, hawks, and falcons — especially kestrels — fly wild in many parts of Saudi Arabia, and falcons are trapped and trained for hunting, which is still to some extent the sport of the ordinary Bedouin as well as of the rich and noble in Arabia. As a conservation measure, the Government of Saudi Arabia has outlawed the hunting of game birds and most animals with firearms, thus limiting hunting to falconry within strict seasonal limits.

Typically, a falcon's quarry includes the houbara bus-

tard (*Chlamydotis undulata;* Arabic *hubara*), the hare, sand grouse, doves, quail, and the stone curlew. Once common in Saudi Arabia, the ostrich was last reported seen in the north of the kingdom in 1938. Ostriches weighed up to 135 kilograms and covered the ground in 2½-meter strides. Though fragments of ostrich eggshells are still sometimes found in the desert, it seems certain that the bird is extinct on the Arabian Peninsula.

REPTILES: Arabia also has some 40 species of land snakes, most of which are harmless. In the Eastern Province the sand boa, Gray's whip snake, the Arabian leaf-nosed snake, the variable sand snake, and the diademed sand snake are common varieties. Also harmless, if alarming to see, is the Moila snake, a rear-fanged species *(Malpolon moilensis)* that is sometimes mistaken for a hooded cobra. Although it belongs to a different

The sand boa is common in sandy areas of the Eastern Province.

family altogether, the Moila snake spreads its neck and raises its head in the way a cobra does. But although its venom is effective against the small rodents, the snake is not considered dangerous to man.

There are some venomous snakes in Saudi Arabia, however. One, fairly common in the Eastern Province, is the sand viper *(Cerastes cerastes)*. Its bite, however, is rarely fatal; nearly all victims recover without special medical attention. This snake is also called the horned viper, but less than half actually have "horns" — raised scales above the eyes. Desert travelers occasionally see this viper's distinctive track in the sand; the trace is discontinuous and consists of series of evenly spaced J-like marks, for the snake is a sidewinder. Because it is a nocturnal reptile, the viper itself is rarely seen. Another species, the saw-scaled viper *Echis coloratus,* is found in central to western Arabia, and in Oman.

One highly venomous reptile is the black cobra *(Walterinnesia aegyptia)*. A hoodless variety of cobra, this 130-centimeter, blue-black to tar-black snake is dangerous, but fortunately it is also very rare or very shy and few specimens have been reported from Saudi Arabia. Bedouin folklore has it that the black cobra can change itself at will into any other shape it chooses — a camel, a woman, a horse — and does so in order to lure humans to their doom. The true cobra, the hooded variety, is found in southwestern Saudi Arabia along with the puff adder.

(Continued on page 110)

Wildlife

Another venomous reptile is the sea snake, several species of which are common to the warm waters of the Arabian Gulf. Sea snake venom is potentially lethal, but sea

Sea snakes, fortunately not aggressive, are often seen in the Gulf.

snakes, although sometimes curious, are shy and tend to avoid swimmers and skin divers. Furthermore, the snake's mouth is too small, and its gape too narrow, to bite humans effectively.

Because Saudi Arabia is an arid land, many lizards can also be found there. None is poisonous. One common type is the *dabb* or spiny-tailed lizard, a heavy-bodied plant-eating species that grows to a length of 50 centimeters. Usually lethargic, but cantankerous when disturbed, it can run very fast and uses its armored tail like a club to defend itself. To avoid capture, the dabb runs to its deep, spiral burrow or sometimes wedges itself into a rock crevice and inflates its body with deep breaths. According to the Bedouins, dabb meat tastes like chicken.

Another type of lizard is the longer but slimmer *waral*, or desert monitor, a carnivore related to the Komodo dragon. The waral is aggressive and very agile, with relatively long legs. Although Bedouins eat the dabb, they will not eat the waral. There is also the fascinating *tuhayhi*, a small agamid lizard which, if chased, can disappear almost magically into the sand by vibrating its body and sinking under the surface. Its blocky head, like that of an American horned toad, and its black-tipped tail identify it. Lacertid

With a nearly translucent body, the gecko has a natural camouflage.

lizards are common too, especially *Acanthodactylus;* with rows of scales projecting from its toes like oars, this 15-centimeter-long lizard can move with the speed of an arrow as it darts from bush to bush.

Another interesting Arabian lizard is the 13-centimeter-long sand-swimming skink, whose smooth submarine-shaped body and tapering snout enable it to swim through the sand as a fish does in water — and with similar movements of its body. Called in Arabic, *dammusah,* it inhabits sand desert, particularly the steep slipfaces of active dunes.

Other lizards are the pale beige or pink, almost translucent geckos, soft-bodied small lizards with fleshy tails and adhesive toes — welcome guests in some houses in Arabia because of their efficiency in hunting insects; and the *nadus,* a reddish, legless lizard (or, technically, *Amphisbaenid*), which lives under the sand surface, its eyes reduced to nearly functionless dots. The nadus is harmless, and may die in a few minutes if exposed to direct desert sun.

AQUATIC ANIMALS: As there are virtually no rivers or lakes and few swamps in lowland Saudi Arabia, freshwater animals are not many and are largely limited to the Caspian turtle and the frog — both seen frequently in oasis irrigation ditches — and the 10-centimeter killifish (*Aphanius dispar*), which is found in springs in the Qatif and al-Hasa oases. Recently a smaller African freshwater fish was introduced into local irrigation ditches because it feeds on mosquito larvae. Several species of other freshwater fish are found in highland streams and pools in 'Asir.

Streamlined, fast-swimming jacks frequent the Gulf coastal reefs.

Life in the seas around the Arabian Peninsula, however, is rich and varied; sports fishing and spearfishing, as a result, are popular pastimes. One of the most common of the Arabian Gulf gamefish is the *hamur,* a dark-colored grouper, slightly speckled and weighing up to 20 kilograms. Its meat is like that of the sea bass and it lives under rock ledges near the coasts. Another is the *kan'ad,* a silver-colored king mackerel which reaches 20 kilograms and is a hard fighter on the line. Others are the *qidd,* the barracuda — which attains a length of 1.2 meters and a weight of 10 kilograms — and the *sayyafi,* the sawfish. Increasingly rare, the sawfish can weigh more than 100 kilograms and is caught near the shore during the spawning season.

Other fish commonly caught by trolling are the *jihabah,* a variety of tuna; and the *subayti,* sometimes caught with a hook but more often by spearing. Scuba divers can sometimes pick up quantities of delicious spiny lobsters or sea crayfish.

Whales and dolphin are occasionally seen in the Gulf too; they enter from the Indian Ocean. There are also several species of sharks that are relatively common.

Altogether, according to a study of the Gulf's marine biology, there are in the Gulf some 4,000 species of marine life — including fish, echinoderms, mollusks, and algae. The study, made over approximately five years by four marine biologists, was sponsored by Aramco, Saudi Aramco's predecessor, as part of a continuing audit of the environment in its area of operations. Its results were published in 1977 under the title *Biotopes of the Western Arabian Gulf: Marine Life and Environments of Saudi Arabia.* Among the findings were some 1,500 species of marine life previously unreported in the Gulf.

INSECTS: Few of the peninsula's insects have economic importance. Wild bees exist in the 'Asir province and some farmers of the Eastern Province keep hives. Some of the most common insects are either pests or plagues. Locusts, for example, periodically devastated crops and wild vegetation over huge areas until the Government of Saudi Arabia, in cooperation with international agencies, began to monitor the growth of locust populations and suppress incipient swarms. Mosquitoes, which spread malaria, and swarms of flies were also major problems until, after World War II, the government launched extensive pesticide-spraying programs in the Eastern Province. One insect commonly encountered in the deserts of Arabia is the dung beetle. Barely able to fly, it is very industrious on the ground, collecting balls of dung in which to lay its eggs. There is also a variety of arachnids: the scorpion, the velvety red rain mite, small jumping spiders, and the hairy *shabath,* a sun spider or solifuge, also known as the camel spider. A nocturnal arachnid, it is harmless but as it is also aggressive and large — up to 15 centimeters in diameter — it is often mistaken for a tarantula.

The fat-tailed scorpion, like other species, moves primarily by night.

Scorpion stings are painful but rarely, except perhaps in cases of very young children, pose a threat to life. The black "fat-tailed scorpion," *Androctonus crassicauda,* is one of the more common species seen in eastern Saudi Arabia. Several other species, yellow or yellow-green in color, may also be encountered. Scorpions are nocturnal and are seldom seen until their sand burrows or under-stone haunts are disturbed. Avoiding barefoot walking at night eliminates most of the risk of being stung.

Watering places in the central plateau are few and far between, though the high elevation draws more rainfall than comes to the regions to the east. The area is sparsely populated except in the north between the depression of Wadi al-Rumah and the southern edge of the Great Nafud. In the region of the mountain ranges of Jabal Shammar lies the great oasis of Hayil with its many dependent towns and villages.

The oasis of Hayil lies just south of the Great Nafud.

THE MOUNTAINS: Along the eastern shore of the Red Sea stretches a coastal plain confined by mountains dropping sharply toward the sea; it is called Tihamah, sometimes divided into Tihamat al-Hijaz, Tihamat 'Asir, and Tihamat al-Yaman. The abrupt western drop of the coastal range is the result of a break or fault in the crust of the earth, between the land masses of what are now Arabia and Africa, into which poured the Red Sea. The Red Sea, as a result, is deep, but as there are numerous shoals near the Arabian shore the coast affords few good natural harbors. The best are at Jiddah, which is the port of Makkah; Yanbu' and al-Wajh in the north; and Jaizan in the south. In southern Yemen the best ports are Aden on the southern coast of the peninsula and, farther east, Mukalla in Hadhramaut.

Between the Gulf of Aqaba to the north and Makkah, the coastal mountains are seldom higher than 2,200 meters and generally do not exceed 1,250 meters. But in 'Asir and Yemen they are much higher. Southeast of Makkah several peaks are more than 2,500 meters high and elevations continue to rise in the south, reaching their highest point — about 3,700 meters — west of San'a, the capital of Yemen. Continuing on, to fringe the southwestern margin of the peninsula, the mountains average 1,800 to 2,500 meters, while eastward the elevations decrease to 920 meters in the Omani region of Dhufar, except for a peak 1,500 meters high north of the port of Mirbat. The mountains of southern Arabia are composed essentially of sedimentary rock dipping gently northward into the Rub' al-Khali basin, with the southern slopes irregular and precipitous.

Terraced fields are cultivated in the mountain valleys of 'Asir.

Between the eastern extension of the mountains of Dhufar and the southern part of the ranges of Oman, there is a plateau some 250 kilometers long that averages 150 meters in height. Then come the mountains of Oman, known as al-Hajar, with conspicuously

The prince of Arabian wildlife is the falcon, trapped and trained for the centuries-old art of hunting in which these spectacular aerial performers catch their prey by sheer speed and often in a power dive. The Saker species above, *Falco cherruq,* is prized for its keen eyesight; next to it is the bustard called hubara, *Chlamydotis undulata,* a plump wide-winged game bird conspicuous in flight but nearly invisible on the ground. Directly above: Tree frog, *Hyla savignyi,* and a colorful species of beetle. At left: Waral, *Varanus griseus,* a powerful, predatory 30-inch monitor lizard. Opposite page: Red Sea coral reef, world famous for the variety of its undersea life; starfish, *Pentaceraster mammillatus,* a gaudy species with no two members quite the same color; green sea turtle, *Chelonia mydas,* which breeds on Arabian Gulf islands; "sand bubbler" crab, *Scopimera scabricauda,* found in millions on tidal sandbanks. Drawings depict the graceful gazelle and larger oryx, both members of the antelope family now rarely seen, and a tiny hedgehog.

The steep north face of the Omani mountains presents a formidable barrier.

rugged topography; elevations reach 3,000 meters in the central massif of the Jabal al-Akhdar (the Green Mountain). Sweeping westward and northward roughly parallel to the Gulf of Oman, the mountains in the area, with steep slopes both seaward and toward the inland plateau, are very unlike the mountains on the western and southern sides of the Arabian Peninsula, which are characterized by steep slopes on the seaward side and gentle slopes toward the interior.

CLIMATE: Weather conditions in Arabia vary almost as much as the terrain. In the mountains that fringe the peninsula on the west, south, and southeast, annual rainfall is about 50 centimeters and there are often torrential downpours and destructive flash floods; yet parts of the Rub' al-Khali may receive no rain for as long as 10 years. Over much of central, northern, and northeastern Arabia, average annual rainfall is five to 15 centimenters and highly variable from year to year.

Temperature and humidity ranges are equally disparate. In the interior the air is dry while on the coasts summer humidity is excessive, particularly at night. In summer, temperatures in some areas may reach 50°C (122°F) in the shade, but in the spring the days are balmy and the nights are clear. In winter temperatures drop below freezing in the central and northern regions and snow sometimes falls in the mountains, the northern plateau, and even in the capital, Riyadh. There are also strong winds such as the prevailing northwest winds along the eastern coast — the winds called the *shamal* — which frequently whip up dust and sandstorms.

The climate of the peninsula, nevertheless, is largely characterized by aridity and heat. As a result the vegetation, wildlife, and domesticated animals of the peninsula share one distinctive feature: a high degree of adaptation to the special demands of life in the desert. The ability of men to adapt and survive in this harsh environment has had important effects on the history of the peninsula and Saudi Arabia.

ARABIA IN THE ISLAMIC ERA: Because of the outward impetus of Islamic expansion in the seventh and eighth centuries, and the overshadowing historical importance of the brilliant civilizations that were established in Syria, Iraq,

Plants

Most of Saudi Arabia is part of the great desert biome that stretches eastward 10,000 kilometers from the Atlantic coast of North Africa to the Indus Valley. Within this zone, which plant geographers call the "North African-Indian desert floristic region," the terrain and climate are relatively uniform — and so, as a result, is the plant life. The southern parts of the Arabian Peninsula — Yemen, the mountains of Oman, and the highlands of 'Asir — because of their higher altitude and different climate, host a different range of species.

Some areas, such as the central and western deserts, support scattered acacia trees, and in the southwestern highlands there are true forests of juniper and wild olive trees — now under government protection. There are also oases and other areas where man has planted such trees as tamarisks for windbreaks. Most plants, however, are small annual herbs or low shrubs that have adapted to the hot, usually dry climate.

Plants have adapted to the Arabian climate in various ways. Some have "learned" to store water in fleshy stems or in roots buried deep in cooler sands. Most limit their surface area by seasonally restricting the number and size of leaves — through which water would evaporate — and some, additionally, secrete a waxy covering to stems or leaves to reduce evaporation. Others have leaf surfaces covered with fine hairs which both reflect heat and cut down on water losses. Some species, known generally as halophytes or saltplants, have evolved a tolerance to salty soils and mineral-rich water that would kill their unadapted relatives.

Many plants produce well-protected seeds that, in some cases, can remain dormant for years if necessary and then germinate and sprout rapidly. Most of them live out their entire life cycle in one rainy season that may be only a few weeks long.

Another characteristic of Saudi Arabia's vegetation is the small number of species that exist. The Eastern Province, for example, has only about 500, while the considerably smaller Lebanon-Palestine area — which borders on several different floristic regions and where the landforms are very varied — has over 3,500 species. This small number of desert species reflects the difficulties faced by plants in adapting to such a harsh environment. A third feature of Arabian desert vegetation is the rarity of what botanists call "endemic species": plants that are peculiar to the area and are found nowhere else. Most of the species apparently did not arise in the

rhanterium eppaposum

acacia

broomrape

calligonum comosum

horwoodia dicksonia

Thompson

area but gradually migrated to it, as they adapted, from neighboring regions.

Of the plants that have adapted to conditions in the Arabian desert, those common enough to be conspicuous are few. In eastern Arabia, a sedge *(Cyperus conglomeratus)* and a perennial grass *(Panicum turgidum)* are common in the coastal sands. Further inland, the *'arfaj* shrublet *(Rhanterium epapposum)* is the dominant species over hundreds of square kilometers, while a saltbush called *rimth (Haloxylon salicornicum)* covers the ground in poorly drained areas where the groundwater is salty. Both *'arfaj* and rimth provide important pasturage to camels, with the saltbush supplying the salt that grazing animals need.

Southwestern Arabia has native cactus-like plants, but these are euphorbias and stapelias, not true cacti, and are not native to the Old World. The few species of prickly pear *(Opuntia)* that are now found in the Middle East — and sometimes in the flowerbeds of the Saudi Aramco communities — were introduced by travelers.

Although some Bedouins formerly ate a score or more different wild herbs in times of famine, none is now important as food. *Samh,* the seeds of a fleshy leaved *Mesembryanthemum,* are still sometimes collected and ground into flour for making bread in northern Arabia. During their brief season desert truffles, not identical to the European species, are considered a delicacy by Saudi Arabs and Westerners alike.

Some desert plants have scientifically proved medicinal value, though claims made for many other folk remedies are yet to be confirmed. *Boswellia* and *Commiphora* trees, whose frankincense and myrrh were transported to the West more than 2,000 years ago from southern Arabia, still grow in Dhufar and Hadhramaut. But the chief economic value of the desert vegetation is as pasturage for grazing livestock.

As if to compensate for the harsh conditions they must survive, many desert plants have flowers of a beauty out of all proportion to their modest numbers and small size. Fragrant bright yellow spikes of broomrape — a root-parasitic plant sometimes called "desert candle" — color the spring roadsides in eastern Arabia, and the delicate desert camomile grows in scattered carpets on the sand. Farther north and west, on the rocky steppes, dense growths of wild iris brighten the depressions where rain

pools have lain; elsewhere tiny yellow lilies grow, as does the mauve *khuzama (Horwoodia dicksoniae),* a member of the mustard family whose scientific name commemorates the botanical work of the late Dame Violet (Mrs. H. R. P.) Dickson of Kuwait. The fruiting of the *'abal* bush *(Calligonum comosum)* marks the height of the spring flowering season on the deep sands. This leafless woody shrub is valued as firewood by desert travelers, but the winter rains bring out pale pink flowers and by late spring the bush is covered with a brilliant trimming of scarlet fruits.

The small annual but woody *kaftah,* or *kaff Maryam* (Mary's hand), known to botanists as *Anastatica hierochuntica,* figures in Arabian folklore. Its small branches drop their leaves at the end of the spring and roll inwards like closing fingers, enclosing and protecting the seeds in a firm brown ball. With the first rain, or when placed in water, the ball opens completely and drops the seeds for germination. It is likened in Arab tradition to the hand of Christ's mother, clenched in the pains of childbirth.

What is probably the most familiar, and valuable, plant on the peninsula, however, is the date palm *(Phoenix dactylifera),* traditionally the most important single cultivated plant of the entire North African-Indian desert region. The limits of distribution of this tree, like those of the single-humped camel, coincide almost exactly with the boundaries of the desert biome. Throughout this zone it provides a staple food as well as valuable wood and fiber by-products still sometimes used for weaving and thatching. It is so much a part of life in Saudi Arabia that it is included as part of the kingdom's emblem.

Other domesticated plants include wheat, barley, sorghum, and millet — the chief grains grown in Arabia — and alfalfa, a common oasis crop grown for feed. Cotton, rice, melons, grapes, pomegranates, oranges, and other tree fruits are also cultivated, and in recent years adapted varieties of a wide range of vegetable crops have been introduced and enthusiastically accepted. On the high terraces of 'Asir, the once profitable coffee plant has virtually disappeared. Coffee is still grown in the mountains of Yemen, but it has been supplanted there to some extent by the more profitable *qat (Catha edulis)* whose leaves, when chewed, are a mild stimulant. In southern Arabia there is also some cultivation of the dyestuffs henna, safflower, and indigo.

Essentially all of a given year's rain in Saudi Arabia —
except in the mountainous southwest — falls between
November and May and not infrequently in a few
violent downpours that may be the season's sum
total. After such rains in early springtime, wadis,
depressions, and flat sandy areas suddenly burst into
color with the bloom of wild annuals in remarkable
variety. Facing page, at top, lavender *Diplotaxis acris*
blooms in profusion. Far right, above: a desert
camomile, or *Anthemis,* and below, *Diplotaxis harra.*
Both the root parasite, *Cistanche tubulosa,* center, and
the fragrant mauve *Horwoodia dicksoniae,* near right,
are native to Arabia as are the *Matthiola*, left, and
Silene arabica, far left. Directly above, from top: the
daisy-like *Senecio desfontainei; Calligonum comosum,* with blossoms resembling fuzzy balls; a dainty
pink cranesbill from the same family as the geranium.
At left above: a lone acacia tree stands amid kilometers of a summer-blooming bush of the family
Compositae, which is characteristic of basins of the
interior.

al-Andalus, and elsewhere, the history of much of the Arabian Peninsula between the death of the Prophet and the rise in the eighteenth century of the reform movement headed by Muhammad ibn 'Abd al-Wahhab is not adequately chronicled. Yet Arabia continued to play an important role during this long period.

In the course of expansion, many tribesmen flocked to the banner of Islam. Soon men born and raised in the peninsula could be found stationed from the borders of China to the shores of the Atlantic. As many tribes moved to the new lands, the political center of the Muslim world shifted from Medina to Damascus and later to Baghdad (see pages 53-76), and the Arabian Peninsula, relatively isolated from the great events occurring in Syria and Iraq, receded in political importance. Makkah and Medina, however, continued to play a paramount role in the life and culture of the Islamic community. Makkah, of course, had been an important entrepôt in the pre-Islamic period. Now, as the goal of the hajj, the Muslims' holy pilgrimage, it also became important in the spread of ideas. Inevitably, pilgrims from every province of the new Islamic empire would discuss recent developments and establish contacts in faraway lands. Medina, first capital of the empire, adopted home of Muhammad after the Hijrah and site of the Mosque of the Prophet, naturally drew thousands of new Muslims eager to see the scene of Muhammad's activities and those of his successors.

The white headdress of a Muslim in the Philippines identifies him as a *hajji*, one who has made the pilgrimage to Makkah.

Medina, consequently, became a center of learning in which the early Muslim scholars gathered and organized the vast body of traditions — the hadith — going back to the Prophet himself. It was a natural impulse to seek as much information as possible about every aspect of life in the city of the Prophet in order to comprehend fully the language and content of the traditions. These gave invaluable testimony on the known practices of the Prophet, and formed the sunnah, which together with the Quran provided the main bases of the shari'ah, the sacred law of Islam (see page 50).

The heartland of the Islamic faith, therefore, did not lose its importance during the Umayyad and 'Abbasid periods. Umayyad caliphs rebuilt the Mosque of the Prophet and the 'Abbasids devoted a great deal of time and money to providing civic amenities in Makkah and Medina and along the pilgrimage route

from Baghdad to the Hijaz. The Caliph Harun al-Rashid and his wife Zubaydah made nine pilgrimages, and Zubaydah saw to the construction of an aqueduct to bring drinking water to Makkah. She also constructed a series of wells and caravanserais along the route from Baghdad to Makkah, which has been known ever since as *Darb Zubaydah* or "the Road of Zubaydah."

The Arabs of the peninsula continued to play an important role in the spread of Islam. Merchants from southern Arabia, for example, brought the new faith into East Africa and islands of the Indian Ocean. Even today, in the tranquil backwaters of Quilon and Cochin on the Malabar coast of India, small communities of Muslims may be found — many proudly tracing their genealogies back to families originally from Oman, Hadhramaut, Yemen, or the Hijaz.

Islam spread to Java and the Philippines in similar fashion. According to tradition, the first Muslims arrived at Jolo Island in the Sulu Archipelago in 1380 on a Chinese junk bound for Canton and began to spread the teachings of the Prophet.

In the intervening centuries, meanwhile, the sharifs of Makkah, direct descendants of the Prophet, had made this city, rather than Medina, politically dominant in the Hijaz. In their early years the sharifs were independent rulers, but from the twelfth century on they became to a degree dependent upon the rulers of Egypt to keep the pilgrimage routes to Makkah and the trade routes of the Red Sea and across the peninsula secure. As a result Makkah continued to be not only the holiest city of Islam but also a prosperous center of trade, especially during the pilgrimage season. The well-known twelfth-century Andalusian traveler Ibn Jubayr visited it and reported that "from all parts produce is brought to it, and it is the most prosperous of countries in its fruits, useful requisites, commodities, and commerce... Not on the face of the world are there any goods or products but that some of them are in Makkah...."

Such prosperity, however, also attracted the attention of the Crusaders, who by then had established themselves at the eastern end of the Mediterranean. In 1181 Reynaud de Châtillon, master of the castle of Karak near the Dead Sea, attacked a pilgrim caravan in northern Arabia. The following year he sent a fleet into the Red Sea, intending to attack Medina. Ibn Jubayr, who arrived in Makkah shortly after this event, says:

> They had then sailed forth to harass (Muslim) pilgrims ... went on to 'Aydhab where they caught a ship coming with pilgrims from Jiddah. On the land they seized a large caravan journeying from Qass to 'Aydhab and killed all in it, leaving none alive. They captured two ships bringing merchandise from the Yemen and burnt many foods prepared on the beaches as provisions for Makkah and Medina — God exalt them. Many infamous acts they committed .... The worst, which shocks the ears for its impiousness and profanity, was their aim to enter the City of the Prophet — may God bless and preserve him — and remove him from the sacred tomb ....

The Crusaders, as it turned out, were repulsed by a Muslim fleet and the raiders were captured. But the expedition, nevertheless, was of greater significance than it might seem. For although the attack was essentially part of the religious conflict of the period, it was also a faint distant harbinger of the coming age of Western colonial expansion.

In 1256 Hulagu Khan, a grandson of Genghis Khan, crossed the Oxus River and entered the lands of Islam. Two years later Baghdad fell before the Mongol hordes and the caliph was put to death. For several years there were no pilgrim caravans from Iraq, but as peace settled over the land the caravans to Makkah resumed.

In succeeding years, Makkah continued to receive visits from the various rulers who came to hold power in the Islamic world. One was Mansa Musa of Mandingo, a new Muslim kingdom far away in West Africa, whose arrival in Makkah in 1325 caused a sensation. His baggage train, it was said, contained a hundred camel-loads of gold, each weighing 300 pounds, and his arrival in Makkah reminded Muslim pilgrims from other lands that despite the Mongols and the Crusades, Islam was still expanding.

Muslim rulers also continued to give financial support for the maintenance of the Holy Places. In the late fifteenth century Qait-Bay, a Mamluk sultan of Egypt, helped the sharifs of Makkah repair and restore the mosques and public buildings of Makkah and Medina. The aqueduct built earlier by Zubaydah was cleaned and repaired, the inner chamber of the Ka'bah was faced with marble, and four centers — one for each of the four canonical schools of Islamic law — were opened, along with a new almshouse for the relief of indigent pilgrims.

CARAVANS TO MAKKAH: It was during this period too that the organization of the caravans to Makkah took its classical form. The caravans were highly organized, for in those days nothing could be left to chance in the deserts of Arabia. They were led by an official called *Amir al-Hajj*, the commander of the pilgrimage, who had the absolute authority of a ship's captain at sea. Under him were a troop of soldiers, outriders, and guides, who were charged with defending the pilgrims from attack; notaries, a judge and secretary; an official charged with the care of the animals and another in charge of provisions; a saddler, a chef with a full staff of cooks; and even an inspector of weights and measures to make sure no pilgrim was cheated along the way.

The caravans marched at night, traveling until sunrise, in order to avoid the heat of the day. During the cooler months they would then rest and set off again later, riding until it was time to say the evening prayer. But during the summer, when travel in the afternoon was impossible, the caravans would move only at night.

The caravan was divided into three parts — an advance guard, the main body, and a rear guard. As many as eight outriders who knew the lay of the land would precede the advance guard, and at night each man in the advance guard carried a torch to guide those who came behind. In the main body of the caravan — which usually traveled half a kilometer behind the advance guard — there would be physicians, spare camels for carrying the sick and infirm, and next, a camel carrying an elaborate litter covered with silk, in which rested a beautifully inscribed Quran, a gift to Makkah. In one of the caravans this was followed by 15 more camels carrying the various sections of the *kiswah*, the embroidered silk covering for the Ka'bah, renewed each year. These camels with their precious burdens were accompanied by a group of

A painting by a Damascus artist depicts pilgrim caravans from Egypt and Syria meeting at 'Arafat in 1905. The sumptuously decorated litters carried equally sumptuous gifts for the Holy Mosque in Makkah.

musicians, who played marching songs throughout the journey.

There were several main caravan routes to Makkah. Pilgrims from Turkey assembled in Istanbul and slowly marched across the Anatolian plateau, gathering pilgrims from every town through which they passed. In Damascus they joined the pilgrims from Syria, then crossed northern Arabia and angled south to Medina and Makkah. This was the same route later followed by the Hijaz Railway from Damascus to Medina, which after its completion in 1908 carried thousands of pilgrims from Turkey and Syria to the Hijaz.

Pilgrims from Iraq assembled in Baghdad and followed "the Road of Zubaydah," which passed near the present pump station of Rafha on the Trans-Arabian Pipe Line (Tapline), crossed Najd directly to Medina, and picked up the main road to Makkah.

A third route, used by pilgrims from Egypt, ran along the Red Sea coast of Arabia, which was more plentifully supplied with water than the routes across north-central Arabia.

The pilgrims coming by these routes converged on Makkah where they joined other pilgrims who had traveled overland across Africa or sailed west from India and the East Indies, thus linking Arabia, the religious heartland of Islam, with Muslims in the rest of the Islamic world.

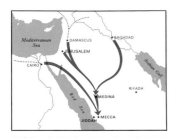

Until the nineteenth century, the main routes of the pilgrim caravans to Makkah were overland from Cairo, Damascus, and Baghdad.

High above the bay at Muscat stands the fort built by the Portuguese in the sixteenth century.

THE FOREIGN INTRUSION: In the early sixteenth century the Muslims of Arabia were feeling a new impact of Western intrusion into the Middle East: by then Vasco da Gama had pioneered the new ocean route from Europe to India (see page 95) and other European powers were beginning to build trading stations and otherwise establish themselves along the Arabian Gulf and the Red Sea.

In Arabia the effect of the European expansion was at first slight. But the Portuguese were well aware of the commercial importance of Jiddah, the port of Makkah. As one traveler wrote to Portugal, "Jiddah... is the most prosperous commercial center in the whole of the Red Sea." They tried, therefore, to secure a foothold on the Red Sea and, when that failed, turned to the far side of Arabia, where in 1507 they captured Hormuz, at the mouth of the Arabian Gulf, and Muscat — and then Bahrain

in 1521. Close on the heels of the Portuguese came the Dutch, the English, and the French, all attempting to establish trading posts and to control ports. In 1649 the Omanis expelled the Portuguese from Muscat, built up a fleet of their own, and restored a degree of Arab control over the Gulf.

The Ottomans (see pages 91-94), who had become the major power in the Islamic world, spared neither money nor pains to beautify Makkah and Medina, and the sister of Suleiman the Magnificent financed another renovation and repair of Makkah's always precarious water supply. Later, under Selim II, the architect Sinan — considered by some authorities to have been as creative an architect as Michelangelo — renovated the Sacred Mosque in an undertaking that would endure until the kings of Saudi Arabia commissioned still more extensive renovations in the twentieth century.

MUHAMMAD IBN 'ABD AL-WAHHAB: In the early eighteenth century, in Najd, the central region of Arabia, the man was born who would in alliance with the House of Sa'ud pave the way for the establishment of the modern Kingdom of Saudi Arabia. His name was Muhammad ibn 'Abd al-Wahhab. The son of a qadi, a religious judge versed in the shari'ah law (see page 50), he was born and grew up in a town called al-'Uyaynah north of Riyadh. Later he studied in Makkah and Medina and then spent some time in Basra and in al-Hasa before returning to Najd. At the end of this period of study Shaykh Muhammad ibn 'Abd al-Wahhab had become familiar with the currents of religious thought and the great political and social problems of his time. By then too he had concluded, from his observations of the world and his wide reading, that reforms were imperative.

Shaykh Muhammad, on his return, was particularly appalled by the superstitions of both the town dwellers and tribesmen of Najd and their devotion to the cult of local saints, which had grown up in many areas of the Islamic world — encouraged perhaps by the isolation of many outlying regions from the centers of Islamic orthodoxy. He began to call for a return to basic principles of Islam as contained in the Quran and the sunnah.

Shaykh Muhammad was the very opposite of an innovator. He was a salafi, a traditionalist, and certainly did not intend to found a new sect of Islam, or even to make a new interpretation of its basic teachings. He believed in a return to Islam as presented in the Quran, a perception of the unimpaired and inviolate oneness of God. Anything that conflicted with that unity was to be denounced as *shirk*, polytheism — that is, unlawful association of anyone or anything with God.

Shaykh Muhammad followed the Hanbali School of Islamic law. The chief influence on him was the thirteenth-century theologian Ibn Taymiyah who, as a modern interpreter has put it, "bowed to no authority higher than the Quran, tradition and the practice of the community and lifted his voice high against innovation, saint worship, vows, and pilgrimages to shrines."

Like most reformers Shaykh Muhammad ran into opposition. He was driven from al-'Uyaynah and forced to take refuge in the town of al-Dir'iyah very close to present-day Riyadh. The old town of al-Dir'iyah was located on high ground above Wadi Hanifah. The ruler of al-Dir'iyah, Muhammad ibn Sa'ud, met him and welcomed him.

THE HOUSE OF SA'UD: It was, for the future Kingdom of Saudi Arabia, a fateful meeting. The two men — one an idealistic reformer bent on bringing the Islamic community back to the purity of its origins, the other an astute chieftain — formed an immediate warm regard for one another's qualities, and established a relationship that links their descendants to the present day.

As Shaykh Muhammad was an eloquent and sincere preacher, his uncompromising reaffirmation of the basic beliefs of Islam soon won many followers. It also alarmed many leaders in Najd, especially those in independent Riyadh, so near al-Dir'iyah. The nascent movement, whose followers called themselves *al-Muwahhidun,* "those who affirm the Unity of God," or "Unitarians," was seen as a threat to established authority.

Muhammad ibn Sa'ud, the friend and protector of Shaykh Muhammad, died in 1765, but under his very able son and successor, 'Abd al-'Aziz, who had been carefully nurtured in the reformist tradition of the Muwahhidun, the movement continued. In 1773 — three years before the American Declaration of Independence — 'Abd al-'Aziz

captured Riyadh and within 15 years controlled all of Najd. Then, in the winter of 1789-90, the Muwahhidun crushed the paramount tribe of al-Hasa, the Bani Khalid, at a famous battle at the hill of Ghuraymil just southwest of present-day Abqaiq. As the Bani Khalid had opposed the rise of the House of Sa'ud from the beginning, the victory at Ghuraymil was very important.

Two years after that battle, Shaykh Muhammad died at the age of 90 and was buried, as he would have wished and in accordance with the sunnah of the Prophet, in an unmarked grave in the town of al-Dir'iyah. But the movement which he had founded — which is sometimes incorrectly referred to as "Wahhabism" — did not die with him. On the contrary, it continued, under the leadership of the House of Sa'ud, to win adherents. In 1801 the reform forces were strong enough to strike inside Iraq and destroy the mausoleums which were built on the tombs of the saints at Karbala.

In 1803 'Abd al-'Aziz was succeeded by his son Sa'ud, but again the movement lost no ground by the death of a leader. By the early years of the nineteenth century, Saudi power extended over most of the Arabian Peninsula, including much of Oman, parts of Yemen, and the holy cities of Makkah and Medina in the Hijaz. Raids were made into Syria and in 1811 a campaign against Baghdad was contemplated.

The historic town of al-Dir'iyah, now restored, provided refuge for the eighteenth-century reformer Muhammad ibn 'Abd al-Wahhab.

By then too the territory controlled by the House of Sa'ud or Al Sa'ud, as it is known in Arabic, was so vast that the peninsula was divided into 20 districts, with a loyal member of the reform movement as governor of each, and a qadi in charge of religious matters and public education. Part of the revenue of each district was consigned to the construction of forts and mosques, and many towns were provided with a simple, unpretentious mosque and a fort.

In Constantinople, meanwhile, the Ottoman sultan had been alarmed at these incursions into what were, nominally, territories of the Ottoman Empire. He therefore ordered Muhammad 'Ali, the governor of the Ottoman province of Egypt, to undertake a punitive expedition against the Saudis and to re-establish Ottoman authority over the holy cities.

Muhammad 'Ali delegated the task to his

Muhammad 'Ali Pasha governed the province of Egypt for the Ottoman Empire.

Family Tree Of The House Of Sa'ud (simplified)

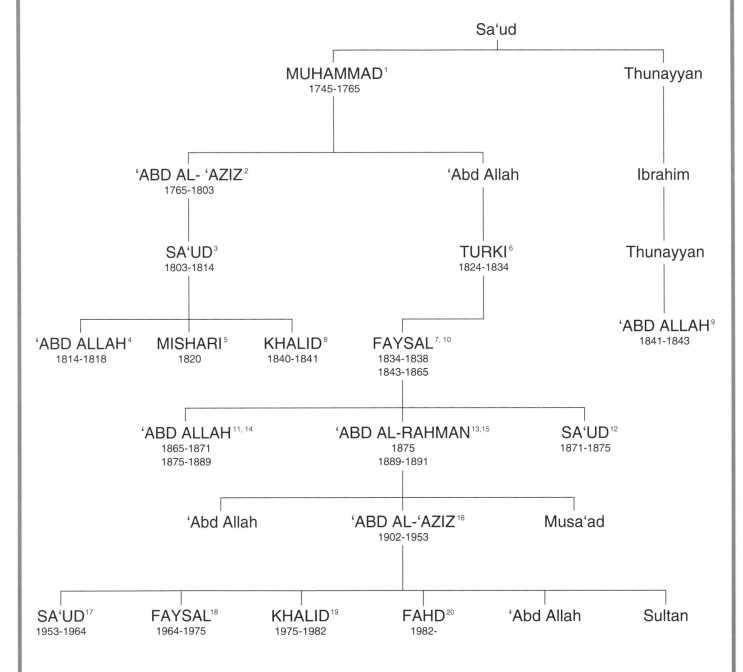

Sa'ud

MUHAMMAD[1]
1745-1765

Thunayyan

'ABD AL- 'AZIZ[2]
1765-1803

'Abd Allah

Ibrahim

SA'UD[3]
1803-1814

TURKI[6]
1824-1834

Thunayyan

'ABD ALLAH[4]
1814-1818

MISHARI[5]
1820

KHALID[8]
1840-1841

FAYSAL[7, 10]
1834-1838
1843-1865

'ABD ALLAH[9]
1841-1843

'ABD ALLAH[11, 14]
1865-1871
1875-1889

'ABD AL-RAHMAN[13,15]
1875
1889-1891

SA'UD[12]
1871-1875

'Abd Allah

'ABD AL-'AZIZ[16]
1902-1953

Musa'ad

SA'UD[17]
1953-1964

FAYSAL[18]
1964-1975

KHALID[19]
1975-1982

FAHD[20]
1982-

'Abd Allah

Sultan

Capital letters denote rulers of the Saudi state and numerals indicate the sequence in which they ruled.

King 'Abd al-'Aziz, the founder of the modern Kingdom of Saudi Arabia, was in the sixth generation in direct descent from Sa'ud ibn Muhammad ibn Muqrin, who died in 1725 and from whom Al Sa'ud and Saudi Arabia derive their names. The simplified family tree shown here extends only as far as the sons of King 'Abd al-'Aziz and of these it includes only those who have ruled as king or who have held one of the top positions in the government of the state, that is, have served as Vice Presidents of the Council of Ministers.

son Tusun Pasha. His forces were badly defeated, although they did succeed in taking Makkah, and Muhammad 'Ali, exasperated by his son's considerable losses of men — both to disease and to the enemy — finally took charge himself. He went to Arabia and initially had considerable success against the redoubtable foe. Then, forced to return to Egypt for political reasons, he sent his second son Ibrahim Pasha into Arabia at the head of a more powerful army.

In 1818 Ibrahim arrived in Najd with the full power of the Egyptian army behind him. Besieging the Saudi ruler 'Abd Allah, who had succeeded Sa'ud, Ibrahim captured al-Dir'iyah, sent 'Abd Allah as prisoner to Constantinople, and leveled al-Dir'iyah. He destroyed forts and other defense works, encouraged local rivalries, and a year later, thinking he had destroyed Al Sa'ud forever, returned to the Hijaz and then to Egypt.

After the departure of the Egyptian forces, a number of mutually antagonistic local amirs in Najd tried to profit from the disruption of the Saudi state. These ambitions, however, were thwarted by Turki ibn 'Abd Allah, a close relative (though not the son) of the dead 'Abd Allah.

TURKI IBN 'ABD ALLAH: During the siege of al-Dir'iyah Turki had taken refuge at a nearby town, but in 1823 he felt the time was ripe for a counterattack. Entering al-Dir'iyah without a fight, he immediately moved against Riyadh, which he took as well. He was the first member of the House of Sa'ud to make Riyadh his capital and it has remained the capital of the dynasty ever since.

Turki reigned for 11 years and by the time of his death had largely restored the boundaries of the Saudi state to what they were before the Egyptian invasions, though he did not recover the holy cities. Having expelled the invaders from Najd, he was to his compatriots a heroic defender of Saudi independence in the face of foreign aggression. But he was also firmly committed to good government in accordance with the Quran and the sunnah of the Prophet, and was concerned about the level of religious education in the provinces. When he recaptured a region, he closely monitored the performance of the five daily prayers, ordered all to follow the precepts of Islam meticulously, and exhorted his people to fear

God, to pray, and to pay the *zakah* (the religious tax). He also stressed the responsibility of the individual not only for his own behavior, but for that of his community, and urged people to avoid the practice of usury — always regarded as contrary to the teachings of Islam. In a letter setting forth his views, he ordered his district governors to standardize all weights and measures to prevent cheating and misunderstandings, stated that any bargain once made was inviolable, forbade the use of tobacco, encouraged study, and ended by asserting that any injustices should be referred to him personally.

The genre of the instructive epistle addressed to the entire community, as in the example just summarized, was very popular during this period. A number of historical works were composed as well — principally chronicles of the Unitarian movement — by scholars such as Ibn Bishr, Ibn Ghannam, Ibn 'Isa, and others. This led, under Turki's successor, to a sharpening of interest in religious literature in Arabia. This revival certainly did not achieve the sophisticated levels of other literary periods although Taha Husayn, one of the foremost twentieth-century critics of Arabic literature, was struck by the poetry written at this time:

> They returned in their poetry to the ancient style and gave us ... that sweet Arabic song, which had not been heard in the memory of living men. This song, whose authors did not imitate the city folk nor force themselves to search for new and strange expressions, was conceived in liberty and borne along by all the greatness, the yearning for the highest ideal, and the strong longing for the revival of ancient glory, which filled its soul.

Turki was succeeded as head of the Saudi state by his son Faysal, who previously had been captured and then had escaped from the Egyptians. Faysal was faced with yet another Egyptian attempt to establish control over Najd. In 1838 he was captured once again, bravely giving himself up to the enemy rather than see his loyal followers slaughtered, and for the second time was taken to Egypt and imprisoned.

For the second time, however, he escaped and made his way back to Najd in 1843, an

Ibrahim Pasha of Egypt led a campaign into central Arabia to subdue the House of Sa'ud.

exploit that added to his popularity among the tribes and also marked the end for a while of Ottoman efforts to quell the House of Sa'ud. Muhammad 'Ali was growing old and the Ottomans, distracted by wars in Moldavia and Wallachia, eventually had to content themselves with exercising titular control of the Hijaz.

Faysal, a just, stern man, continued to reign until his death in 1865. He was described by Lewis Pelly, the British Resident in the Gulf, as generous to the poor and a scholar who held regular discussions on the works of Ibn Taymiyah and Muhammad ibn 'Abd al-Wahhab. According to Pelly, Faysal was also interested in the economic problems of Arabia. Pelly said that Faysal spoke "very sensibly on the physical and political position of Arabia, explaining its great want to be that of rain. If only rain would fall agriculture would be possible, and the tribes might then be rendered sedentary." He was also, Pelly went on, "successful in curbing the predatory habits of his tribes; and...was desirous of inculcating among them more settled habits, and turning their minds towards agriculture and trade."

Both agriculture and trade were limited at that period, but during Faysal's rule a market for Arabian horses developed in British India. For centuries Arabia had bred fine horses, but in the 1860s demand picked up and exports soared. Horses bred by the famous Shammar tribe were exported through Kuwait and such Saudi ports as Qatif; according to one estimate, 600 horses were exported through Kuwait in 1863. The demand, in fact, eventually exhausted the supply and a moratorium was placed on the export of horses for four years. Other than Arabian horses, however, the only exports of any consequence were dates and pearls. According to one estimate some 750,000 tons of dates a year were exported during this period — two thirds of which came from the oasis of al-Hasa and the rest from Qatif. The trade of the period, although limited, still required some form of currency, and there were several to choose from: coins minted by the Ottomans, the British, and the Indians, and the most common currency — the Maria Theresa thaler or dollar. There were also at that time coins called in Arabic *tawilahs* ("long ones"). Minted in al-Hasa, they were made of silver, copper, or lead and consisted of a strip of the metal folded in half to form what looked like tweezers and stamped on the side with an Arabic inscription in kufic script. At one time tawilahs — called *larins* outside Arabia — were accepted as currency from the Gulf to Ceylon.

THE BIRTH OF 'ABD AL-'AZIZ AL SA'UD: With the death of Faysal in 1865 this relatively brief period of peace and prosperity came to an end. First, rivalry over the succession weakened Saudi unity. Then, taking advantage of the Saudis' internal conflict, the Ottomans occupied much of the eastern seaboard and the oasis of al-Hasa. Landing in 1871 at Ras Tanura, the present site of a Saudi Aramco oil refinery and a major shipping terminal, the Turks took Qatif, then marched 160 kilometers to Hofuf, the main town of al-Hasa, where they overcame stubborn resistance put up by the Saudi governor and occupied the fortress. The Turkish commander was awarded a sword, studded with diamonds, with the word "Najd" engraved on the blade — although no Turkish troops had entered the central region of Arabia.

In the story of the House of Sa'ud, however, two other events overshadow the Turkish invasion. One was the birth of a son to 'Abd al-Rahman ibn Faysal, the man who had emerged as the reigning head of the House of Sa'ud. The son — who would later create the Kingdom of Saudi Arabia — was named 'Abd al-'Aziz ibn 'Abd al-Rahman Al Faysal Al Sa'ud, known to the West as Ibn Sa'ud and to his subjects simply as 'Abd al-'Aziz. The other event was the rise to power of a dynasty called Al Rashid or the House of Rashid, which led to a conflict that would occupy the Saudis, and young 'Abd al-'Aziz, for decades.

Based in north-central Arabia, with the town of Hayil as capital, the House of Rashid had taken advantage of the uncertainties over the Saudi succession to install a deputy governor in Riyadh in Saudi territory. In response, 'Abd al-Rahman, when he emerged as head of the House of Sa'ud, attacked Riyadh and recaptured it. Muhammad ibn Rashid, in turn, marched on Riyadh and, finding the defense too strong for a direct assault, besieged it. During the siege he cut down a large number of date palms on which the townspeople depended for sustenance, a common practice in Arabian warfare in that period. After 40 days of this harsh but indecisive activity, Ibn Rashid proposed negotiations with the defenders. In the Saudi delegation was the young boy 'Abd al-'Aziz, making his debut on the stage of history.

The truce that was arranged as a consequence of these negotiations was short-lived. Muhammad ibn Rashid soon led his men to the area of the Qasim, in northern Najd, where he attacked and routed the Saudis at the battle of al-Mulayda on January 21, 1891. Isolated and bereft of allies, 'Abd al-Rahman sent his women and children to the protection of his friend the ruler of Bahrain, while he himself, with no hope of returning to Riyadh, took to the desert in the south where he had friends among the tribes. For a time 'Abd al-Rahman and a handful of loyal followers roamed the fringes of the Rub' al-Khali, but later they moved on to Qatar and then to Bahrain. Finally they took refuge in Kuwait, where they spent the better part of a decade.

For 'Abd al-'Aziz the years he and his father spent in Kuwait as guests of the ruler, Mubarak al-Sabah, provided valuable insights into international politics. Mubarak was an able politician and Kuwait, strategically placed at the head of the Arabian Gulf, was then the focus of Western activity in the area. The British had special treaty relations with Kuwait, Bahrain, Oman, and the shaykhdoms on the Trucial Coast; the Russians, as part of their centuries-old search for a warm-water port, were probing for outlets in the Arabian Gulf; and both the Germans and the Turks were looking toward Kuwait as the possible terminus for the Berlin-to-Baghdad railway, an attempt to challenge Britain's hegemony in the Gulf.

As he observed the international negotiations of Mubarak al-Sabah, however, 'Abd al-'Aziz also kept an intent eye on the Rashidis, now allied with the Turks in al-Hasa and led by a nephew of Muhammad ibn Rashid whose name was also 'Abd al-'Aziz and who had a reputation for being weaker and more rash than his formidable uncle.

In 1901 'Abd al-'Aziz Al Sa'ud, then about 20 years old, decided that it was time for the House of Sa'ud to win back the lands wrested from it by the Turks and Ibn Rashid. As the entire force the Saudis could muster at this time consisted of only 40 men, 'Abd al-Rahman tried to dissuade him. But 'Abd al-'Aziz would not be dissuaded; the House of Rashid had just mounted an attack on Kuwait and Shaykh Mubarak was willing to let 'Abd al-'Aziz create a diversion. Toward the end of 1901 therefore, 'Abd al-'Aziz set off for Najd with his 40 men in the first phase of a campaign to rebuild the fortunes of the House of Sa'ud.

THE ATTACK ON RIYADH: The Saudi force did not attack immediately. Instead, it spent several months on the northern fringes of the Rub' al-Khali, hoping to win reinforcements from the tribes camped near the wells of Yabrin and Haradh, which is now one of the kingdom's largest agricultural projects. 'Abd al'Aziz was able to recruit only 20 more warriors but, reluctant to wait any longer, he decided to attack Riyadh anyway, and with only 60 men behind him he set off.

At a distance of one and a half hour's march from Riyadh, 'Abd al-'Aziz left one-third of his force with the camels, instructing them to return to Kuwait if no message was received from him within 24 hours. Then, with the rest of his men, he advanced on foot — to be less conspicuous — until he reached the outskirts of the city. There 'Abd al-'Aziz waited for night to fall.

At last it was time. 'Abd al-'Aziz stationed his brother Muhammad in the groves with 33 men to act as a back-up force and quietly scaled the walls with the others. Inside the walls, they knocked on the door of the house of a cattle dealer, who fled. His daughters, recognizing 'Abd al-'Aziz, were gagged and locked up. Next 'Abd al-'Aziz sent a messenger back to tell Muhammad, waiting outside the walls, to advance with all possible stealth into the city. Finally, by standing on one another's shoulders, 'Abd al-'Aziz and his men entered the house of 'Ajlan, Ibn Rashid's governor, silenced the servants, and searched the house. Learning that 'Ajlan was in the custom of spending his nights in the fort of al-Masmak in the city, they decided to wait for morning when the gates of the fort would be opened.

It was difficult to wait. As 'Abd al-'Aziz recalled in later years, they "slept a little while...prayed the morning prayer and sat thinking about what we should do." But at last the dark desert sky lightened and they prepared for action.

Mubarak al-Sabah, ruler of Kuwait, was host to the young exiled member of the House of Sa'ud.

A building boom has changed the face of Saudi Arabia in recent years, with new office towers and apartment blocks dominating the skylines of such established cities as Jiddah, which is endowed with a number of architecturally distinguished buildings, above right. Across the peninsula, the fast-growing city of al-Khobar on the Arabian Gulf boasts blocks of modern office towers, bottom right. Riyadh too has witnessed extraordinary growth, epitomized in its international airport, center, a facility that incorporates Islamic architectural details in its advanced design concept. New residential areas, often constructed in conjunction with the kingdom's growing industrial areas, are another component of Saudi Arabia's recent development. The industrial city of Yanbu' on the Red Sea, above left, lies nestled at the foot of the Hijaz mountains, only minutes away from the area's refining and shipping facilities, while the residential area of Jubail, lower photo at right, is within easy commuting distance of the community's manufacturing and oil refining sector, seen in the background. In contrast, minarets of the Sacred Mosque loom behind comfortable apartment buildings in the holy city of Makkah, above, a magnet for pilgrims and religious scholars.

'Abd al-'Aziz reestablished the Saudi Dynasty in the heartland of Arabia with a daring assault on Riyadh, whose city walls, seen at left above, he with a small band of men scaled stealthily by night, using a palm trunk for their ladder. As al-Masmak fort came to life at dawn, the outnumbered attackers sprang towards its entrance and with gunfire and rocks hurtling about them pursued the Rashidi governor through the postern gate seen above — killed him, captured the garrison, and sent criers forth to proclaim that the House of Sa'ud had retaken the town.

Originally they had planned to take 'Ajlan prisoner as soon as he left the fort and entered the house. But as the sun rose and the gates of the fort opened, they saw that 'Ajlan was not alone; he walked out of the gate accompanied by 10 bodyguards. Instantly 'Abd al-'Aziz and his followers sprang to the attack, leaving four men in the house to cover them with rifles.

At the sudden appearance of 'Abd al-'Aziz, 'Ajlan's bodyguards bolted, leaving 'Ajlan facing the Saudi onslaught alone, with only a sword for defense. Darting forward, 'Abd Allah ibn Jiluwi, a cousin of 'Abd al-'Aziz, threw a spear at 'Ajlan but missed; the spear went into the gate of the fort where the steel point, imbedded in the wood, can still be seen today.

No coward, 'Ajlan lunged at 'Abd al-'Aziz, who later reminisced: "He made at me with his sword, but its edge was not good. I covered my face and shot at him with my gun. I heard the crash of the sword upon the ground and knew that the shot had hit 'Ajlan, but had not killed him. He started to go through the postern gate, but I caught hold of his legs. The men inside caught hold of his arms while I still held his legs. His company were shooting their firearms at us and throwing stones upon us. 'Ajlan gave me a powerful kick in the side so that I was about to faint. I let go of his legs and he got inside. I wished to enter, but my men would not let me. Then 'Abd Allah ibn Jiluwi entered with the bullets falling about him. After him 10 others entered. We flung the gate wide open, and our company ran up to reinforce us. We were 40 and there before us were 80. We killed half of them. Then four fell from the wall and were crushed. The rest were trapped in a tower; we granted safe-conduct to them and they descended. As for 'Ajlan, Ibn Jiluwi slew him."

Such is the epic story, as related by King 'Abd al-'Aziz, of how Riyadh was taken on January 16, 1902, as the sun was rising over the desert and the city was just coming to life. The capture of Riyadh marked the dawn of a new era in the history of Arabia and a turning point in the fortunes of the House of Sa'ud.

In Riyadh, as the news swiftly spread, the people welcomed the new ruler joyfully; they had suffered under the rule of Ibn Rashid. 'Abd al-'Aziz, knowing his small band could never hold the city if the Rashidis were to counterattack, immediately set about repairing the defenses of Riyadh. The Rashidis did not attack, however, and during the next six months 'Abd al-'Aziz completed his defenses, sent for his father in Kuwait, and handed over the city to him while he took the field. This close relationship between father and son was to last until the death of 'Abd al-Rahman nearly 30 years later. It was a relationship of mutual respect, of profitable consultation on all important matters, of willingness on the part of the son to cede to his father the place of highest honor on every public occasion.

Meanwhile, the inhabitants of the area south of Riyadh, learning that the House of Sa'ud had retaken the city, hastened to acknowledge the suzerainty of 'Abd al-'Aziz; his exploit in capturing Riyadh with so small a force had won the admiration of the Bedouins, who rode into Riyadh from the desert to join him.

Simultaneously 'Abd al-'Aziz began to prepare for a large-scale offensive against the Rashidis. For although the taking of Riyadh was a master stroke, he knew that his situation was still precarious and that he still faced a long, arduous campaign against the House of Rashid and its Turkish allies. 'Abd al-'Aziz had demonstrated his daring with the capture of Riyadh. Now, as he would for the next quarter of a century, he would demonstrate the rarer qualities of knowing when to advance and when to retreat, when to conciliate and when to punish.

'Abd Allah ibn Jiluwi in later life became Governor of the Eastern Province.

Riyadh in 1950, with its flat-topped roofs and crenelated parapets, looked much as it did in the early days of 'Abd al-'Aziz Al Sa'ud.

COUNTERATTACK: At Hafar al-Batin, meanwhile, Ibn Rashid continued to plan his campaign against Kuwait. He apparently regarded the young Saudi as no great threat to his power and felt he could deal with him at a later date. Knowing that 'Abd al-'Aziz Al Sa'ud had few resources, he assumed that he could offer few tangible advantages to the villagers and tribesmen of Najd, wholly absorbed as they were in their hard struggle for existence.

Ibn Rashid's assumption turned out to be wrong. In another of the shrewd moves for which 'Abd al-'Aziz later would be famous, he and his brother Sa'd set about winning the hearts of the people of al-Kharj, al-Aflaj, and Wadi al-Dawasir — all regions over which the authority of the House of Rashid had been at best nominal.

The Rashidis finally struck back. In the fall of 1902 Ibn Rashid headed south. Bypassing Riyadh, now strongly fortified, he made a raid into al-Kharj hoping to subdue the district before 'Abd al-'Aziz had prepared for its defense. Again, he miscalculated. 'Abd al-'Aziz, by means of a forced night march, had reached the village and date groves of al-Dilam earlier and had prepared an ambush. It caught the Rashidis unaware, and after a fierce, daylong battle Ibn Rashid was forced to retreat all the way to his northern capital of Hayil.

This second victory gave another tremendous boost to the prestige of 'Abd al-'Aziz Al Sa'ud and consolidated his popularity with the people. He had proved himself in battle and could now count on the support of the inhabitants of Najd from Riyadh to Wadi al-Dawasir.

THE STRUGGLE FOR NAJD: Ibn Rashid, however, was not going to let Najd slip from his fingers without a fight, and for the next few years the two sides battled almost continually for mastery. 'Abd al-'Aziz attacked the heartlands of Al Rashid in the area of Jabal Shammar and in 1904 attacked the province of the Qasim, capturing the two key towns of 'Unayzah and Buraydah. He won over to his cause the people of the Washm and Sudayr as well. By the spring of the year he was master of central Najd.

At that point Ibn Rashid, unable to contain the Saudi advance himself, appealed to his Turkish allies for help and when he renewed the contest toward the end of May 1904 he was accompanied by eight battalions of Turkish troops equipped with artillery.

The Saudi forces confronted the combined troops of Ibn Rashid and the Turks in the province of the Qasim, outside the town of Buraydah. The battle was running in favor of the Saudis when 'Abd al-'Aziz was wounded by shrapnel from a Turkish cannon shot and forced to retire. It was, of course, a defeat, but by the time he had recovered from his wounds and returned to the fray some months later, the fierce heat of an Arabian summer had taken its toll on the Turkish troops. In the next encounter 'Abd al-'Aziz, seeing some of his men begin to waver, personally led an attack on the Turkish artillery. The Turks gave ground, Ibn Rashid's forces broke,

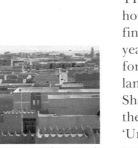

Buraydah, a trading center in Najd, was seized by 'Abd al-'Aziz Al Sa'ud at an early stage.

Coffee making is the very essence of Bedouin hospitality.

Life In The Desert

In the arid lands which make up most of Saudi Arabia, men, as well as plants and animals, have adapted to the stringent conditions imposed by terrain and climate. This is particularly true of the Bedouins, the nomadic people who have inhabited large regions of the peninsula for centuries.

The word "Bedouin" comes from the French version of the Arabic word *badawi* (plural, *badu*) which means simply desert dweller. It is an accurate term but used only by townsmen. They refer to themselves, simply and proudly, as "Arabs." Bedouin life evolved from the demands of a harsh environment. The constant and compelling need for water and pasturage, for example, led the tribes to divide much of the Arabian Peninsula into tribal areas — called *dirahs* or ranges — some of which were very large. Within each dirah were the wells used by the tribe, often marked with the tribe's camel brand. Use of another tribe's wells without permission was forbidden, and unlawful crossing of another tribe's dirah could lead to war.

The need for water and pasturage, indeed, shaped every aspect of Bedouin life. When, for example, desert

130

A Bedouin's life is one of independence and self-reliance.

Wool from Bedouin's animals goes into heavy tent fabric, woven by women of the tribe.

thunderstorms lighted the horizon in the evening, the tribe would head for the area that had received the downpour so that they could pasture their herds on the quick-growing vegetation that resulted. When little or no rain fell, the tribes would cluster about their wells, drawing water — often from considerable depths — to keep their animals alive. And when there was a drought, the desert would often erupt in tribal clashes, as tribes were forced to move from their own dirahs to those of others. Until the introduction of firearms, the conflicts — often swift raids on enemy encampments — were fought with swords, daggers, and long cane spears. Raiding parties sometimes rode by camel to enemy territory, then mounted their horses for the attack.

The demands of life in the desert also shaped codes of hospitality. As any traveler would die without water, food, and shelter, every man — even an enemy — could claim three days of hospitality from any other man. It would not, furthermore, be granted grudgingly; the host might find himself in the same predicament as his guest at another time. Similarly the host would offer travelers not only hospitality, but also protection.

The shaykhs or leaders of Bedouin tribes were usually chosen from a few specific families, but as tribal society was basically democratic their authority was derived from the consent of their fellow tribesmen, who were free to oppose it at any time. Sometimes, it is true, large tribes split into factions because of opposition to a shaykh, but these factions would almost always unite against a common enemy; collective responsibility was a powerful force for order within the tribe.

Bedouin society accorded women a position somewhat superior to that of many women in towns. Married women were often consulted on important decisions, and in the old days some women accompanied the tribes to the battlefield and cheered on the warriors. Their work, furthermore, was vital: to raise the children, to gather fuel — a time-consuming occupation in itself — fetch water, milk the animals, cook, spin yarn, and weave the tent cloths and blankets that constitute a good part of the wealth of most Bedouin families. As a result they were accorded great respect.

One of the jobs given women was to pitch the tent, the "house of hair," which is symbolic of Bedouin life and which suggests the adaptability of the Bedouin to his demanding environment. The black tents had to fulfill stringent requirements. They had to be easy to erect and dismantle, light and portable, easy to maintain and repair, airy, and resistant to wind and rain, yet provide insulation from the sun and protection against the cold.

Essentially, the black tent consists of long strips of cloth woven from black or brown wool and goat's hair, which are sewn together to provide a roof and to make sides, which are attached to the roof with wood or pins, all of this being supported by ropes and poles and pegged to the ground. The open side faces away from the winds, and the interior is usually separated by vertical curtains into three sections: the men's section, where guests are received, the family section, and the kitchen. The floor area is covered with rugs, either woven by the women of the tribe or purchased in market towns, and the owner's sword or rifle is hung from the tentpole in the men's section, in front of which is the hearth, with coffeepots and other coffee-making implements ranged around.

To settled Arabs, the Bedouins have traditionally been regarded as the repositories of manly virtues: they were proud, independent, resourceful, courageous, loyal, hospitable, and generous. In addition, they were thought to speak the purest Arabic, uncontaminated by contact with foreign elements in the cities. These views, no doubt, were and are often exaggerated; but Bedouins nonetheless did develop a distinctive way of life and their contributions, historically, have been important as well as interesting. In the perpetual hunt for scarce meat (their herds were kept primarily for milk) the Bedouins painstakingly developed the art of falconry and bred the saluki, a handsome greyhound. Bedouins also developed the Arabian horse, one of the more famous breeds in the world, and learned to make use of the camel, the animal which enabled them to master the desert and control overland trade routes for millennia.

The camel, curiously, seems to have originated in North America as a small Eocene mammal which probably migrated and eventually made its way to central Asia, the Middle East, North Africa, and the Sahara Desert, evolving into today's one-hump and two-hump camels. Another group of ancestral camels migrated to Central and South America where they evolved into today's llamas, alpacas, guanacos, and vicuñas.

(Continued on page 132)

Life In The Desert

The camel, despite its awkward appearance, is a remarkable example of successful adaptation to environment. It can traverse arid lands in which no other large mammal can survive and has enabled man to penetrate into and survive in these regions as well.

Domesticated centuries ago, the camel enabled nomadic peoples to conquer the desert.

The most striking adaptation of the camel to desert life is its ability to conserve water — although not, as was once believed, in its hump or stomach. Instead, the camel has evolved a number of mechanisms which allow it to absorb and retain large quantities of water. Its drinking capacity naturally varies with the seasons of the year, weather conditions, and the level of the animal's dehydration, but is truly large. It may range from only some 64 liters (16 gallons) to as many as 200 in a drinking session of two watering episodes separated by only a few hours. Two hundred liters (50 gallons) would be the equivalent of about one-half of the animal's dehydrated weight. The camel's body temperature, moreover, can rise as much as three degrees centigrade (six degrees Fahrenheit) during the day without its losing precious body moisture in wasteful cooling through sweating. The camel can also eat low-protein desert shrubs and convert them to body tissue and drink water high in salts and minerals and convert it, along with salty desert shrubs, into milk (a primary element in the Bedouins' diet). The camel's peculiar kidney physiology, finally, allows it to recycle its own urea, a process which not only provides an additional source of protein but also assists in water conservation.

The camel has remarkable eyelashes — two sets for each eye — which effectively screen its eyes from sand and dust. It also possesses large, soft, padded feet which work on the snowshoe principle and allow it to walk over drift sand without sinking in. The long neck of the camel allows it to nibble low-growing desert herbs comfortably as well as leaves from trees, and also serves to balance the animal's rather unwieldy body as it runs.

During the winter, camels are ordinarily not watered at all; they receive enough moisture from what they eat. As the weather grows warmer, however, they are watered every seven to nine days and in full summer every two days. The top speed of a good riding camel is around 22 kilometers an hour. They are capable of many different paces: from a walk that covers about six kilometers an hour to a long-distance stride with which a camel can, relatively easily, cover 160 kilometers in a day.

When, and exactly how, the camel was domesticated is not precisely known but it is easily conceivable that the process of domestication got under way between 3000 and 2500 B.C. in south Arabia. In due course men started riding camels — the stage of development that allowed the Bedouins to venture deeper, and for longer periods, into the large areas of arid land of the Middle East.

As a beast of burden, the camel had no peer in the arid regions; it was less expensive to maintain, had a longer working life, and could carry loads in areas where oxen or donkeys could not. For all these reasons, the early Arabs were able to monopolize the ancient trade routes between the East and the civilized towns of Mesopotamia, Egypt, and the Mediterranean and were able to establish mercantile cities and kingdoms in northern and central Arabia. The camel's efficiency, in fact, impeded the development of roads; wheeled vehicles drawn by oxens, for example, could not compete with the low cost of camel transport.

As a means of transport, the camel is no longer as important as it used to be in Saudi Arabia. Statistics are scarce but the age of the camel seems to have passed into history. Traditional caravan routes are falling into disuse and many wells on these routes, untended for years, are filling up with sand and vanishing. The camel is still used for milk and meat, but rarely for transport.

The days of the camel caravan are fading into history.

The life of the Bedouin has been changing too. Even the Bedouins who cling to the nomadic, pastoral way of life are relying increasingly on trucks to haul water to their herds and are abandoning some of the spartan ways of the past for the more comfortable ways of an industrialized world. Many Bedouins still range through the deserts but today they are more often than not at the wheel of heavy-duty trailer trucks rather than at the head of a column of camels. Others have left herding to work in factories and oil installations and not a few, under Saudi Arabia's extensive educational program, have enrolled and been graduated from universities abroad. In short, present-day Bedouins are adapting to the exigencies of modern life as, millennia ago, they adapted to the demands of the arid steppes and deserts.

the retreat became a rout, and the Saudis captured stores of equipment and a large sum of Turkish gold.

A few months later, while 'Abd al-'Aziz was dealing with problems in the south, Ibn Rashid tried yet once more to reassert his authority over the province of the Qasim. The people of the province, however, sent a delegation to 'Abd al-'Aziz in Riyadh, pleading with him to return. He did so, and caught Ibn Rashid by surprise. While Ibn Rashid was desperately trying to rally his troops, he was shot down in his camp near Buraydah in April 1906. The contest between 'Abd al-'Aziz of the House of Rashid and 'Abd al-'Aziz of the House of Sa'ud, which had lasted since the capture of Riyadh four years and three months before, was at an end.

CONSOLIDATION:

CONSOLIDATION: After the defeat of the Rashidis, 'Abd al-'Aziz began the difficult task of consolidating his victory. But within two years he was confronted with still another adversary: the Sharif Husayn ibn 'Ali.

In 1908 the Sharif Husayn, appointed by the Ottoman Government, became Amir of the Hijaz. Husayn had been born in Makkah, but had spent most of his life in Istanbul and was nearly 60 years old. He was a man of great culture, charm, and integrity, but also extremely ambitious and committed to the preservation of the Hijaz as a fief of his family, the Hashimites.

Husayn, therefore, was delighted when, in 1908, the Turks completed the Hijaz Railway linking Damascus to Medina. The Hijaz was now tied to the Ottoman Empire by a more substantial bond than the formal recognition of Ottoman suzerainty by the sharifs. The Turks could move troops expeditiously toward the very heart of Arabia if need be.

To make matters worse Husayn had apprehended 'Abd al-'Aziz Al Sa'ud's brother Sa'd during a mission to enlist the aid of the 'Utaybah tribe against the Rashidis. Husayn offered to release Sa'd if 'Abd al-'Aziz would recognize his hegemony over the Qasim and pay him tribute. Although torn between his political aims and his love for his brother, 'Abd al-'Aziz agreed to meet Husayn's terms and Sa'd was released.

Another difficulty was the continual shifting of allegiances by the Bedouin tribes. Depending on the personality of their shaykhs,

tribal feuds, and complex family relationships, the Bedouins would support first one ruler and then his opponent. To 'Abd al-'Aziz this seemed to be an even greater stumbling block to the unification of his realm than the threat of outside intervention.

In pondering this problem 'Abd al-'Aziz concluded that until order had been established in Najd his ambition to unite the provinces of Arabia into a single kingdom would be doomed to failure. He concluded too that order would depend on providing security for his subjects and a better life for the tribes; in 1910 a drought had killed thousands of camels and sheep and spawned unrest in both Najd and Kuwait. After long consultation with the members of his family and his religious advisors, therefore, 'Abd al-'Aziz decided upon a plan to settle the Bedouins in agricultural communities, while still preserving their ability as fighting men. At the same time he dispatched teachers to the tribes, summoning them to take part in a revival of pure Islam.

Securing the loyalty of the Bedouins was a critical task facing the young 'Abd al-'Aziz.

Those who answered the call were called *Ikhwan* — "Brethren" — and grouped into settlements which were at first based primarily upon common beliefs and practices, though later they came to be associated with particular tribes or tribal groups. The settlers were given land, agricultural implements, seeds, money for their everyday needs, and arms for defense. They were also given money for the construction of mosques to which religious teachers were assigned.

The first of the Ikhwan settlements was started at the wells of al-Artawiyah in northern Najd in 1912 and in a very short time numbered 10,000 men. Soon there were scores of Ikhwan settlements with thousands of men at strategic points throughout Najd, the most famous being at Ghutghut at no great distance west of Riyadh. These settlers would become the elite troops in 'Abd al-'Aziz's growing forces.

Ikhwan settlements established by the future king became the backbone of his forces.

EXPANSION:

EXPANSION: 'Abd al-'Aziz now turned his attention toward eastern Arabia where the Turks, who had occupied much of it in 1871, were still in control. At this time — 1913 — the Ottomans were plagued with difficulties both in Turkey itself, where the Young Turks were agitating for much-needed reforms (see page 94), and in their European provinces,

133

al-Qurayyat

Sākaka
al-Jawf

Tayma

Khaybar

Medin

Yanbu'al-Nakhl
Yanbu'al-Bahr

Rabigh

Jiddah   al-Hi   iya
Ma

RED   al-L

SEA

GROWTH OF THE MODERN SAUDI STATE

Having recaptured Riyadh from the Rashidis in 1902, 'Abd al-'Aziz Al Sa'ud spends the next 30-odd years dealing with threats to his position and in reestablishing, consolidating, and extending Saudi control. The first few years bring expansion to the south toward Wadi al-Dawasir and to the north and west into Sudayr and the Qasim as far as the towns of Buraydah and 'Unayzah. In 1913 the Turks are expelled from al-Hasa and the coastal towns on the Arabian Gulf, and 'Abd al-'Aziz gains control of most of what is now the Eastern Province of Saudi Arabia. The next great period of expansion comes after World War I when the Saudis come into conflict with the Hashimites, expel them from Makkah, Medina, Jiddah, and the rest of the Hijaz, and push on to the northwest as far as the borders of Trans-Jordan. The final period of growth comes in the following decade when the Saudis move south to establish control over Jaizan and the district of Najran.

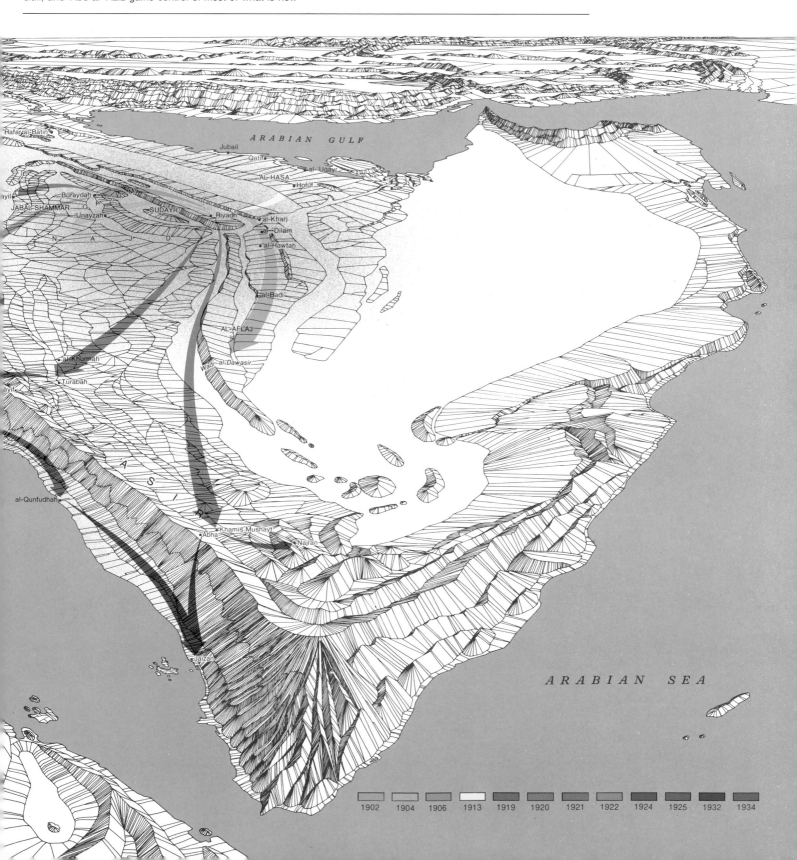

ARABIAN GULF

ARABIAN SEA

| 1902 | 1904 | 1906 | 1913 | 1919 | 1920 | 1921 | 1922 | 1924 | 1925 | 1932 | 1934 |

The fort at Hofuf was the first of the Turkish garrisons to surrender as 'Abd al-'Aziz pushed his authority into the eastern areas.

Ruins of the fort at al-'Uqayr on the Gulf recall the days of Ottoman occupation.

British recognition of 'Abd al-'Aziz followed contacts with the Saudis by Captain Shakespear, British political agent in Kuwait.

where independence movements were challenging their supremacy. It seemed a propitious time for action and 'Abd al-'Aziz, with his newly formed corps of Ikhwan behind him, struck off on a moonless night in April 1913. At the head of 600 men, he crept up to the fortifications of Hofuf, the Turkish headquarters in al-Hasa, scaled the wall on ropes and the trunks of palm trees, and, before the sleeping garrison could gather its wits, captured the inner fortress and accepted the Turkish governor's surrender. To press his advantage 'Abd al-'Aziz moved quickly on other Turkish garrisons in al-'Uqayr and Qatif, chivalrously allowing the Turks to depart without surrendering their weapons.

The consolidation of al-Hasa completed, 'Abd al-'Aziz then sought an agreement with the Turks to stabilize the situation in the Arabian Gulf. The Turks, however, while negotiating with 'Abd al-'Aziz, made an agreement to deliver arms to the supporters of the House of Rashid, who were still hostile. 'Abd al-'Aziz, therefore, laid the groundwork for an understanding with the British, and in the winter of 1913 Captain W. H. I. Shakespear, the British Political Agent in Kuwait, visited 'Abd al-'Aziz in Riyadh. The discussions between 'Abd al-'Aziz and Shakespear laid the foundations of the Anglo-Saudi treaty of 1915 whereby Britain recognized Ibn Sa'ud as hereditary ruler of Najd, al-Hasa, Qatif, Jubail, and their dependencies in return for support in World War I.

In Europe, by then, the war had begun. One of the many causes of the conflict was international rivalry for control of the Middle East — partly because, with the opening of the Suez Canal (see page 95), the Middle East had become a key link between Britain and her Indian Empire, between France and Indochina, and between the Dutch and the East Indies. Other European nations, furthermore, had strategic and commercial interests in the Middle East which they wished to expand, and many European governments hoped to profit from what they regarded as the inevitable dismemberment of the Ottoman Empire.

In the Middle East, at the same time, national movements for independence from the Ottomans were growing, while sentiments of pan-Arabism were everywhere in evidence. The Arab world, after centuries of Turkish rule, was rediscovering its identity.

When the powers of Europe went to war in 1914 control of the Middle East was divided between two of the greatest colonial powers in the world: Great Britain and the Ottoman Empire. The Ottomans still exerted a form of authority over most of the Red Sea coast as far south as Yemen, and still actively ruled Palestine, Syria, and Mesopotamia. The British controlled Egypt and the Suez Canal, their lifeline to India, as well as Aden. As they considered the Arabian Gulf lay within their sphere of influence, they had announced in 1903 that the establishment of a naval base or fortified port in the Gulf by any other power would be regarded as "a very grave menace to British interests" and one which Britain would certainly resist with all the means at its disposal.

Britain's first objective, after the outbreak of war, was to wrest control of Mesopotamia — now Iraq — from the Turks. As part of their campaign they sought Saudi help against the House of Rashid, an ally of the Ottomans, while at the same time encouraging the Sharif Husayn to expel the Turks from the Hijaz.

'Abd al-'Aziz did not take a decisive part in World War I. He had domestic troubles enough: first, as always, with the House of Rashid, and second with a rebellion by the tribe of the 'Ujman during which he was twice severely wounded and lost his beloved brother Sa'd, whom he had ransomed from his captivity by the Sharif Husayn.

Toward the end of World War I, however, 'Abd al-'Aziz found himself faced with the open hostility of the Sharif Husayn, whom he had refrained from attacking during the war, despite provocations, because of their common wish to see the last of the Turks leave the peninsula. But now Husayn was armed and supported by the British, with whom he had concluded a secret agreement under which they committed themselves to support the establishment of an independent Arab state comprising most of the Arab lands east of Egypt.

At the end of the war the Allies recognized Husayn not as King of the Arab Countries — the title he chose for himself — but only as King of the Hijaz. Britain and France themselves, by mandates, took over a large part of the territories the sharif thought he had been promised: Palestine, Iraq, and Syria.

If disappointed by the Allies' betrayal, however, Husayn was still determined to retain at least the Hijaz, and launched three attacks upon the oasis of al-Khurmah, which was strategically located on the route between the Hijaz and Najd. In the first attacks the people of al-Khurmah succeeded in holding off Husayn's troops, but as the situation became desperate they appealed to 'Abd al-'Aziz for help. At first 'Abd al-'Aziz simply warned Husayn to desist, but when the sharif, ignoring the warning, sent 4,000 men armed with machine guns and artillery against al-Khurmah, 'Abd al-'Aziz sent his troops into action. In May 1919, in the middle of a dark night, the Saudi Ikhwan attacked from all sides and inflicted a crushing defeat upon the sharif's expedition at the nearby town of Turabah.

With al-Khurmah and Turabah safely attached to the Saudi realm, the ways to the Hijaz and to 'Asir, the mountainous region south of Makkah, were now open and in 1920 the Saudis captured Abha, the capital of 'Asir. Two years later 'Abd al-'Aziz's son Faysal, later to become the third king of Saudi Arabia, led an army to 'Asir to suppress a serious rebellion there. Since then 'Asir has remained a tranquil part of the Saudi state.

Meanwhile the Rashidis were once again at war with Al Sa'ud and to end this constant threat to stability 'Abd al-'Aziz besieged the Rashidi headquarters in Hayil, long a thorn in the Saudi side. This time his efforts met with success, and after three months Hayil capitulated and the surviving members of the House of Rashid were brought to Riyadh, where they remained his honored guests. The long rivalry between the two houses was finally at an end.

In 1922 'Abd al-'Aziz extended his expanding domains to the west and north by annexing the oases of Khaybar and Tayma. On his northern frontiers, he took the important Wadi al-Sirhan and the oasis of al-Jawf.

Events were moving toward a climax. The last act of the drama began in September 1924 when a detachment of the Ikhwan appeared before the city of Tayif in the highlands above Makkah. The force probably had no intention of attacking the town, for the commander of the detachment was a few hours' march in the rear. But the sight of the Ikhwan created a panic and 'Ali, a son of the

Sharif Husayn, withdrew with his troops toward Makkah. The townsmen, left defenseless, agreed to surrender the town peaceably.

When the gates were swung open, however, the Ikhwan were reportedly fired upon. Thinking themselves betrayed, they attacked, and fierce fighting ensued before their commander arrived and restored order. This was the only time during the Hijaz campaign when the Saudi forces overreacted; normally, they kept strict discipline.

With the Saudis in Tayif the days of the Sharif Husayn were obviously numbered, and prominent citizens of the Hijaz began to bring pressure to bear on him to abdicate in favor of his son 'Ali. For a long time the aged ruler refused, but finally he yielded and in October 1924 he boarded a ship in Jiddah and sought refuge in Transjordan, from which the British later moved him to Cyprus, where he spent his last years.

One of the first acts of his son 'Ali, now king, was to evacuate Makkah, considered indefensible once Tayif had fallen, and to concentrate his troops in Jiddah. In October 1924, therefore, after Saudi advance forces had entered Makkah, 'Abd al-'Aziz entered the holy city — not as a conqueror but in the humble garb of a pilgrim.

For the sharifs of Makkah the end was now in sight, and in 1925 it came. In the first days of December, Medina surrendered, and two weeks later Jiddah capitulated. 'Ali boarded a ship for Iraq, where his brother Faysal reigned, and on January 6, 1926, the leading citizens of Makkah offered allegiance to 'Abd al-'Aziz as successor to the title King of the Hijaz. Soon after, in a ceremony of investiture in the Sacred Mosque of Makkah, the new King swore that he would rule his new domains in accordance with the sacred law of Islam.

REORGANIZATION: These domains were now extensive. Reaching from the borders of Iraq in the north to the far side of the Rub' al-Khali in the south, and from the Arabian Gulf in the east to the Red Sea in the west, they comprised an area three times the size of France and included the holy cities of Makkah and Medina.

Fully aware that the addition of the Hijaz to his dominions brought unprecedented responsibilities as protector of the holy cities, 'Abd al-'Aziz convened an Islamic Congress

Tayif, a summer resort from pre-Islamic times, still attracts local tourists with well-appointed hotels.

Countless pilgrims once passed through the old Makkah Gate, now demolished, in Jiddah.

The foundations of this mosque at al-Jawf, an important center on the caravan route, reportedly go back to the Caliph 'Umar ibn al-Khattab.

Much of Saudi Arabia's land is cultivable and with added nutrients can be productive wherever water can be made available. Deep, prolific aquifers make agriculture by irrigation profitable in many areas, and water-management studies enhanced by experimentation with new crops, machinery, and farming techniques continue to add to the success of new ventures. At left: Vegetable production is increased with use of hydroponic greenhouses and, in the field, by plastic sheeting to help retain soil moisture. Continuing clockwise: Fat-tailed sheep, once the Bedouins' traditional source of meat, milk, and wool; a government farm in al-Hasa oasis, scene of experiments with crops and livestock; a citrus grove in Tayif, a major supplier of many varieties of fruit for both local consumption and export; an al-Hasa water project where hundreds of kilometers of concrete irrigation and drainage channels web across more than 20,000 hectares. Directly above: A field of barley, an important grain crop along with wheat, millet, and sorghum, adjacent to a view of a project at Wadi al-Sahba where some 4,000 hectares of reclaimed desert have been turned over to a commercial firm for production of meat, poultry, and dairy products.

in Makkah after the pilgrimage in the summer of 1926. The purpose of the congress was to reassure Muslim opinion throughout the world and to give all Muslims an opportunity to suggest measures regarding the pilgrimage and regulate other religious matters of common interest. Thus reassured, Muslims from all over the world during the following years flocked to the holy cities for the hajj season.

Every Bedouin chief was made accountable to the King for the actions of his tribe.

Remaining in the Hijaz for a time, 'Abd al-'Aziz devoted his attention to organizing the administration of the province and to finding a way of controlling the fiercely independent Bedouins. This was not a simple problem; historically the Bedouin tribes had opposed the imposition of any central authority. But the King's solution was simplicity itself: the chief of every tribe was told that he was responsible for any crime committed by his people and that if for any reason he failed to control them the King would intervene. As 'Abd al-'Aziz (who understood the Bedouins well) dealt with tribal problems with tact and firmness, his system was successful. In talking to the various shaykhs he made it clear that the land could only resist foreign intervention if it were united under one rule and that the alternative was the anarchy he had spent much of his lifetime bringing to an end. By assisting the tribes in periods of drought and distress he removed the basic incentive for raids upon neighboring tribes and passing caravans.

'Abd al-'Aziz Al Faysal Al Sa'ud was to become famous in the West simply as King Ibn Sa'ud.

In January 1927 'Abd al-'Aziz was officially proclaimed King of the Hijaz and Najd and its Dependencies, with Riyadh and Makkah as his two capitals, and in May of the same year Great Britain in the Treaty of Jiddah formally recognized the kingdom as a fully sovereign state. On September 22, 1932, the country was renamed the Kingdom of Saudi Arabia.

In 1928, however, as 'Abd al-'Aziz was still organizing his kingdom, some of the Ikhwan leaders who had helped him conquer the Hijaz attacked an Iraqi border fort, in defiance of his express orders. As this was in direct violation of the boundary agreement he had concluded with the British at al-'Uqayr in 1922, 'Abd al-'Aziz was, reluctantly, forced to take action; he could not tolerate insubordination which threatened the existence of his young state.

The Ikhwan were motivated by a number

Stamps

Saudi Arabia has been a member of the Universal Postal Union since 1927 and of the Arab Postal Union since 1948, and through arrangements sponsored or supervised by these organizations it maintains regular postal contact with almost all countries of the world. Since 1970 the Saudi Arabian Ministry of Telegraphs, Posts, and Telephones has more than quadrupled the number of post offices in the kingdom from 108 to more than 450 and as of the late 1970s the national airline, Saudia, and other carriers provided daily service between Saudi Arabia and other countries in Europe, Asia, Africa, and America. Postal service in Saudi Arabia today compares favorably with similar services throughout the developed world.

In the 1930s when the first geologists of the California Arabian Standard Oil Company (precursor of Aramco and Saudi Aramco) came ashore at Jubail, however, postal services were very limited, especially in the eastern part of the kingdom. The only international mail service in what was then called al-Hasa Province was by boat between al-'Uqayr and Bahrain, and the small post office at al-Khobar was unable to handle the quantities of mail involved in the oil operations. The mail pouches, accordingly, were taken down to al-Khobar pier and put aboard a company launch, accompanied by a Saudi postal official, and sent to Bahrain. Saudi stamps were not used; Bahraini stamps were applied, and return addresses in this period show "CASOC, Bahrain Island." Late in 1941 the al-Khobar post office was expanded and supplied with the large quantities of stamps required, and from this time on the company's mail began to show Saudi stamps. The first mail-carrying commercial airline flight (flown by TWA) landed in Dhahran on July 6, 1946, and in 1962 the first jet air service to and from Dhahran was inaugurated — an event the Saudi Post Office marked with the issue of a commemorative set of five stamps.

The first Saudi air stamps appeared in 1949; a set issued in 1963 marks first jet service.

The postage stamps of the Kingdom of Saudi Arabia and its forerunners offer a fascinating field for the stamp collector, from the sometimes crude provisional issues of the early days before the establishment of the kingdom to the colorful and professionally produced issues of recent times. The first stamps issued for use in what is now Saudi Arabia date back only to 1916, but the postal history of Arabia goes back well before that. In fact the first postal service in the modern sense was provided by an Egyptian post office which operated in Jiddah from June 1865 until it was closed at the end of June 1881. This service was maintained principally for the use of Egyptian businessmen in Jiddah and for Egyptian pilgrims on their way to or from the holy cities, but it was available to all who wished to make use of it, and the records of the Egyptian post office indicate that the Jiddah office was a successful and profitable one. Egyptian stamps used in Jiddah can be identified by the cancellation, which reads "Gedda" or "Djeddah."

The postmark indicates an Egyptian stamp showing the Sphinx was used in Jiddah.

The Hijaz at this time was under the suzerainty of the Ottoman Empire, and more or less contemporaneously with the operation of the Egyptian Post Office the Turks began to extend their postal system into the Hijaz and southward into Yemen. Turkish post offices were opened in Jiddah and Makkah in 1871 and by 1892 had reached at least as far south in what is now Saudi Arabia as the town of al-Qunfudhah. Stamps used at these offices were the contemporary Turkish issues and here again collectors can identify them by their cancellations. Most of the Ottoman post offices in the Hijaz closed in 1916 when the Sharif Husayn, with British backing, declared himself King of the Hijaz, but Turkish forces held out in Medina until January 1919, and the Turkish post office in that city was the last to close. The Turks also maintained post offices in Qatif, al-Hasa, and Jubail in eastern Arabia; these offices were closed when Saudi forces captured this area and expelled the Turks in 1913. Collectors know that stamps bearing cancellations from these three offices are very scarce.

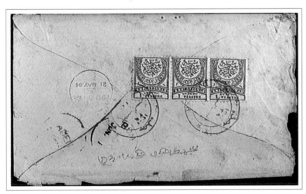

A letter bearing Turkish stamps mailed from an Ottoman post office in the Hijaz.

The first adhesive postage stamps of Arabia appeared in 1916-1917 under the auspices of the Sharif Husayn and were inscribed "Hijazi Postage" and "Makkah." The designs, selected by Colonel T. E. Lawrence ("Lawrence of Arabia") and Sir Ronald Storrs of the British High Commission in Cairo, were based on decorative carvings on mosques and other historical buildings in Cairo and on pages from famous copies of the Quran.

This 1/8-qirsh stamp of 1917 was one of the first postage stamps of the Hijaz.

In 1921-1922 Husayn, relying on secret commitments given him by the British during World War I and wishing to indicate that his domains were not restricted to the Hijaz alone, reissued these stamps with an Arabic overprint reading "Hashimite Arab Government" and the Hijrah date 1340. A short time later a set of locally printed stamps appeared with the same inscription as well as the arms of the Hashimite family, and in 1924 this set was reissued with an overprint commemorating the proclamation of King Husayn as caliph.

Arms of the ruling Hashimite family are shown on this 1/8-qirsh Hijaz stamp.

In July 1925, after the abdication of Husayn, a further set of stamps was issued, based on stylized arabesque designs and inscribed "Hijazi Arab Government." The central feature of each design was an elaborate calligraphic rendering of the name 'Ali ibn al-Husayn, and these were the last stamps issued by the Hashimites; five months after they appeared the Saudi forces captured Jiddah and the Hashimites withdrew from the Hijaz.

Stamps issued in 1925 bear the name of 'Ali as King of the Hijaz.

The Saudi Government in Najd and eastern Arabia at first had no formal apparatus of public postal services, and individuals who had a need to correspond with the world outside had to arrange for private travelers to carry letters either westward to the Hijaz and Egypt or eastward to Bahrain or Kuwait. As the Saudis moved into the Hijaz in 1924 and 1925 and took over the existing postal system, however, there was a need for new stamps to represent the

(Continued on page 142)

Stamps

new status quo. The immediate solution was to improvise by using a series of overprints reading variously "Najdi Postage" or "Najdi Sultanate Postage" which were applied to whatever stamps the retreating Hashimites had left behind them — including not only regular Hashimite postage issues but also Hijaz Railway tax stamps and various other Hijazi revenue stamps and even a few Turkish stamps left over from a still earlier period.

Saudi forces entering the Hijaz issued overprinted Hashimite and Turkish stamps.

The first definitive issue of Saudi stamps, printed in Egypt and consisting of five denominations from one-quarter to five qirsh inscribed "Hijaz and Najd," appeared in February 1926, shortly after 'Abd al-'Aziz was named King of the Hijaz and Sultan of Najd. These have been followed by a number of later issues. The first regular issue with which Aramco employees became familiar as they moved into the eastern part of the kingdom was the so-called *tughra* or monogram issue (so named after the main feature of the design, a highly stylized monogram of King 'Abd al-'Aziz) which appeared piecemeal between 1934 and 1957. Between 1960 and 1975 there appeared a very long series of over 225 stamps in three designs depicting the dam across Wadi Hanifah near Riyadh, Aramco's Buqqa gas-oil separator plant in the Abqaiq oil field, and a Saudia airplane. A so-called "tourist" issue showing various local scenes such as the Ka'bah in Makkah, an Arabian horse, and the rock tombs of Madain Salih appeared between 1968 and 1975; and another definitive pictorial set depicting the Ka'bah, a drilling rig in the offshore Khafji oil field, and other scenes began to appear in 1976.

Regular-issue Saudi postage stamps have featured Islamic themes as well as oil facilities.

Commemoratives mark the once-endangered oryx and Saudi participation in the 1994 World Cup for soccer competition.

The commemorative stamp issues of Saudi Arabia form a fascinating and often colorful and attractive record of some of the major events in the kingdom's history. The first Saudi commemoratives consisted of five stamps overprinted and issued in July 1925 to mark the first pilgrimage to take place after the reestablishment of Saudi rule over Makkah, and other overprinted issues appeared later in the same year to commemorate the extension of Saudi authority to Medina and Jiddah after the withdrawal of the Hashimites.

A 1985 commemorative marked a Saudi astronaut's participation in a scientific space shuttle mission.

A set of commemoratives issued in January 1934 to mark the proclamation of King 'Abd al-'Aziz's son, Sa'ud, as Crown Prince was the first set of Saudi stamps to be inscribed "Saudi Arabia" instead of the earlier "Hijaz and Najd," and these have since become some of the scarcest and most valuable stamps of Saudi Arabia. Later commemoratives have honored various international events and anniversaries, such as World Refugee Year, the centennial of the World Meteorological Organization, and the hundredth anniversary of the invention of the telephone, as well as milestones in the history and development of Saudi Arabia including the fiftieth anniversary of the capture of Riyadh, the expansion of Dammam port, the completion of the Dammam-Jiddah highway, the thirtieth anniversary of Saudia, and the expansions of the holy mosques. The handsomest of all the commemorative stamps of Saudi Arabia is probably the set issued in July 1975 as a memorial to King Faysal and showing a classically simple portrait of the late king in the act of performing his prayers.

The memorial set issued in 1975 shows King Faysal at prayer.

of factors in their rebellion. They were disturbed by 'Abd al-'Aziz's political realism; they saw no reason to honor arrangements with any powers that did not belong to the reform movement of the Muwahhidun. After the conquest of the Hijaz, furthermore, many of them had come in contact with a more worldly urban life than they had known, and they looked with suspicion and mistrust on 'Abd al-'Aziz's readiness to accept some aspects of this life. In 1925, during the siege of Jiddah, for example, the more fanatical among the Ikhwan had opposed the use of the telephone, and 'Abd al-'Aziz had hesitated for some weeks before installing a line between Makkah and his headquarters at the front. At that time he reportedly had won over the Ikhwan by having them listen to verses from the Quran read over the telephone, thus convincing them that an instrument that carries the word of God cannot be working for the devil.

Doubters remained, however, and at a conference of the Ikhwan in 1926 at al-Artawiyah, the first and greatest of the Ikhwan settlements, 'Abd al-'Aziz was criticized for making use of the telegraph, the telephone, the radio, and the motorcar — all of them described as "inventions of the devil." In January 1927, therefore, 'Abd al-'Aziz convened a conference of Ikhwan leaders and ulema or religious leaders in Riyadh to discuss these and other matters. During the conference the ulema issued a *fatwa* or religious opinion declining to make a decision on the telegraph, since there was no precedent either for it or against it in the shari'ah (Islamic law). They also refrained from taking a stand on the radio, the telephone, and other modern means of communications.

Faced now, in the attack on the Iraqi border, with outright disobedience of his orders on the part of the Ikhwan, 'Abd al-'Aziz moved against them. As they were, by the King's own design, elite forces, the fighting lasted for a year and a half. Finally, at a battle near al-Artawiyah in the spring of 1929, 'Abd al-'Aziz defeated them, brought the rebel leaders to Riyadh, and eventually broke the power of all the rebellious Ikhwan settlements.

In the 1930s the King was confronted with another, less tractable, problem: the economy of the kingdom was still based on stock rearing and traditional subsistence farming in the fertile areas. As more than 40 per cent of the population was engaged in agriculture, Saudi Arabia's revenues, essentially, stemmed from customs duties and levies on pilgrims. Now, however, as the worldwide depression of the 1930s worsened, the numbers of pilgrims dropped drastically: in 1929 a hundred thousand pilgrims visited Makkah; in 1931, only 40,000 came.

In 1931, at the request of 'Abd al-'Aziz, an American mining engineer named Karl Twitchell came to visit the King in Jiddah and to discuss a survey of the potential resources of the kingdom — particularly untapped sources of water in the Hijaz. Twitchell was unable to discover new water sources there, but he did find the ancient tailings of five gold mines (one of which may have been worked during the time of King Solomon) and in 1934 the Saudi Arabian Mining Syndicate, Ltd., was formed to work Mahd al-Dhahab, "the Cradle of Gold," 80 kilometers south of Medina.

Karl Twitchell spearheaded early efforts to tap Saudi Arabia's natural resources.

The originally controversial radio system proved its usefulness in a brief war against Yemen which broke out in 1934 when the Imam Yahya of Yemen disputed the Saudi title to the 'Asir region north of Najran, and encouraged local tribes to revolt. 'Abd al-'Aziz sent his oldest sons, Sa'ud and Faysal — both later to become kings — to deal with the threat and kept in touch with them via radio. Sa'ud led a column into the mountains, while Faysal followed the coast and captured the port of Hodeida in Yemen. Yahya abandoned his claim to the disputed area and signed a peace treaty with the Government of Saudi Arabia in May 1934 at Tayif.

The Saudi victory and 'Abd al-'Aziz's generous terms to his defeated foe — he made no attempt to add Yemen to his kingdom — created a favorable impression throughout the Arab world. The war also led to an ambitious road-building project — a response to the introduction of modern vehicles to transport troops.

In the meantime 'Abd al-'Aziz, still in need of funds to finance the development he thought necessary, had opened discussions with respect to petroleum exploration in Saudi Arabia. In economic terms it was the most important decision of his reign. In 1933, after a concession had been granted to the Standard Oil Company of California, two American geologists landed at Jubail on the east coast to

Protruding stone slabs prevent rainwater from dissolving the characteristic mud-brick buildings in 'Asir.

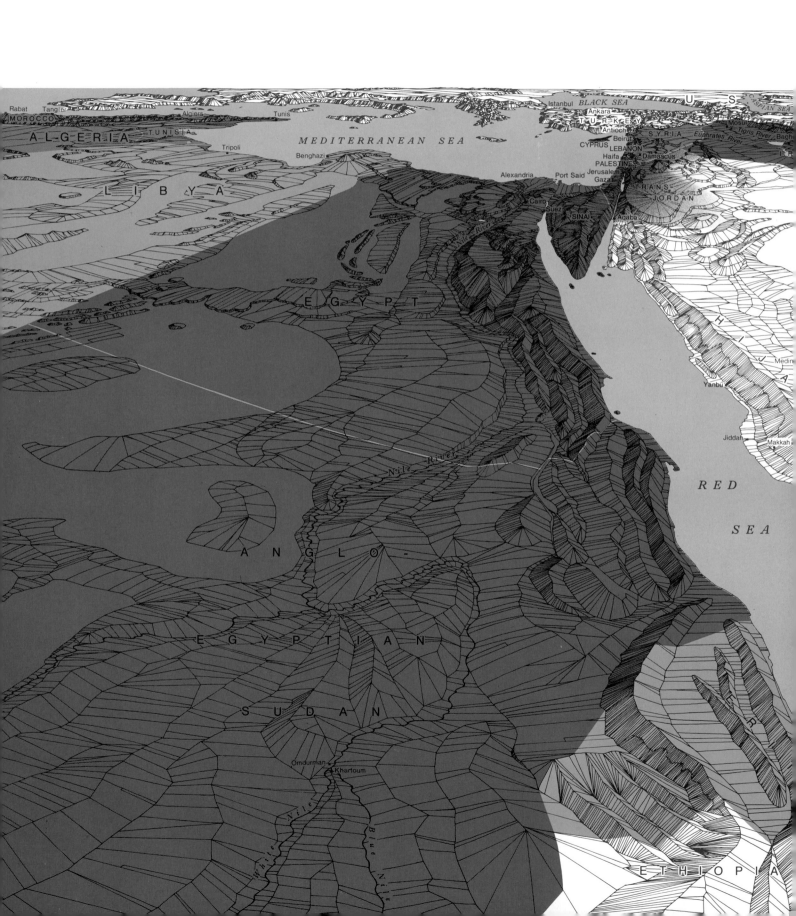

THE MIDDLE EAST AFTER THE FIRST WORLD WAR

With the final defeat of the House of Rashid and the expulsion of the Hashimites from the Hijaz, the Saudi state under King 'Abd al-'Aziz has reestablished itself and extends its control over much of the Arabian Peninsula, and its independence is given *de jure* recognition by Great Britain with the Treaty of Jiddah in May 1927. Simultaneously, however, European colonial control over the other lands of the Middle East and North Africa has reached its peak. Algeria is ruled as a part of metropolitan France, Morocco and Tunisia are French protectorates, and Libya is an Italian colony. The British protectorate over Egypt formally ended in 1922, but British influence remains strong and British troops are stationed in the country. Except for Yemen and the territories under 'Abd al-'Aziz, all of the Arabian Peninsula is under some degree of British influence or control. To the north, Iraq is a British mandate, France has been given control over Syria and Lebanon, and the British have mandates over Palestine and Transjordan.

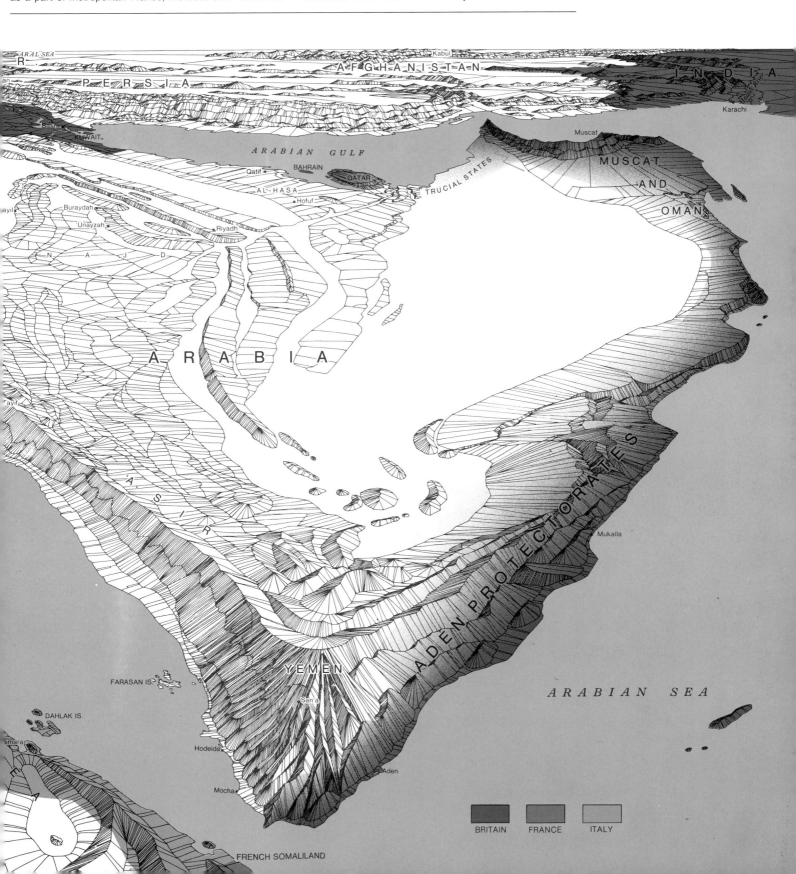

BRITAIN   FRANCE   ITALY

begin a search for petroleum that would eventually lead to the discovery of the world's largest oil reserves (see page 244) and transform Saudi Arabia from a simple desert kingdom into a power in world economics and politically one of the most important countries in the Middle East.

These effects, of course, were not felt until after World War II; for although oil in commercial quantities was found in 1938, the war held back the development of production facilities. In the interim, because the war prevented many pilgrims from reaching Makkah, and because severe drought decimated the Bedouins' herds, economic conditions worsened.

At first 'Abd al-'Aziz met his essential expenditures by advances against future oil royalties. Later Great Britain assumed responsibility for aid to Saudi Arabia (since existing U.S. legislation precluded direct American financial aid to the kingdom), and eventually in the last years of the war Saudi Arabia became eligible for lend-lease and other assistance from the United States.

Prince Faysal headed his country's delegation to the founding of the United Nations in San Francisco.

FOREIGN RELATIONS: During most of World War II, the King maintained a position of benevolent neutrality towards Britain and the United States. In return for U.S. military aid, however, he allowed the Americans to construct and use an airfield at Dhahran for maintenance, repair, and other technical services for Allied planes. In 1942 an American legation was opened in Saudi Arabia and in 1944 a Saudi Arabian legation was opened in Washington — both later raised to the rank of embassies.

In February 1945 the King agreed to go to Egypt, where he met first with President Franklin D. Roosevelt aboard an American cruiser anchored in the Great Bitter Lake in the Suez Canal Zone, and then with British Prime Minister Winston Churchill and Foreign Minister Anthony Eden at the Fayyum oasis south of Cairo.

This historic meeting between the American President and the Saudi King confirmed Saudi Arabia's adherence to the Allied cause and in March 1945 Saudi Arabia entered the war on the Allied side. 'Abd al-'Aziz and Roosevelt also exchanged letters in which the king stated that all countries should help and receive the Jewish victims of Nazi Germany,

The Suez Canal Zone was the setting for a historic wartime meeting between King 'Abd al-'Aziz and President Roosevelt.

and that Palestine could not be expected to do this alone. Roosevelt, for his part, recalled in his letter the assurance he had personally given the King at their meeting that the U.S. Government would make no change in its policy toward the Palestine question without full consultation with both the Arabs and the Jews and that he would undertake no actions which might prove hostile to the Arab people. Two weeks later, Roosevelt reported to the American Congress on his meeting with the King of Saudi Arabia in these words: "Of the problems of Arabia, I learned more about the whole problem, the Muslim problem, the Jewish problem, by talking with Ibn Sa'ud for five minutes than I could have learned in the exchange of two or three dozen letters."

In March 1945 Saudi Arabia joined Egypt, Iraq, Lebanon, Syria, Transjordan, and a representative of the Palestinian people in signing a Pact of Union of Arab States, the foundation of the Arab League.

During the war too a number of the King's sons visited the United States and one of them, Prince Faysal, later headed the Saudi delegation at the founding of the United Nations in San Francisco — the first of several assignments he was to undertake as his role in forming the foreign policy of Saudi Arabia expanded.

INTERNAL DEVELOPMENT: At the end of the war Saudi Arabia's situation in the world had already begun to change. As its oil resources had attracted U.S. Government support, oil revenues were starting to mount at a gratifying rate, and at last 'Abd al-'Aziz could begin to move more rapidly toward fulfilling his dream of transforming a tribal and regional society into a modern, centrally administered state.

One of the country's vital needs was, and still is, education. Immediately upon unifying Saudi Arabia in 1926 'Abd al-'Aziz founded a new school in Makkah. In 1945 he instituted an extensive program of school construction and by 1951 had raised the number of schools in the kingdom to 226, with 1,217 teachers and a combined enrollment of 29,887 pupils, all supplied with free books and tuition. This was the beginning of a long campaign that reached a climax with the government devoting one-sixth of its $143 billion second five-year Development Plan to education.

Another need to which the Saudi Government gave attention was agriculture, a reflection of 'Abd al-'Aziz's earlier perception that agricultural settlements were vital to a country where less than one percent of the land is arable, and where the food supply, as a result, was in earlier years sometimes problematical. Agriculture, in fact, has always been important in Arabia. Throughout the kingdom the evidence of ancient expertise in the arts of irrigation and water conservation is unmistakable, and remains of ancient dams that once spanned wadis in the western part of the country can still be seen today. Near Tayif, one of the best preserved — which today looks like the wall of a giant terrace — displays on its downstream face an inscription dating it to the first century of Islam.

During the long years of decline and tribal warfare, unfortunately, much of the ancient knowledge of irrigation techniques had been lost, and 'Abd al-'Aziz therefore invited a U.S. agricultural mission to visit the country in 1942 and report on its agricultural potential. He also laid the groundwork for pilot projects set up in al-Kharj and elsewhere — projects that became the nucleus of the later government farms and breeding stations. Lastly the King strongly backed development of the vast underground supplies of water of the eastern part of the kingdom which, today, are being tapped for government irrigation projects.

In July 1947 the government announced that it intended to spend some $270 million to build roads, schools, and hospitals, to install electrical power generation plants, and to foster agricultural irrigation projects. In the early 1950s, as state revenues from the postwar boom in oil production approached the $100 million mark, Saudi Arabia was able, with American technical assistance, to launch these and other development programs. Construction was begun of a system of paved roads tying together the major cities of the kingdom, an air service linking the two coasts was established, and modern harbor facilities were built in Jiddah so that the ancient port could take larger ships. A powerful radio station was put into operation in Jiddah, and authorization was given for development of plans for a telephone network which culminated in the late 1970s with the construction of a cross-country telephone and telex communications system based on coaxial cable, microwave transmission stations, and satellite ground stations. The year 1951 saw the completion of a project begun in the late 1940s which was of particular importance to 'Abd al-'Aziz — a railroad linking his capital, Riyadh, with the port of Dammam on the east coast.

Steady growth and industrialization have stimulated today's use of modern cargo-handling techniques at Dammam port.

Health and medicine were other interests close to the King's heart. On conquering the Hijaz one of his first acts was to improve medical facilities for the pilgrims to Makkah, an act which earned the gratitude of Muslims everywhere; and in 1947, even as plans for hospitals were being discussed, he ordered four surplus field hospitals from the American Army to be used in the inevitable interim between the planning and construction of permanent ones.

These hospitals were actually the first elements in a building program that, in the mid-1970s, brought forth the King Faysal Specialist Hospital and Research Center in Riyadh — one of the most modern medical facilities in the world — and scores of other hospitals throughout the country. As a result of the King's interest in health and medicine each administrative district in the kingdom was helped to establish its own medical infrastructure, and today there are over 3,400 dispensaries and health centers, as well as mobile medical units which periodically visit the more isolated parts of the country. Another result is that such diseases as malaria, trachoma, and smallpox, all once endemic, have been virtually eliminated.

A specialist works with an electron microscope at King Faysal Specialist Hospital and Research Center, a superbly equipped institution for high-technology medicine.

King 'Abd al-'Aziz established the basis of the country's present administration. In 1953 he sanctioned the formation of a Council of Ministers to act under the presidency of his son, Crown Prince Sa'ud. King 'Abd al-'Aziz, however, did not live to see the profound effect that such projects were to have on his kingdom. On November 9, 1953, with many of them barely launched, he died after 51 years as leader and king of a country from which he had been exiled as a youth but later had reconquered, pacified, and finally developed. His life, by any standards, was remarkable. In 1900, when he was about 20 years old, the Arabian Peninsula was one of the least known and most isolated areas in the world. When he died Saudi Arabia, because of his efforts, was well on the way to becoming a world power. His achievements indeed were

Jiddah's bustling harbor is the largest on the Red Sea coast.

Saudi Arabia's commitment to education suited to the kingdom's specific needs is reflected in institutions across the country. Above left: Inspired by the architecture of Najd, the elegant King Sa'ud University in Riyadh houses almost 1,500 classrooms and labs supporting studies in the sciences, medicine, business, education, and the arts. Facing page, top and center: The King Fahd University of Petroleum and Minerals atop the *jabal* at Dhahran is an architectural masterpiece as well as a source of graduates for the professional needs of the country's principal industry. Directly above: Professor and students at King 'Abd al-'Aziz University in Jiddah. At left: Pupils at a boys intermediate school in Yanbu' and, at far left, opposite page, students in an elementary girls school.

King 'Abd al-'Aziz had been leader of his country for more than half a century when he died in 1953.

almost unparalleled, and due, by common consent, to a powerful intelligence, unflagging courage, exceptional vision, and above all a profound faith in God.

KING SA'UD IBN 'ABD AL-'AZIZ: 'Abd al-'Aziz was succeeded by his son Sa'ud who, 20 years before, had been designated as his successor. Sa'ud was born in Kuwait in 1902, the year his father captured Riyadh. After the conquest of the Hijaz in 1925, he was put in charge of the affairs of Najd, with a special responsibility for tribal affairs. While president of the Council of Ministers he also worked with his younger brother Faysal in extending the administrative structure of the state — through establishment of ministries and autonomous agencies which eventually modified or eliminated most elements of the direct, personal rule that had characterized the reign of 'Abd al-'Aziz. In 1933 he was proclaimed his father's heir and in November 1953, in accordance with the established procedure for the succession, he became king.

During his reign King Sa'ud broadened Saudi Arabia's relations with its neighbors by undertaking a series of trips and official visits to friendly states. But his reign covered troubled years. On October 17, 1955, he signed a treaty of defense and friendship with Egypt and in the following year, when the French and British attacked Egypt following the nationalization of the Suez Canal Company, Saudi Arabia broke its diplomatic relations with Britain and France. In 1958, with the creation of the United Arab Republic, and Saudi concern over the growing influence of the Soviet Union in the Middle East, relations with Egypt cooled swiftly and did not improve significantly until the early 1970s.

Relations with Jordan, still ruled by the descendants of 'Abd al-'Aziz's former opponents the sharifs of Makkah, improved, and Saudi Arabia also supported Kuwait, its small neighbor to the north, against external threats. In 1958 Sa'ud and the ruler of Bahrain agreed on a boundary in the waters between their two countries.

During Sa'ud's reign too the kingdom was plagued with financial difficulties. Despite increasing oil revenues, spending had outstripped income and, in order to finance development projects, large loans against future oil income had been drawn and the

Food

The Bedouins have a saying that translates to "He makes coffee day and night." Among a people justifiably famous for their hospitality, it is a way of describing a generous man. In the old custom, the preparation of Arab coffee — flavored with aromatic cardamom seed but unsweetened — was indeed man's work, and the ring of the brass mortar in which he pounded fresh-roasted beans to powder was music to the ears of expectant guests. If the guest was a man of standing, the host would be obliged to kill a sheep, and the coffee would be followed by an offering of trays heaped high with rice and chunks of mutton or, on rare occasion, rice-stuffed baby lamb rubbed with spice and roasted whole — all surrounded by circles of flat, unleavened bread. The guest would know it was time to go when his host passed the censer trailing the filmy smoke of frankincense or scented wood, for by tradition this signaled the end of the "Arab feast."

Loyalty to custom and tradition is the essence of all Middle East food and feasting, and many of the finest dishes of the Arabs' heritage are centuries old. Arab poets of the Middle Ages celebrate some — like the Damascene sweets still relished today — in detailing the banquets of the caliphs of Baghdad, and both peasant fare and court cuisine spread with the marching armies of Islam. Part of the table of present-day Saudi Arabia comes from the traditions developed by the peninsular Arabs themselves, part from this age-old culinary pool.

Typically Levantine, for instance, are two purees eaten with bread as appetizers: *hummus* made from ground chick peas, and baba ghannouj *(baba ghannuj)* made from mashed eggplant, both mixtures rich with tahini *(tahinah)* or sesame-seed paste, and pungent with lemon juice and garlic. Typically Turkish are tomatoes, peppers, zucchini, or eggplant stuffed with a spicy mixture of rice, pine nuts, and currants that may also be the center for a wrapping of grape leaves in a version more typically Greek. *Burghul*, or cracked wheat, mixed with finely chopped parsley, onions, tomatoes, and mint, is the main ingredient for the well-known salad called tabbouleh *(tabbulah)* in Lebanon and Syria; mixed with ground lamb and baked in a variety of shapes it is the basis for a meat dish called kibbeh *(kubbah)*. The wheat is *jarish* in Saudi Arabia, where it is a

favorite rice substitute in central Najd and the oases of the Eastern Province. Everywhere one finds members of that vast family of skewered meats and ground-meat kebabs. And everywhere there is *laban,* the yoghurt of the Middle East, and labneh *(labnah),* the rich cream cheese made from *laban* drained through cheesecloth.

For centuries the date was the Arabs' universal staple. Today in Saudi Arabia dates and coffee may still be offered to a caller; dates stuffed with almonds are a popular confection; dates baked into tiny, sugared cookies known as *ma'mul* are essential to the proper celebration of 'Id al-Fitr, the festival that comes with the close of Ramadan. Dates are the basis for *hunaynah,* a hearty, nearly stiff porridge of ground dates cooked with butter and semolina and flavored with cardamom that is a classic breakfast dish of Najd.

Cardamom is the flavoring, too, for *saliq,* the best known evening rice dish of the country — a hot pudding for which the rice is first half-cooked with meat or chicken broth, then usually simmered with milk until soft. Essential here is a hint of *mustaka,* gum arabic, the aromatic resin of the mastic tree which, like frankincense, recalls the ancient incense trade. A spice mixture of cumin, cinnamon, turmeric, and black pepper which the housewife buys ready-mixed flavors a favorite dish of the Eastern Province made with the tender trout-size *subayti,* a local fish that abounds in the Arabian Gulf along with fat-fleshed grouper, king mackerel, and succulent shrimp.

To sample the food of Saudi Arabia in its greatest variety, one must look to the Hijaz, where trade and pilgrim caravans have long lent a cosmopolitan air. Paradoxically, the best time to sample this rich table is during Ramadan when, after the dawn-to-sunset fast, the evening meal is lavish.

The Ramadan fast is broken with a few sips of water, then a few dates and a thick drink prepared from sheets of dried, pressed apricots chopped and pureed with water. The meal itself begins — invariably, in the Hijaz — with a thick, nourishing soup made from soaked wheat and lamb stock, rich with chunks of lamb and spices: cinnamon, cardamom, and the dry, curled leaf of tree wormwood.

Then come the brown Egyptian *ful,* broad beans cooked with tomato, onion, and oil, and next the much-loved *sanbusak,* paper-thin pastry made up in triangular shapes stuffed with ground meat, onion, the pungent leaf of coriander, or the long, spike-like local member of the garlic family called *kurrath.*

The meal may continue with eggs gently cooked on a bed of fried onion, green pepper, and tomato, followed by one or two main dishes. There might be *kabsah* — chicken or lamb sauteed with garlic, onion, tomato, grated carrot, and grated orange rind, served atop rice that has simmered in the meat sauce and all garnished with raisins and almonds. *'Aysh abu lahm* is a local speciality the Hijazis describe as "something like pizza." A leavened dough, egg-rich and flavored with seeds of fennel and black caraway, is baked in the shape of a thick-bottomed pie shell, filled with fried mutton and chopped kurrath or spring onion, and topped with a sauce made from tahinah. There might be a large grouper laid open and covered with onion, tomato, garlic, hot pepper, and cumin, baked and served with lemon, or fish baked with a sweet-sour sauce made from dried tamarind.

The customary way to end such a banquet is with a sweet, preferably one that is soft and cold like the elegant *muhallabiyah,* a delicate pudding of rice flour and milk, lightly flavored with orange blossom or rose water and decorated with almonds and pistachios. For special guests many sweets would be served, and there are many identified especially with the Ramadan season: *luqmat al-qadi,* spoonfuls of soft, light dough sometimes lightly spiced with cardamom and saffron, fried and then dipped in syrup; *qatayif,* store-bought pancakes stuffed at home with nuts and cheese, then fried and covered with syrup; *basbusah,* semolina cooked with sugar syrup, baked into squares and sometimes served with a topping of *qishtah,* the local answer to an Englishman's clotted cream; or *kunafah,* top and bottom layers of pastry that resembles shredded wheat with a middle layer of goat cheese, butter, and pine nuts, and over all, once again, a sweet syrup scented with rose water. The well-known sweet tooth of the Arabs is indulged in Ramadan, and never are rewards better earned.

country was plunging deep into debt. In 1958, therefore, Sa'ud granted his brother, Crown Prince Faysal, full executive powers in financial, internal, and foreign affairs.

In view of the gravity of the problems, Prince Faysal moved with dispatch. He introduced political and economic reforms, developed the cabinet system of government, granted greater freedom to the Saudi press, and, in addressing the financial crisis, insisted on strict austerity, reducing both government expenditures and those of the Royal Family. He also published a state budget for the first time.

Sa'ud's reign was characterized by many positive achievements, particularly in the field of education. King Sa'ud University in Riyadh was founded, government girls' schools were opened, the number of students sent to study abroad was increased, and within the kingdom many schools and religious institutes were established. In the field of social welfare, the government set up a grievance board to insure impartial hearings on citizen complaints, granted interest-free loans to farmers, and in accordance with Islamic law devoted proceeds from the zakah to helping the poor and needy. During Sa'ud's reign the government built a huge housing complex in Jiddah for pilgrims to use free of charge during the hajj, renovated and extended the Sacred Mosque in Makkah and the Prophet's Mosque in Medina, and paved roads used by the pilgrims.

Sa'ud reigned until November 1964, when, because the King's health was declining, the Royal Family and religious leaders persuaded Faysal to replace him.

KING FAYSAL IBN 'ABD AL-'AZIZ: Born in Riyadh in 1906, Faysal was educated by his maternal grandfather, 'Abd Allah ibn 'Abd al-Latif Al al-Shaykh, a noted religious scholar and a descendant of Muhammad ibn 'Abd al-Wahhab. In 1918 at the age of 12 he accompanied his father on his military campaign against Hayil and in 1922 he led the campaign in 'Asir. Later, in 1934, he also commanded one of the Saudi columns in the war with Yemen.

In 1926, after the conquest of the Hijaz, Faysal was appointed Viceroy of the Hijaz as well as chairman of the new Consultative Council, and in 1930 he became Minister of Foreign Affairs. While holding those posts

The Natural History Museum of King Sa'ud University in Riyadh belongs to the oldest and largest university of the kingdom.

Saudi Arab women have their own well-equipped institutions of higher learning.

Dress

As in all countries, dress in Saudi Arabia is influenced by climate, utility, and custom. With a few minor regional exceptions (usually slight variations or additions to the basic garment) traditional clothing is uniform across the peninsula, both among residents of towns and cities and among those who dwell in the oases, mountains, and deserts. Differences tend to be found only in the quality of materials used and in the fineness of tailoring; and fashion, for the most part, exercises an influence only on small details, although it does play a role in women's clothing and in the dress of some young men in urban areas.

Boys and men wear the *thawb,* a loose-fitting, ankle-length, usually white shirt which allows free circulation of air though covering almost the entire body. This was traditionally made of cotton, but today artificial fibers are also frequently used. The thawb is slightly more close fitting than the similar garment, called the *dishdashah,* worn in neighboring Gulf states. Both differ from the extremely full galabiya *(jallabiyah)* of Egypt, which is often of polished cotton and has broad, open sleeves, and the similar garment — usually of wool and less frequently of cotton — which is called djellaba *(jallabah)* in North Africa. The sleeves of the thawb close at the wrist. It also buttons at the collar, and matching sets of decorated buttons and cuff links are worn for formal occasions. Breast or side pockets are individual options.

In the cool months, warmer cloth, including wool worsted, is used for the thawb and the color may vary from white and cream to subdued shades of light gray, brown,

and sometimes blue or green. It is not at all unusual to see a dark Western-style suit jacket worn over the thawb during the winter, but the more traditional, flowing floor-length cloak called the *bisht* in the eastern and western provinces and *mishlah* elsewhere in the kingdom is still more common. Extremely handsome, with a sweeping air of dignity, the cloak is made of finely woven wool or camel hair, and is worn over the shoulders like a cape, enclosing the arms but open in the front. Small slits in the seams allow the hands to extend when desired. It comes in natural colors of black, brown, and more rarely tan or cream, and is trimmed or edged in black cord or even — usually in the more expensive versions — in gold braid.

Worn on the head are the ubiquitous *ghutrah* and *'iqal*, the headcloth and double ring of black rope or cord to hold it, which to many in the West and elsewhere most represent the image of the adult Arab male. Today the ghutrah and iqal are seen worn not only with the thawb, but also with work clothes, military and boy scout uniforms, and even academic gowns.

The square headcloth comes in plain white and in red-and-white or black-and-white checks. The color is a matter of personal or regional preference, although the checks seem to be somewhat more common in the interior and rural regions of the peninsula and the white in the more urbanized areas. Folded diagonally to make a triangle, the headdress is draped with the fold across the forehead and the points hanging down the back of the head and shoulders. One or both ends or the tail of the triangle can also be casually tossed forward over the shoulders, or wrapped loosely around the neck, across the ears and lower face, or even around the top of the head, offering the wearer varying degrees of protection from wind, dust, smoke, sun, heat, or cold. The way in which the cloth is wrapped and the angle or position in which the cord sits on the head are very much a matter of the wearer's personal preference.

Further north, in the Arab countries of the Eastern Mediterranean, the ghutrah is called *hattah* or *kufiyah* by those who wear it. In Saudi Arabia kufiyah is the name for the small, brimless, usually white cap worn by younger boys and by many men in informal situations, and frequently worn beneath the ghutrah to help position the ' iqal.

Sandals complete the traditional male dress, although today Western-style shoes are equally common. All kinds of sandals are worn, but the traditional type has a broad, ovalshaped natural leather strap decorated with geometric patterns of colored or metallic stitching across the top of the foot.

Other variations in the national dress are few. In the winter, herdsmen in the northern plateau regions along the Jordanian and Iraqi frontiers may wear very heavy cloaks, or *farwahs,* lined with sheepskin and covered in thick red, brown, or black canvas decorated on the back with bold open diamond patterns in black braid. Tribesmen or travelers in remote areas sometimes belt the usually

(Continued on page 154)

Dress

free-falling thawb with a wide leather strap in which they can carry a knife or money. In the extreme southwest, both men and women sometimes wear broad-brimmed conical straw hats.

In the coastal regions, workers in traditional jobs requiring freedom of movement, such as sailors, or farmers climbing date palms or planting rice, often wear a thin knit T-shirt with a brightly patterned cotton sarong called *izar* or *futah* tucked around the waist and reaching to mid-calf. In modern industries where men work close to heavy vehicles or machines with moving parts, Western-style trousers, shirts, safety shoes, and hard hats are replacing flowing thawbs, sandals, and ghutrahs, at least during working hours.

Little girls may wear frilly party dresses much like their counterparts in Europe and America, while their older sisters and their mothers slip the floor-length 'abayah over their heads and around their shoulders when going out in public, drawing it closed in front with their hands. Beneath the somber black 'abayah or when it comes off at home or at school or while visiting female friends, dresses are invariably bright and cheerful. Traditional dresses are basically colorful versions of the thawb, often decorated with glittering sequins or metallic thread. Many women in the larger towns today wear the latest Western fashions

beneath their 'abayahs and watch the rise and fall of the hemline as closely as their sisters abroad. Both women and girls, even those from families of modest income, often wear great amounts of gold or silver jewelry — which doubles as ornament and as investment. Traditional wedding dresses are known for their intricate and colorfully embroidered fronts.

Like the 'abayah, the veil is generally worn by women and older girls outside the immediate family circle. Bedouin women and farm women of the 'Asir mountains have traditionally not worn the veil while performing daily chores, and tend rather to pull their 'abayahs modestly across their faces when strangers approach. This custom is also seen more frequently today in the larger cities. Veil styles range from a stiff fabric, cut mask-like with narrow eye slits, to a sheer piece of black chiffon or an ornamental draping with embroidery, beads, or rows of coins. But throughout the peninsula, and certainly among older women, the veil in some form is as typical and ubiquitous as is the ghutrah of the men.

On Thursdays and Fridays in the marketplaces of the major cities these days, one sees national costumes from the world over. Western expatriates in Saudi Arabia dress much as they would for similar climate and circumstances at home except that clothing, both in offices and on social occasions, is usually somewhat more casual. Revealing or tight-fitting clothing is considered inappropriate for either men or women, however, and even in sports situations a degree of modesty is a thoughtful sign of consideration and respect for local tradition.

Faysal traveled extensively in Europe, including the Soviet Union in 1934, and the Middle East. In 1939 he attended a conference on Palestine in London, and in 1943, with his brother Khalid, who would succeed him as king, he made a state visit to President Roosevelt in Washington. He also represented Saudi Arabia at the United Nations Conference of 1945 and, during the 1947 UN debate on the Palestine problem, delivered a historic speech opposing partition. Later, when King Sa'ud succeeded his father, Faysal was appointed president of the Council of Ministers and then, in 1964, became king.

As he had before his accession, King Faysal devoted much of his attention to foreign affairs, particularly, in the first years of his reign, to the problems arising from civil war in Yemen. To strengthen ties among the Muslim states of the world, he also traveled widely, particularly in Africa and the Middle East. As the spokesman and interpreter for Muslims everywhere, he was a spiritual force as well as a political leader.

On the domestic front Faysal continued to press forward with fiscal and administrative reforms. In keeping with his own expressed views on development — "change, but change slowly" — the reforms came gradually. The effect, nevertheless, was more centralization of government and administration along with an increased emphasis on the economic development of the country.

MODERNIZATION OF THE KINGDOM:
Under Sa'ud, and after him Faysal, the efforts of 'Abd al-'Aziz continued to bear fruit in all fields and particularly in transportation, communications, and industry.

Cars and trucks were introduced into Saudi Arabia in the 1920s. Because of the nature of much of the terrain, and because of the lack of adequate roads, their use was at first largely restricted to the bigger towns — though Twitchell and geologists of the California Arabian Standard Oil Company (Casoc) crossed the peninsula by car a number of times in the early 1930s. Road construction on a large scale did not get under way until after World War II, but by 1992 the government had built more than 33,000 kilometers of paved main highways including the 1,530-kilometer transpeninsular highway linking Dammam with Riyadh and Jiddah and the 1,400-kilo-

meter road running from Dammam to the frontier with Jordan, as well as more than 92,000 kilometers of feeder roads. In the late 1940s 'Abd al-'Aziz had also pushed construction of a 580-kilometer railroad from Dammam to Riyadh, completed in 1951.

To link the distant provinces of a country the size of all of Western Europe, Saudi Arabia also established a national airline in the 1940s. Now called Saudia, the airline grew rapidly into one of the major airlines of the Middle East. By the late 1970s its fleet of jets provided domestic flights daily throughout Saudi Arabia and international flights to many Middle East terminals as well as to major cities in Asia, Africa, Europe, and North America.

During Faysal's reign too the government encouraged private development of agriculture and those sectors of the economy other than petroleum. In earlier times the pillars of the economy were date farming, livestock, pearling, and the export of such simple goods as *bishts* (handwoven woolen cloaks) and traditional jewelry. In the 1960s the kingdom's entrepreneurs began to turn to the processing of plastics, steel, cement, and tiles. As encouragement the government began to provide low-cost industrial sites and special tax and customs incentives.

Many of these steps forward in the development of the resources and economic infrastructure of the kingdom were achieved under the first five-year Development Plan, inaugurated in 1970, under which the economy of the country grew very rapidly, with real national income increasing at a rate of almost 45 percent per year.

In 1969-1970, when the plan was being drawn up, Saudi Arabia's oil production was relatively low (less than 3.5 million barrels a day), the financial resources of the kingdom were limited, and there were definite restrictions on government spending. The plan was drawn up with these restrictions and limitations in mind.

The unforeseen rise in oil production in the years immediately after the plan went into effect (see page 209) and the subsequent very large rise in government income meant that these restraints were no longer required. The period 1970-1975 turned out to be one with a very high rate of economic growth and accomplishment in which most of the goals of the Development Plan were reached or surpassed. One of the

Casoc geological parties were first to use cars in the Eastern Province.

This Saudi pilot flies for Saudia, the national airline of Saudi Arabia, the largest in the Middle East.

Thousands of kilometers of modern transportation lines today link the provinces of Saudi Arabia and help the kingdom support its growing commercial traffic and increase the mobility of its people. This page, above: Port installations on the Gulf at Jubail, a fast-developing industrial site in the Eastern Province. Top: A Tri-Star of the fleet owned by Saudia, the national airline of Saudi Arabia and the biggest carrier in the Middle East. Opposite page, beginning at top: The LPG tanker *Al-Berry*, the first vessel of this type to fly the Saudi flag; terminal buildings at Jiddah airport, successor to an outgrown facility; the Jiddah-Tayif highway, a showpiece of modern road-building techniques, cuts through the mountains of the Hijaz, while the Saudi Government Railroad carries passengers and freight across almost 500 kilometers of desert between Riyadh and Dammam.

SAUDI ARABIA: ROADS, PORTS AND AIRPORTS

From its early days the Saudi Arabian Government has seen the development of internal transportation as one of the most effective ways of unifying and thus strengthening the kingdom. When George Sadlier, probably the first Westerner to cross the Arabian Peninsula from coast to coast, made his journey in 1819, it took him almost three months to travel from Qatif to Yanbu', and when Philby made the trip from al-'Uqayr to Jiddah a century later it took him almost a month.

Today the coasts of Saudi Arabia are only hours apart by air, and its town and villages are linked by a network of 33,000 kilometers of paved highways. The country is served by a national airline that in 1994 carried over 9 million passengers on its domestic routes alone, a railroad running between Dammam and Riyadh, and ports that have been enlarged and modernized to handle the huge volumes of cargo needed to support its development plans.

MEDITERRANEAN SEA

al-Qurayyat

JORDAN

Tabuk

Duba   Tayr

al-'Ula

al-Wajh

Yanbu'

Rab

Jid

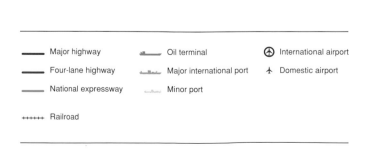

——— Major highway   ◢▬ Oil terminal   ⊕ International airport

——— Four-lane highway   ◢▬▬ Major international port   ✈ Domestic airport

——— National expressway   ◢▬ Minor port

++++++ Railroad

King Faysal, monarch from 1964 until 1975, earned renown as a spiritual as well as political leader.

Ruler from 1975 until 1982, King Khalid presided over a key phase of the kingdom's development.

The Abu Hadriyah highway was constructed as part of a massive infrastructure-development program.

general goals of the plan was reducing the country's dependence on the oil industry for its income, but although there was a very substantial absolute increase in the importance of the non-oil sectors of the economy, their contribution to the gross national product was actually reduced in proportionate terms because of the very great increase in income from the oil industry. Nevertheless the kingdom's planners gained valuable experience which they were subsequently able to put to good use in drawing up and implementing the second five-year Development Plan.

During the period of the first Development Plan, Saudi Arabia emerged as a pivotal source of the world's petroleum supplies and King Faysal's government was faced with a financial problem that was the reverse of his father's. Instead of worrying about how to make ends meet, the government had to cope with a monetary surplus that some Western economists said could upset finances and currencies around the world. The economists were quickly proven wrong. Although Saudi Arabia invested a good share of its surplus funds abroad, a substantial percentage was earmarked for aid to other, less fortunate, nations and in addition the kingdom began to channel billions of dollars into the ambitious second five-year Development Plan to expand the industrial infrastructure of the country.

THE DEATH OF FAYSAL: In March 1975 King Faysal was assassinated by a deranged young relative as he sat in his office in Riyadh. To the world at large his death was a shocking development and a serious loss; for Faysal, to the West as well as to the East, to Christians as well as to Muslims, had emerged in his last years as a moderating influence politically and economically throughout the Middle East. At a time when the political and economic impact of the Middle East was increasingly affecting the entire industrialized world, it was Faysal more than any other leader who spoke for balance and reason.

In Saudi Arabia, of course, the loss of Faysal was felt even more deeply. Like his father, he had ruled with the consent and blessings of his subjects, had governed them in the spirit of Islam, and had devoted his rule to providing a better life for his people and a better future for his country.

To some foreign observers the smooth transition of power and the lack of hysteria surrounding Faysal's death were surprising; to those who knew the country, however, they were not. Although Faysal had been an exceptional man and an effective leader, the Kingdom of Saudi Arabia, for which he so deeply cared, was not dependent for its stability on the rule of any one man.

KING KHALID IBN 'ABD AL-'AZIZ: Born in 1912 in Riyadh, Faysal's successor King Khalid was exposed to the problems of the country at an early age. After the conflict with Yemen in 1934 he was his country's representative at the peace negotiations at Tayif. With Faysal, he had attended a number of international meetings on behalf of Saudi Arabia — in particular, the 1939 conference on Palestine held in London, the official visit to President Roosevelt in 1943, and the founding of the U.N. in 1945. In October 1962 Khalid was named Vice President of the Council of Ministers and in 1965 he was designated Crown Prince.

When Khalid assumed the responsibilities of the kingdom the senior members of Al Sa'ud chose Fahd, a younger brother, to be crown prince. Assisted by Fahd, Khalid began to address himself to the problems of a kingdom that had changed almost beyond recognition since 'Abd al-'Aziz formally named it the Kingdom of Saudi Arabia almost 50 years before. To a large extent the change in the country had come about because of Saudi Arabia's emergence as the single most important source of petroleum in the world. With oil reserves equal to about one-fourth the combined reserves of the entire world, Saudi Arabia at the time of Khalid's accession was playing a decisive role in world affairs economically and politically.

On the domestic front, Khalid's reign will be remembered as a time of massive infrastructure building. By the 1970s, the kingdom's fast-growing economy was outstripping its support systems of roads, ports, electricity supply, and some social services. The second Development Plan, which got underway in 1975, and the third Development Plan, launched in 1980, were full-scale attacks on these bottlenecks. Quickly expanding oil revenues provided the finances for this expensive task, and each of the two plans was funded at more than $200 billion. The effort

Money

These quarter-, half-, and one-riyal silver coins of Saudi Arabia appeared in 1936.

The first money of the House of Sa'ud was issued shortly after the capture of Makkah in 1924 and consisted of copper half-qirsh and quarter-qirsh coins bearing the *tughra* or monogram of King 'Abd al-'Aziz and inscribed "Minted in the Mother of Villages" (a traditional name for Makkah). These coins circulated along with the gold dinar, silver riyal, and copper qirsh struck by the Hashimite government of the Hijaz, as well as the Turkish silver majidi, the Austrian Maria Theresa dollar, and various other foreign coins. They were followed a year later by quarter-, half-, and one-qirsh coins issued in the name of 'Abd al-'Aziz as King of the Hijaz and Sultan of Najd.

Gold Hashimite dinars and silver Maria Theresa dollars circulated before Saudi coinage.

On January 24, 1928, 'Abd al-'Aziz established an independent monetary system based on the Saudi riyal, a silver coin about the size of a U.S. silver dollar and valued at one-tenth of a British sovereign. In 1936 a new riyal coin about the size of a U. S. half dollar was issued. This coin weighed 11.66 grams and was eventually valued at approximately 3.75 Saudi riyals to the U. S. dollar. New half-riyal and quarter-riyal silver coins and one-qirsh, half-qirsh, and quarter-qirsh cupronickel coins were issued. By the late 1950s the value of the silver content of the silver coins had become considerably more than their face value, and

The large silver riyal of 'Abd al-'Aziz was inscribed "King of the Hijaz and Najd."

they disappeared from circulation. They were officially demonetized in December 1959. The cupronickel coins were withdrawn from circulation in 1958 and demonetized in the following year. They were replaced by new cupronickel coins in one-qirsh, two-qirsh, and four-qirsh denominations.

For many years Saudi Arabia issued no paper currency. Major transactions had to be carried out with gold or large quantities of silver riyals and occasionally foreign banknotes such as Indian rupees, U. S. dollars, and Egyptian pounds. The expansion of oil production, with its impact on payments of all kinds, so increased the demands for cash, however, that the handling of silver and gold coins in the quantities needed became almost impossible. To pay oil royalties to the government, for example, Aramco for many years had to arrange for kegs of foreign gold coins to be flown to Jiddah for delivery to the Ministry of Finance. To pay its employees (whose biweekly wages in silver coins might weigh as much as 4½ kilograms) the company had to transport, guard, and count some 40 tons of silver coins every month.

In 1951, therefore, the government, realizing that it needed a more workable monetary system, invited a team of foreign financial experts to draw up a program of monetary reform. Out of the recommendations of the experts came the formation in the following year of the Saudi Arabian

Saudi Arabia issued these gold sovereigns in 1952 (left), and 1958.

Monetary Agency (SAMA), a form of central bank established to strengthen and stabilize the country's currency, to centralize government receipts and payments, and to control disbursements authorized under the government's budget.

The monetary agency was forbidden by its charter to issue paper money, but in the year it was established it issued the Saudi sovereign, a gold coin of the same weight and approximate fineness as the British sovereign and with a par value of 40 Saudi riyals. Following the discovery of a large number of counterfeit Saudi sovereigns, a new sovereign of the same size and weight but of a different design was issued in January 1958. Both types of sovereigns were later demonetized.

(Continued on page 162)

Money

Despite the explicit prohibition in its charter against the issue of paper money, SAMA in 1953 won government permission to issue "pilgrim receipts," a sort of traveler's check designed to meet the seasonal needs of foreign pilgrims for riyals. It was announced that the receipts were issued primarily to relieve pilgrims of the necessity of carrying large sums in heavy silver coins, but the receipts (fully backed by silver riyals and eventually issued in 10-, five-, and one-riyal denominations) were an immediate success and soon circulated throughout the kingdom. Within a short time more than 150 million riyals worth were in circulation, rapidly replacing both gold sovereigns and silver riyals, and they soon gained the virtual status of official paper currency. A new series of pilgrim receipts was issued beginning in 1954.

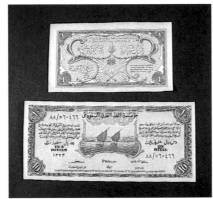

Pilgrim receipts, first issued in 1953, were forerunners of official paper currency.

In 1953 the financially cautious King 'Abd al-'Aziz died and government spending, previously held in check by him, began to increase substantially. Before long foreign exchange payments were exceeding income and debts had begun to accumulate. At the same time, the government abandoned the requirement that the receipts must have 100 percent hard-money backing — a move that proved disastrous when, between 1956 and 1958, severe inflation swept the country and the value of the riyal sank in relation to both goods and foreign currencies.

In 1956 additional problems developed. In the wake of the second Arab-Israeli war, oil production fell sharply and government revenues dropped from $341 million in 1955 to $290 million. On top of unchecked government spending, excessive borrowing, and the flight of capital, this drop in revenues weakened the riyal even more and by early 1958 the country was facing uncontrolled inflation, exhaustion of its foreign exchange reserves, and nearly half a billion dollars in debt.

At this point Crown Prince Faysal asked two financial experts, Anwar Ali of Pakistan and Ahmad Zaki Saad of Egypt (both on loan from the International Monetary Fund), to design and implement a program of fiscal and monetary reform. The objectives — to balance spending and income, pay the national debt, curb inflation, and stabilize the riyal — were challenging, particularly as the decisive measures needed had to be introduced cautiously and without imposing unenforceable controls. Yet the experts, with Faysal's firm and active backing, succeeded. By clamping curbs on unessential imports, establishing a dual rate of exchange for the riyal, reducing government expenditures drastically, and devaluating the riyal, they reversed the decline in a year and a half.

At the beginning of 1960 the Government of Saudi Arabia revalued the riyal, fixing its value in terms of gold at the equivalent of 4.5 riyals to the U. S. dollar and insuring stability of the riyal by providing 100 percent backing in gold and foreign currency. At the same time it was announced that henceforth the riyal would contain 20 qirsh rather than 22 and that the qirsh would be divided into five halalah. Thus the currency, in effect, was put on a decimal basis for the first time.

By the end of 1960 the inflationary spiral had been stopped, the international exchange value of the riyal had been stabilized, and a start had been made on retiring the national debt. Saudi Arabia had joined the International Monetary Fund in 1957, and in March 1961 it accepted the Fund's obligation to make the Saudi riyal a fully convertible currency.

In June 1961 the Monetary Agency issued an official Saudi paper currency in denominations of one, five, 10, 50, and 100 riyals. A 500-riyal note was added later. The pilgrim receipts remained in circulation alongside the new paper money until 1963. Since 1976 Saudi banknotes (printed in Arabic on one side and English on the other) have borne not only crossed swords and a palm tree, the national emblem, but also portraits of the reigning monarchs. There are coins of five, 10, 25, and 50 halalahs, as well as one-riyal (100-halalah) cupronickel coins. There is also a bronze one-halalah coin, but it is rarely seen in circulation.

New coins in halalah denominations appeared after decimalization of Saudi currency.

By the mid-1970s Saudi Arabia had developed a sophisticated and responsible fiscal system capable of handling both the country's expanded internal money supply and its large international reserve holdings. The kingdom, furthermore, was a key contributor to multimillion-dollar international aid programs, and had become an important source of capital for the World Bank and the International Monetary Fund. In 1978 it became entitled to appoint its own director to represent Saudi Arabia on the Executive Board of the IMF, a reflection of its position as one of the largest creditors of that organization. Saudi Arabia in the 1990s continued to enjoy the advantages of a sound and stable monetary system supporting a national economy that had become the largest in the Middle East.

was a resounding success; within a few years port and road congestion was overcome with the construction of a world-scale national highway system, marine facilities, international airports, and expanded telecommunications. In the 10 years covered by the plans, paved highways quadrupled in length, port tonnage increased some 10 times, and power generation increased from under 2 billion to over 44 billion kilowatt-hours. The number of schools jumped from 5,600 to over 15,000, with an average completion rate of more than two schools a day. The number of hospitals nearly doubled, to 176.

A major start was also made on another development phase, which aimed at diversifying the kingdom's economy through the development of heavy industry — much of it emphasizing petrochemical production based on the fuel and feedstocks supplied by the kingdom's Master Gas System, also under construction (see page 219). Planners decided to concentrate the new industries in two massive complexes, one at Jubail on the Gulf coast, the other at Yanbu' on the Red Sea — each linked by pipeline to the Master Gas System. A new, independent organization, the Royal Commission for Jubail and Yanbu', was established to coordinate these mixed government and private ventures and provide their special infrastructure requirements.

King Khalid, during the latter years of his reign, played leading roles on the regional and international political scenes. He hosted the Third Islamic Summit Conference in Makkah and Tayif in 1981. In May of the same year he made his nation a founding member of the Cooperation Council for the Arab States of the Gulf, generally known as the Gulf Cooperation Council (GCC). Saudi Arabia was instrumental in establishing this economic and cultural organization, which also includes Bahrain, the United Arab Emirates, Kuwait, the Sultanate of Oman, and Qatar. Looking forward to a long-term goal of a "Gulf Common Market," the GCC soon achieved major steps toward freer transit of people and goods, the reduction of customs barriers, and greater security for citizens of its member countries.

KING FAHD IBN 'ABD AL-'AZIZ: Crown Prince Fahd ibn 'Abd al-'Aziz was proclaimed king in June 1982 following the death of King Khalid. Prince 'Abd Allah ibn 'Abd al-'Aziz, a younger son of 'Abd al-'Aziz and Head of the National Guard, then became Crown Prince and First Vice President of the Council of Ministers. By this time Fahd was already a veteran of many years of distinguished government service, having been appointed the kingdom's first Minister of Education with the establishment of that ministry in 1954. In 1962 he became Minister of Interior and served in that capacity until he was named Crown Prince and First Deputy Prime Minister in 1975. During this period he worked on developing and implementing many of the kingdom's major projects. He served as chairman of a number of high-level committees and councils, including the Supreme Council for Petroleum and Minerals, the Supreme Council for Education, and the Royal Commission for Jubail and Yanbu'. In 1986 King Fahd expressed his wish to eschew royal titles and honorifics and to be called only "Custodian of the Two Holy Mosques." Only that title, replacing the traditional "His Majesty" before the still-used descriptor "King," has since been considered the proper form of address or reference to Fahd as head of state.

King Fahd's reign will be marked in the kingdom's history as the period witnessing completion of the first great wave of macro-scale infrastructure construction and the coming to fruition of industrial and other developments dependent on them. Perhaps most notably major strides have been, and are still being, taken to diversify the kingdom's sources of income across a growing and profitable industrial base and to achieve self-sufficiency in many primary manufactured and agricultural commodities.

INDUSTRY AND INFRASTRUCTURE: By the mid-1990s the highly successful Royal Commission industrial complexes at Jubail and Yanbu' had already passed through a capacity expansion phase and had taken a six percent share of the world petrochemical market in competition with the long-dominant plants of North America, Europe, and Japan. They were also providing a major share of the kingdom's domestic needs for such manufactured staples as construction steel, plastic resins, and chemical fertilizers. Secondary industries were making use of these raw materials to manufacture consumer items. Total investment in Saudi Arabia's industrial enterprises, numbering more than 2,000 in 1995, was estimated at over $34 billion.

Custodian of the Two Holy Mosques King Fahd ibn 'Abd al- 'Aziz.

His Royal Highness Crown Prince 'Abd Allah ibn 'Abd al- 'Aziz.

His Royal Highness Prince Sultan ibn 'Abd al- 'Aziz.

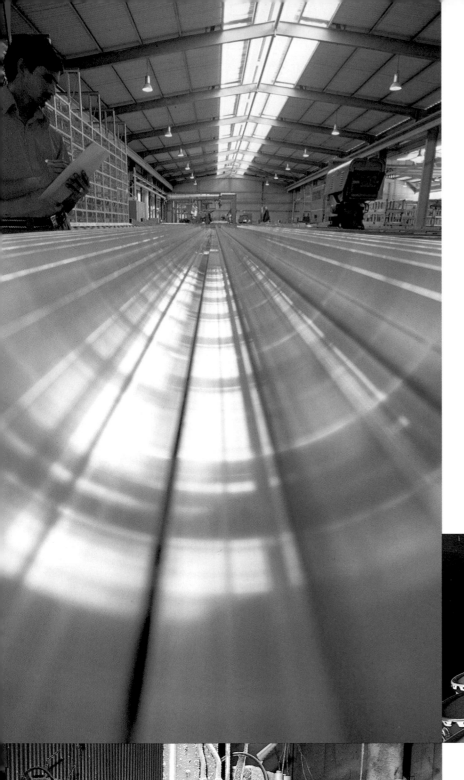

Saudi Arabia's massive development program of the 1970s and 80s — one of the most ambitious ever undertaken by any nation — has achieved industrial diversification as well as a balance of development across the Arabian Peninsula. Industries not even envisioned 20 years ago are today thriving in regions throughout the kingdom. Much of the heavier industry is centered in the new industrial cities of Yanbu' on the Red Sea, and Jubail on the Gulf. Facing page at top: A Yanbu' petrochemical company produces both polyethylene and ethylene glycol from ethane. Bottom, Jubail industries include manufacturing plants for steel rebar and water-proofing materials. At lower right, engineers for another Jubail firm make a progress check on a DGA stripper column being constructed for the oil industry. Older, established cities have also developed a strong economic base. Among Jiddah industries are a computerized steel rolling mill and an automotive assembly plant, both seen above. Dammam industries include a fertilizer plant, at left, and a paint factory, directly above.

Some 26 kilometers long, this soaring causeway connects Saudi Arabia with Bahrain, the island nation to its east.

The Ghazlan power plant, near Ju'aymah on the Gulf coast, forms a hub of the Eastern Province's electricity transmission system.

The Jaizan dam in southwestern Saudi Arabia stores 51 billion liters of rain water (13.47 billion gallons).

Private enterprise has always been a cornerstone of Saudi industrial development policy. Even the huge petrochemical plant investments of the Saudi Basic Industries Corporation (SABIC), funded largely by the government because of their great scale, had a 30 percent private equity component. One of the major objectives of the five-year Development Plans is to give private enterprise a primary role in the national development effort.

Other development achievements of the 1980s and 1990s provided additional public service infrastructure. By 1991 the number of public and private hospitals in the kingdom had grown to 274, supplemented by 3,150 clinics. Virtually every village had been reached by the electric power grid and all regions had been linked by a highway network running throughout the kingdom. Among special road projects was one that can only be described as an international engineering masterpiece: the 26-kilometer-long causeway and bridge system over the sea linking al-Khobar in the Eastern Province with the neighboring island state of Bahrain. Funded and maintained by Saudi Arabia and known as the King Fahd Causeway, the new highway link was opened in 1986 and by 1993 was carrying 1,800,000 vehicles annually.

AGRICULTURE AND WATER: Agriculture has been another area of recent tremendous development. Historically dependent on imports for a large proportion of its foodstuffs, the kingdom is now virtually self-sufficient in the production of such staples as poultry, eggs, dairy products, and bread grains. The government-subsidized production of wheat using modern center-pivot irrigation methods was successful beyond expectations, with production far outstripping domestic needs. Output peaked at over 4 million tons in 1992 and in 1993 Saudi Arabia was the sixth largest wheat exporter in the world. Emphasis shifted thereafter to barley production, with the aim of achieving self-sufficiency in this important livestock feed-grain supplement.

The Saudi Arabian Government has long been conscious of the importance of developing the water resources of its arid lands. Large-scale agriculture here depends largely on the exploitation of non-renewable "fossil" water through the use of pumped wells penetrating deep aquifers. Some of these aquifers are of

enormous extent but may range widely in quality, from sweet to highly saline. Starting in the late 1970s the desalination of seawater began to be an important supplement to the kingdom's water supplies for its fast-growing cities and towns. Desalination plants — some of them among the largest in the world and fueled by the nation's abundant oil and gas resources — were completed on both the Gulf and Red Sea coasts. Even cities deep inland, such as Riyadh, are served partly by desalinated water transported from the Gulf coast in enormous concrete-lined pipelines. In the mid-1990s the kingdom's desalinated water production accounted for 30 percent of the world's total, turning out more than 1.9 million cubic meters (500 million gallons) per day with projects under way to increase this to over 3 million cubic meters (800 million gallons).

GOVERNMENT ORGANIZATION: In 1953 King 'Abd al-'Aziz, recognizing that the simple direct methods of tribal administration would no longer suffice in a kingdom the size of Saudi Arabia, had established the Council of Ministers under a president who was in effect the prime minister. Sa'ud continued his father's efforts to improve Saudi Arabia's government system, and the kingdom's administrative machinery was modernized further by King Faysal, who had served as president of the Council. One step was to invite experts from such institutions as the United Nations and the Ford Foundation to help with administrative organization and training. Another was to invest heavily in education and send thousands of young Saudi students to colleges and universities, both at home and abroad, in an effort to meet the country's need for trained personnel.

During the reign of 'Abd al-'Aziz the *amirs*, or governors, of the provinces — because of the size of the country and the absence of modern communications facilities — governed their areas semiautonomously, although in the King's name. Generally, even local officials of the central government were subordinates of the amirs. Later, under Faysal's gradual centralization of administration, this changed: Most of the amirs of the provinces began to report to the Ministry of Interior, and local representatives of the central government were

Calendar, Clocks

In the seventh century, according to the historian al-Biruni, the Caliph 'Umar ibn al-Khattab, after conquering Persia and driving the Byzantines from Syria, found that administration of the new Islamic empire required extensive correspondence with his far-flung regional governors and generals. That, in turn, involved him in the problem of calendars; in attempting to date the correspondence he learned that the various systems of dating then current were both complicated and linked to other religions and states. As they were therefore for one reason or another both impractical and unacceptable, it was decided to establish a new calendar system based on the advent of Islam.

In establishing the new system 'Umar decided, after consulting with his companions, to use the year of the Hijrah, the Prophet's emigration from Makkah to Medina, as the starting point of the Islamic era, and to start the new calendar from the day of Muhammad's departure from Makkah on July 26, 622. Year One of the Islamic calendar thus corresponds to A.D. 622-623. The Western method of designating Islamic dates is by the abbreviation A.H., for Anno Hegirae or "Year of the Hijrah."

Relying on a number of passages in the Quran (surahs ix: 36-37, and x: 5), 'Umar established the new Islamic calendar, in contrast to most other calendars, on the basis of a strictly lunar year. One lunar month — which is the cycle between two new moons — contains 29 days, 12 hours, 44 minutes, and 2.8 seconds. A lunar year of 12 months, therefore, contains 354 days and 11/30 of a day. This fraction adds up to 11 days in every cycle of 30 years, and these are inserted into the calendar by establishing leap years. Therefore every period of 30 years has 11 leap years of 355 days. The intercalated day is always added to the month of Dhu al-Hijjah, the last month of the year.

The difference between the 354- or 355-day lunar year and the astronomical solar year of 365 and 1/4 days accounts for the difficulty of converting dates from the Islamic (*Hijri* or "Hijrah") calendar to the Gregorian calendar and vice versa. One method of conversion is to use the following equation on the basis that every 32 Gregorian solar years are approximately equal to 33 Muslim lunar years.

(A.D. = Gregorian year, A.H. = Hijrah year):

$$\text{A.D.} = 622 + \left(\frac{32}{33} \times \text{A.H.}\right) \text{ or A.H.} = \frac{33}{32} \times (\text{A.D.} - 622)$$

It should be remembered that these formulas give only the year in which the corresponding year began. For example, A.H. 1416 began on May 30, 1995, almost halfway through the Gregorian year of 1995.

Some other guides for approximate conversion are: a Gregorian century equals 103 Hijrah years; and 100 Hijrah years equal 97 Gregorian years. The 12 months of the Islamic year are: Muharram, Safar, Rabi' al-Awwal (Rabi' I), Rabi' al-Thani (Rabi' II), Jumada al-Ula (Jumada I), Jumada al-Akhirah (Jumada II), Rajab, Sha'ban, Ramadan, Shawwal, Dhu al-Qa'dah, and Dhu al-Hijjah.

In theory the months contain 30 or 29 days alternately, with the odd-numbered months generally containing 30 days and the even-numbered months generally containing 29 days. But the actual length of each month depends on when the new moon is sighted, as the sighting of the new moon marks the beginning of the month and the sighting of the next new moon marks its end.

The lunar year is completely unrelated to the seasons because each year begins 10 or 11 days earlier in the Gregorian year than the previous one. Thus any one of the months of the Islamic year may occur in any season. If your birth date is Rajab 15, you might celebrate the occasion one time in July and 15 years later in February. Schools in Saudi Arabia, as a result, close for the summer holiday in any three consecutive months that may fall in that season; and Ramadan, the month of fasting (see page 51), may occur in winter, spring, summer, or autumn, making a complete cycle every 33 Gregorian years.

Similarly, Muslim festivals, unlike Christmas or the Fourth of July, may occur in any season. But in addition to the difference in timing, there are major differences with regard to festivals. In Western countries holidays celebrate a variety of religious, historic, social, and military events. But in Saudi Arabia, a country that strictly follows the doctrines of Islam, the main holidays that are recognized and observed are the two religious feasts of '*Id al-Adha* and '*Id al-Fitr.*

'Id al-Adha or "the Festival of the Sacrifice," falls on the tenth day of the twelfth month, Dhu al-Hijjah, and marks the end of the hajj (see page 52). It is celebrated with the sacrifice of a sheep or goat — or sometimes, if the family can afford it, a camel or an ox — which is then given to the poor. Public worship is also performed by the whole community. In most Muslim countries the duration of the festival is three days, during which most people dress in new clothes, exchange presents, visit friends and relatives, and may serve special foods and sweets.

The other feast day, 'Id al-Fitr or "The Festival of the Breaking of the Fast," occurs on the first day of the month of Shawwal and marks the end of Ramadan, the month of fasting. Like 'Id al-Adha, it is celebrated by public worship and by the exchange of gifts and visits, but there is no

(Continued on page 168

Calendar, Clocks

sacrifice. It is also usually celebrated for three days. As it marks an end to the austerities of Ramadan, 'Id al-Fitr is a particularly joyous festival.

Both festivals are primarily religious occasions — designed for expression of communal solidarity as well as individual satisfaction and fulfillment at having carried out two major duties of the faith — and are characterized by the giving of alms to the poor and generous gifts (generally of money) to one's younger relatives.

Although other Muslim countries have introduced a number of additional public holidays, such as the Prophet's Birthday, New Year's Day (the first day of Muharram), and various national Independence Days, Saudi Arabia observes only the two religious festivals as public holidays. National Day, though not a public holiday, commemorates the unification of the country under the name of the Kingdom of Saudi Arabia on September 22, 1932. It follows neither the Hijrah nor the Gregorian calendar, but rather the solar zodiacal calendar, and falls on the first day of the month of Libra, corresponding more or less to September 23 in the Gregorian calendar. That date was picked for the convenience of foreign governments and their diplomats. It is customary for them to send formal congratulatory messages to the Government of Saudi Arabia and for Saudi diplomats abroad to hold receptions on that day. The solar zodiacal calendar is also used in Saudi Arabia for defining the government's fiscal year. Beginning with 1989, it was decided that each fiscal year should commence on 11 Capricorn, corresponding to January 1, and end on 10 Capricorn of the following year (December 31). Thus, since January 1, 1989, fiscal years have corresponded exactly with Gregorian calendar years.

The adoption of the lunar month also had effects on Saudi Arabia's system of keeping time. As the month begins with the sighting, just after sunset, of the new moon and ends with sighting of the moon of the next month, the Muslim system followed ancient convention in which each day was considered to begin at sunset. For many centuries this system of starting the "day" at sunset, dividing it into 24 equal parts — 12 for daylight and 12 for nighttime — worked well.

But in more recent times the element of confusion caused by the fact that in each city and town clocks were set daily to correspond with actual local sunset, and the additional problems caused by the introduction of modern time zones and daylight saving time, caused frequent difficulty and misunderstanding. At one point, a popular gift to newcomers to Saudi Arabia was a watch with two dials, one giving Arabian time, the other whatever time the newcomer himself was using. Eventually the confusion caused by the use of two systems of time became too great, and when the requirements of international telecommunications and such factors as airline schedules began to demand the use of a single system, Saudi Arabia in the 1960s adopted Greenwich Mean Time plus three hours (GMT+3) as the standard time for the entire country.

Jumada II 1416 — NOVEMBER 1995 — Rajab 1416

SUN	MON	TUE	WED	THU	FRI	SAT
			1 / 8	2 / 9	3 / 10	4 / 11
5 / 12	6 / 13	7 / 14	8 / 15	9 / 16	10 / 17	11 / 18
12 / 19	13 / 20	14 / 21	15 / 22	16 / 23	17 / 24	18 / 25
19 / 26	20 / 27	21 / 28	22 / 29	23 / 1	24 / 2	25 / 3
26 / 4	27 / 5	28 / 6	29 / 7	30 / 8		

Rajab 1416 — DECEMBER 1995 — Sha'ban 1416

SUN	MON	TUE	WED	THU	FRI	SAT
					1 / 9	2 / 10
3 / 11	4 / 12	5 / 13	6 / 14	7 / 15	8 / 16	9 / 17
10 / 18	11 / 19	12 / 20	13 / 21	14 / 22	15 / 23	16 / 24
17 / 25	18 / 26	19 / 27	20 / 28	21 / 29	22 / 30	23 / 1
24 / 2	25 / 3	26 / 4	27 / 5	28 / 6	29 / 7	30 / 8
31 / 9						

Sha'ban 1416 — JANUARY 1996 — Ramadan 1416

SUN	MON	TUE	WED	THU	FRI	SAT
1 / 10	2 / 11	3 / 12	4 / 13	5 / 14	6 / 15
7 / 16	8 / 17	9 / 18	10 / 19	11 / 20	12 / 21	13 / 22
14 / 23	15 / 24	16 / 25	17 / 26	18 / 27	19 / 28	20 / 29
21 / 1	22 / 2	23 / 3	24 / 4	25 / 5	26 / 6	27 / 7
28 / 8	29 / 9	30 / 10	31 / 11			

supervised by their respective ministries. Local leaders such as the heads of municipalities, towns, and villages were attached to the Ministry of Municipal and Rural Affairs.

Saudi Arabia's government and administrative framework has since grown into a huge, complex, and modern system. King Fahd, upon his accession, became King of Saudi Arabia, Supreme Commander of the Saudi Arabian Armed Forces, and President of the Council of Ministers. Assisting him were the Crown Prince, 'Abd Allah ibn 'Abd al-'Aziz, First Vice President of the Council of Ministers and Head of the National Guard; and Prince Sultan ibn 'Abd al-'Aziz, Second Vice President of the Council of Ministers, Minister of Defense and Aviation, and Inspector General.

Reflecting the complexity of government, the Council of Ministers in 1995 was made up of the Ministers of Agriculture and Water, Commerce, Communications, Defense and Aviation, Education, Finance and National Economy, Foreign Affairs, Health, Higher Education, Industry and Electricity, Information, Interior, Islamic Endowments and Guidance, Justice, Labor and Social Affairs, Municipal and Rural Affairs, Petroleum and Mineral Resources, Pilgrimage, Planning, Post and Telecommunications, and Public Works and Housing, as well as other officials bearing the title of Minister of State.

ADMINISTRATIVE CHANGES:

Legislation introduced by Royal Decrees of King Fahd in 1992 and 1993 further modernized and defined the system of government while broadening the bases for decision making. The Basic System of Government, issued in March 1992, has many characteristics of a Western-style constitution but is never referred to by that term. Its first article, in fact, stipulates that the constitution of the kingdom is the Holy Quran and the *sunnah* of the Prophet (see page 50). The Basic System, among other provisions, defines eligibility for accession to the throne, establishes the judiciary as an independent authority subject only to the shari'ah, or religious law, and sets forth general rules for the state financial system. Another series of provisions acts, in effect, as a bill of citizens' rights, guaranteeing rights of private property, the inviolability of the home, and protection from arbitrary penalties.

One important provision of the Basic System was for reactivation of an advisory body called Majlis al-Shura, or the Consultative Council, an institution first set up in the early years of the Saudi state but which had hitherto played only a small role in public affairs. The new council, as defined by Royal Decree in 1993, is a body of 60 members appointed for four-year terms with the function of advising the Council of Ministers on matters of economic and social development, international agreements, and the interpretation of legislation. King Fahd opened its first session on December 29, 1993, the members representing a broad spectrum of Saudi leaders with extensive experience in the academic world, public administration, and various technical fields. The majority hold advanced academic credentials, many from well-known Middle Eastern and Western universities.

PROVINCIAL ADMINISTRATION:

Another major component of these legislative developments, issued in March 1992, was a formal and more detailed definition of the kingdom's system of provincial administration. This included the establishment of provincial councils, with members of recognized stature and expertise appointed from both the private and public sectors. The primary function of these bodies is to advise the executive branch on regional development matters.

Originally the Kingdom of Saudi Arabia was divided, for administrative purposes, into broad geographical regions. The main divisions were Najd, which covers a large part of the interior and includes the capital Riyadh; the Hijaz, which contains the holy cities of Makkah and Medina and also Jiddah, the kingdom's main port on the Red Sea; 'Asir, a mountainous district ruled from Abha and including Najran, a fertile valley close to the border with Yemen; and al-Hasa (later called the Eastern Province), named after the oasis of al-Hasa but also including Dammam, the Qatif oasis, Jubail, the principal oil fields, and the major oil company communities of Dhahran, Abqaiq, and Ras Tanura.

Today, however, the kingdom is divided into 13 administrative districts, generally called provinces, which do not necessarily correspond with the earlier geographical regions. These are: 'Asir, al-Bahah, the Eastern Province, Hayil, al-Jawf, Jaizan, Makkah, Medina, Najran, the Northern Frontier, al-Qasim, Riyadh, and Tabuk (see map, page 170).

King Fahd chairs a session of the Council of Ministers.

The terraced fields of 'Asir province in southwestern Saudi Arabia are a source of agricultural produce.

Jiddah, long an important Red Sea port, is one of several key commercial centers in Saudi Arabia.

A number of modern hotels are located in Medina province, an important center of pilgrimage.

Each is headed by an appointed amir. Each province is divided into smaller units called *muhafazat,* or districts, often centered on smaller towns. Journalists and other writers sometimes also, for convenience, divide the kingdom into "central, northern, southern, western, and eastern provinces." Except for the Eastern Province, however, these units have no official administrative status.

Administrative areas of Saudi Arabia

LAW IN MODERN SAUDI ARABIA: King Fahd and his predecessors, while carrying out administrative reform and modernization, continued to adhere strictly to the fundamental law of the kingdom. This was, and is today, the shari'ah, the sacred law of Islam. Rooted in the Quran and the practices and sayings of the Prophet, it is the shari'ah, rather than a man-made constitution or legislative acts, that is the heart of the legal system of the kingdom. Although essentially religious in nature, the shari'ah extends far beyond religious matters. In Saudi Arabia, as it once did throughout the Islamic world, the shari'ah regulates virtually all matters covered in Western countries by

civil and criminal law. Because, furthermore, the shari'ah is considered as divine law, it is interpreted and administered by the *qadis* (shari'ah judges) and ulema, or religious scholars; as a result the judiciary system in the kingdom has a considerable degree of independence from other branches of government.

In matters not expressly prohibited or enjoined by the shari'ah, however, the king, through the Council of Ministers and a number of commissions and councils, administers a relatively large body of regulations covering such areas as public health, customs, commerce, and labor — all increasingly important as the process of modernization moves ahead.

THE PILGRIMAGE TODAY: The modernization of Saudi Arabia has also been extended to the kingdom's role as custodian of the holy cities of Makkah and Medina. To the House of Sa'ud this role has always been highly important. Immediately upon extending his rule to the Hijaz, 'Abd al-'Aziz took steps toward eliminating the payments extorted from pilgrims by tribes along the caravan routes. Later, when the revenues from petroleum permitted it, he abolished all taxes on pilgrims. He also assigned his son Faysal to improve conditions for the pilgrims who came to Saudi Arabia each year to make the hajj, one of the most significant spiritual events in the lives of Muslims. The pilgrimage has required from the kingdom, the physical as well as historic heartland of Islam, a logistical achievement of impressive proportions. In recent years it has involved the handling of about a million and a half pilgrims, most of whom have to be provided with food, water, shelter, and medical care. Also — in great throngs and on a precise schedule — they have to be transported, guided, and instructed during the trips to and between the holy places and back, generally within three weeks.

For the better part of 14 centuries Muslims from around the world had been making the holy journey, some on foot, some in caravans, and others, in recent years, by steamer and rail. But starting after World War II — when someone converted a handful of war-surplus bombers into passenger planes and chartered them to pilgrims — more Muslims began to come by air. By the 1960s, when airlines began to build fleets of modern aircraft, Muslims from such distant centers of

170

Islam as Indonesia and the Philippines to the east and Nigeria to the west were flying to Saudi Arabia. By the 1970s most major airlines of Muslim countries were offering such service and some were assigning the biggest jets they owned to carry pilgrims — sometimes at the rate of 300 flights a day. One of these was Saudia, the national airline of Saudi Arabia. As a result, the numbers of foreign pilgrims increased sharply — from 316,000 in 1967 to 996,000 in 1994, mostly from the neighboring Arab countries and Iran, with smaller numbers coming from the rest of Asia and Africa, and a few thousand from Europe and from North, South, and Central America.

Traditionally, many of the pilgrims' logistical problems were left in the hands of a unique institution: a corps of expert guides called *mutawwifs*. For generations the mutawwifs had arranged transport and accommodations, provided food and water, guided the pilgrims to Makkah and the other holy places, and offered help with regard to the rituals. Today all pilgrims must be registered with official, government-supervised guide establishments. Each establishment is responsible for pilgrims from a specific part of the world throughout their visit and through their departure. This is partly because the visas issued to pilgrims are special visas; they restrict the pilgrims to Jiddah, holy places traditionally included in the pilgrimage and, of course, the *haram*, the sacred "area of the pilgrims," which is strictly forbidden to non-Muslims. The Ministry of Pilgrimage also sets strict health and safety standards for the housing and other services provided for pilgrims.

With the swift rise in the numbers of pilgrims, however, Saudi Arabia has had to do even more. At Jiddah, for example, the government has built huge transit centers and in the haram has constructed a network of high-technology roads and interchanges capable of handling many thousands of vehicles simultaneously. King 'Abd al-'Aziz International Airport in Jiddah, completed in 1981 and winner of an international prize for design, has become the pilgrims' primary gateway to the holy places.

Improvements to the physical facilities of the holy places themselves have been a continuing high priority beginning with the reign of King 'Abd al-'Aziz. Each Saudi monarch has taken a special personal interest in a long series of works that have increased the public area of the Holy Mosque in Makkah from an original 28,000 square meters to the 309,000 square meters completed under King Fahd in the 1990s. This increased the capacity for simultaneous worship nearly 15 times, to 695,000 persons. The area and facilities of the Prophet's Mosque in Medina has likewise been expanded and improved from an original capacity of 28,000 persons to accommodate 257,000 worshippers.

The government has also reorganized the tent cities in which the pilgrims live, deployed thousands of police, Boy Scouts, and National Guardsmen to direct traffic and assist the aged and the ill, and taken steps to eliminate or reduce the danger of infectious diseases. In addition it has equipped its pilgrimage staffs with such modern tools as two-way radios, closed-circuit television, and helicopters. By the early 1990s Saudi Arabia was spending over $480 million annually to ensure that a faithful Muslim anywhere could, at least once in his or her lifetime, make this holiest of journeys.

Although the kingdom, under 'Abd al-'Aziz and his successors, has emerged as a prosperous nation and an important power in world affairs, its rulers and people have never forgotten the long struggle waged by the House of Sa'ud to adhere to and promote the faith revealed to the Prophet 14 centuries ago.

The Holy Mosque, Makkah.

The many faces of Saudis of different generations at work, play, school, and repose.

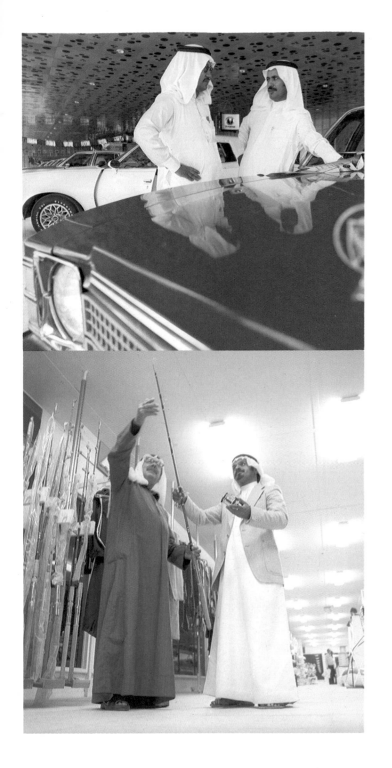

OIL AND
SAUDI ARAMCO

Overleaf: On any drilling rig, the work goes on around the clock.

Men of ancient civilizations, for example the Sumerians, used natural asphalt to cement blades to handles.

Asphalt, or bitumen, also served as mortar in ancient structures of sunbaked brick.

The Venetian traveler Marco Polo wrote about the "fountains of oil" he had seen at Baku on the Caspian Sea.

Noah, preparing the Ark for the Flood, seems to have been one of the first men to have realized the value of petroleum. He used it to caulk the seams of his primitive vessel "within and without." But even before Noah, prehistoric man undoubtedly had seen, smelled, and tasted this strange blackish-green substance that seeped out of the earth into wells, pools, and streams. For a time ancient man probably had feared it too; strokes of lightning occasionally ignited escaping jets of gas and started fires that in some instances burned for centuries.

Precisely when, and to what degree, ancient man first began to grasp the potential of petroleum are still matters for archaeological investigation; but by 4000 B.C., when the Sumerians were beginning to keep records, he had already started to put it to use. The Sumerians — and later the Assyrians and Babylonians — used natural asphalt to attach blades to handles, treat the sick, and serve as mortar for their buildings.

These civilizations, grouped on the banks of the Euphrates River in what is now Iraq, obtained their asphalt from the "fountains of pitch" in Hit, about a 160 kilometers west of Baghdad, where large seepages of petroleum existed. But Hit was by no means the only source in ancient times. Other seepages existed along the Dead Sea and near the Yellow River in China.

According to the available evidence, ancient peoples knew more about asphalt than other forms of petroleum. In the Bible there are references to "slime" and "pitch" — that is, bitumen or asphalt, the substance left after the gases and liquids of crude oil evaporate. The basket in which Moses floated in the Nile was waterproofed with asphalt, and the famous wall on which Belshazzar saw the handwriting was mortared with it. Nor was liquid petroleum unknown. There are numerous early references to *neft* (an Old Persian word for petroleum from which the English "naphtha" is derived) and to "burning waters," and in the fifth century B.C. the Greek historian Herodotus described how both bitumen and oil were obtained.

As the centuries rolled on, less benign uses of petroleum were recorded. Byzantine commanders introduced the dreaded Greek Fire, a compound similar to napalm that the Greeks poured on enemy troops beneath city walls or sprayed on enemy fleets through tubes mounted on the prows of ships. It was difficult to extinguish and is credited with being a key weapon in Byzantium's defense when Arab fleets attacked Constantinople in the seventh and eighth centuries.

The Arabs had discovered petroleum too. In the ninth century the scientist al-Razi produced a handbook in which he recorded a description of how to distill neft and in about 950 a group of Arab scholars in Basra worked out a theory on the origin of oil. Arabic literature of the period has references to "Eternal Fires," presumably caused by natural gas, seeping from the earth, which had somehow been ignited and burned for many years. One repeated reference was to the oil of Baku, a city on the Caspian Sea that was once part of Persia and is now the center of oil production in Azerbaijan. The 'Abbasid caliph al-Mu'tamid granted the revenues from naphtha springs at Baku to the people living in the vicinity.

The oil of Baku — where one of the most prolific fields in history was found in the nineteenth century — was undoubtedly known thousands of years earlier, but clear references to it did not appear until the seventh century. In the following centuries, however, Baku attracted considerable attention. The Arab writer al-Mas'udi, who visited Baku about 915, told of a place where "fire…ceases not to burn." Marco Polo, who visited the area about 1271, wrote that he had seen "a fountain from which oil springs in great abundance, insomuch that a hundred shiploads might be taken from it at one time." He also noted that "people come from vast distances to fetch it, for in all the countries round about they have no other oil."

THE NEW WORLD: Similar if less spectacular indications of petroleum were noted in many places in Europe — in Italy, Poland, Germany, and what are now Slovakia and Rumania — and later in the New World. Francisco Pizarro, the Spanish conqueror of Peru, found the Indians there producing asphalt, and Sir Francis Drake later used asphalt to coat his ships' hulls during his far-ranging raids on Spanish shipping in the Caribbean. In 1543 Hernando de Soto, on his way to the Mississippi, found asphalt in a Texas creek; and in 1595 in Trinidad Sir Walter Raleigh found asphalt which, he wrote, "melteth not with the sunne as the pitch of Norway." In 1627 a missionary found oil in a spring now called Cuba Lake near Buffalo, New York; it was the first of many such springs that French and British explorers discovered as they pushed into the North American wilderness.

The Indians had collected petroleum from natural springs for centuries and prized it highly for medicinal and other purposes. Physicians followed their example, and petroleum became a familiar item on the shelves of drugstores. It was also used to a small extent for lighting and for lubrication and by the early nineteenth century had been refined experimentally.

During the nineteenth century the development of machines and factories created a large demand for lubricants and lighting oils. Mutton and beef tallow, lard, and castor oil served as lubricants. The demand for lubricants and lamp oil gave rise to a whaling industry so successful that whales were almost exterminated and sperm oil became scarce and expensive.

By the middle of the century scientists in France, Great Britain, Canada, and the United States had begun to develop substitutes. Between 1845 and 1850 a French company began to make illuminating oil from shale, and in 1850 a process to produce oil from coal and oil shale was patented and a geologist named Abraham Gesner demonstrated how to make kerosene from coal. Not long after, in 1855, a Pittsburgh druggist named Samuel Kier, who had been bottling and selling petroleum as a medicine called "Kier Rock Oil," built a distillation system that produced what he called "carbon oil." It was possibly North America's first real oil refinery. Out of these processes came grease for nineteenth-century machinery and the fuel known as kerosene, which provided a fine heating fuel and a clearer, brighter light for living rooms, carriages, and locomotives.

DRAKE'S WELL: By then, of course, it had occurred to more than one alert businessman that oil might be far more useful than anyone had suspected. It had also occurred to them that there might be more oil in the earth than was seeping to the surface and that there must be better methods of obtaining petroleum than skimming it off the surface of the

so-called oil springs. From such conclusions came the experiments that eventually led to the digging of a 15-meter well in Canada which produced enough oil to justify a small refinery that made lamp oil. This was in 1857, two years before a former railroad conductor, "Colonel" Edwin L. Drake, decided that he could drill for oil the way others drilled for water. Pursuing this idea, Drake hammered an iron pipe nine meters into the ground and, inside the pipe, drilled a well beside Oil Creek in Pennsylvania, near what is now called Titusville. The well was about 21 meters deep. It produced up to 35 barrels a day and launched the petroleum industry as it is known today.

At first the search triggered by the Drake well focused on Pennsylvania, which dominated oil production for the next 25 years. But throughout the United States men began to take a closer look at any area where streams and wells had ever shown traces of oil. As a result, discoveries were made in Colorado, Wyoming, Ohio, Illinois, California, Texas, and Kansas before the end of the century.

Drake's discovery quickly led to experiments and innovations in refining and transport. In June 1860, a year after the discovery at Titusville, the country's first commercial refinery was opened in Oil Creek Valley in Pennsylvania. In 1861 railroads began to transport oil in vertical wooden tanks mounted on flatcars. Four years later the first successful oil pipeline was built — five-centimeters in diameter and eight kilometers long — and in 1886 a precursor of the modern oil tanker, the *Glückauf*, was launched. Toward the turn of the century oilmen were even beginning to experiment with offshore drilling. In 1894 they drilled the country's first offshore well at Santa Barbara, California. Drake's well also gave impetus to exploration and development in Burma, Java, Sumatra, Borneo, Venezuela, Rumania, and Russia.

ROCKEFELLER: In those first exuberant years the petroleum industry was turbulent, oilmen wasteful and disorganized. Lacking both knowledge and supporting technology, they first drilled wells haphazardly, worked them inefficiently, and sold kerosene, lubricants, and other rudimentary products in ruinous boom-and-bust cycles of glut and

Sir Walter Raleigh reported huge pools of asphalt on the island of Trinidad.

North American Indians discovered in oil seeps a substance useful as medicine and war paint.

Samuel Kier converted petroleum to lamp oil with a simple distillation system.

Drake's Pennsylvania well gave birth to an industry when it brought oil to the surface by methods earlier used only to drill for water.

shortage, wealth and bankruptcy, until in 1870 John D. Rockefeller, a young Ohio businessman, began to impose a wholly integrated structure on refining and transport throughout the United States. Rockefeller, a grain merchant, started his career in oil with a small 500-barrel-per-day refinery in Cleveland, but by 1872 had bought or annexed nearly 80 refineries in the oil regions and in Pittsburgh, Philadelphia, and New York. By 1880 his monolithic organization controlled close to 95 percent of all refining in America.

Rockefeller's methods provoked outraged opposition that plagued the oil industry for decades, and in 1911 the U.S. Supreme Court upheld legislation that dismantled the organization. But it is also true that he and his associates created the disciplined, coordinated structure that standardized the wildly fluctuating cost, distribution, and quality of petroleum products. Their reorganization of the industry contributed enormously to the soaring growth of American industrialism and the national prosperity that industrialism created. It also enabled the United States to hold its own in the strong competition for international petroleum markets that developed among British, Dutch, Russian, and American companies in the 1880s.

This competition arose as European oil interests began to develop fields in such places as the East Indies, Rumania, and especially Russia — where oilmen in 1874 had tapped the historic potential of Baku, establishing Russia as a world leader in oil production. By 1900 Russia was producing some 80 million barrels a year compared to America's 65 million.

John D. Rockefeller imposed a disciplined structure on a previously disorganized and wasteful oil business.

SPINDLETOP: In that year, however, a former Austrian naval officer named Anthony F. Lucas drilled a well in a 4.5-meter hillock of land in Texas 40 kilometers from the Gulf of Mexico; a few months later, in one of the most colorful episodes in the history of oil, he brought in Spindletop, the biggest oil field of the period.

For years before the discovery people in the nearby town of Beaumont had said that you could poke a cane into the hillock and light the gas that escaped. But as most experts (including one of Rockefeller's top geologists) dismissed the possibility of oil, no one but a self-taught geologist named Patillo Higgins

The infant U.S. oil industry reached maturity with the gigantic Spindletop discovery near Beaumont, Texas, in 1901.

The Search For Oil

Until the nineteenth century man rarely searched for oil methodically. In some places he detected it seeping into pools or glistening on the surface of streams. Elsewhere he could scarcely miss it — for example, in the "pitch lake" in Trinidad, which is estimated to have held some 200 million barrels of heavy oil and asphalt that could simply be dug by hand.

In the nineteenth century, however, geologists noted that oil seeps often seemed to originate in or near the downward-folded layers of rock known as "anticlines" and speculated that oil was trapped in these structures. An anticline is one kind of "structural trap" in which the strata are bent downward and away from the center to form a solid rock structure that resembles an upside-down bowl. Over millions of years petroleum that has formed from the decomposition of plant and animal remains flows through porous rock layers and into the anticlinal trap. The oil is held in the trap because the structure is "capped" by a layer of impermeable rock, usually a shale. There are several kinds of structural traps, but anticlines are generally the most important in the search for oil as they are easier to detect and are likely to contain more oil than other types.

The folds of rock that form traps do not necessarily reveal themselves on the surface and the search for oil must rely on a variety of methods to deduce what lies beneath the surface. The basic tool in this search is an understanding of the earth itself — how it was formed, what it is made of, and how it was modified through time. The search begins with a study of surface features, starting with the compilation of maps based on aerial photographs and, more recently, on satellite images processed to emphasize selected geological features. The geologist uses his knowledge of such disciplines as the study of the origin, composition, and distribution of rock strata (stratigraphy); the classification and occurrence of rocks (petrology); and the fossilized animal and plant remains from ancient geological periods (paleontology) to construct geologic maps of the earth's subsurface. These maps, when used by specialists knowledgeable about the processes that lead to the formation and accumulation of oil, are important tools in locating potential oil-bearing traps.

Since the 1930s exploration methods based on the relatively new science of geophysics (the study of the physics of the earth) have played an important role in the search for oil. The three primary methods are magnetic, gravity, and seismic surveying, which can be used alone or together, often with considerable accuracy, to map the shape and size of subsurface structures.

Surveying with the use of the magnetometer is historically the oldest of these three techniques. The magnetometer detects minute variations in the earth's magnetic field. Because sedimentary rocks are practically nonmagnetic while igneous rocks have strong magnetic effects, differences in the local magnetic field are used to determine the size, shape, and thickness of sedimentary strata, which may possibly be oil bearing. Magnetic surveys are often made today by specially equipped aircraft that gather data continuously over wide geographical areas. A related technique used on the surface, magnetotelluric (MT) surveying, measures the magnetic and electric fields of the earth to determine the depth and thickness of subsurface geologic units.

The gravity meter measures minute differences in the earth's gravitational field. Variations between the gravity values of the sedimentary rocks and the underlying basement rocks enable the explorationist to determine the shape, size, and distribution of subsurface sedimentary structures.

While the magnetometer and gravity meter can play important roles in oil exploration, advances in geophysical technology have made the seismic survey the most widely used and perhaps the most important method for studying subsurface strata. The traditional seismograph is a device that records and measures vibrations in the earth produced by earthquakes. In oil exploration instruments working on similar principles are used to record and measure smaller-scale, man-made vibrations or shock waves generated by artificial means. The detonation of explosive charges placed in a series of shallow drill holes, or "shot points," can be used to generate the shock waves. The most common

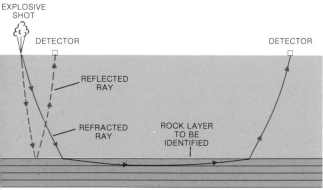

Information obtained from induced shock waves helps identify underground strata.

method now used by Saudi Aramco makes use of a heavy vibrating metal pad mounted on the underside of specially equipped trucks. The truck, moving along surveyed lines, stops at regular intervals and lowers the pad to the ground to generate a series of shock waves. The vibrations formed by either method travel downward through the earth and, as they encounter layers of rock, are bounced back to the surface. There they are recorded by highly sensitive microphones called geophones.

Shock waves transmitted through the earth may return to the surface as reflected rays, which are bounced off the surface of some rock layers, or as refracted rays that travel laterally along the layers before angling back to the surface. Of the two types, the reflected signal provides the most accurate information because it returns to the geophone by a direct route. The time that a shock wave takes to travel from the vibration source to a rock layer and back to the geophone is measured with an accuracy of up to a thousandth of a second. Shock wave data recorded by the geophones are processed to make a visual representation of

the layers of rock beneath the earth's surface, along what is called a seismic line. A series of intersecting seismic lines forms a grid that comprises the complete survey. The geoscientist with an understanding of geophysics and geology can interpret the visual display of the earth's strata seen on seismic lines. After all of the data in the seismic survey has been analyzed, he can prepare detailed maps of the shape of the subsurface strata down to depths that are beyond even the reach of modern drilling equipment.

At one time seismic data was recorded as graphs on long strips of photosensitive paper, but since the mid-1960s when digital computer processing came into use, the data from the geophones has been recorded on magnetic tape, translated by a field computer into numerical values, and then sent to a computer center for further processing. In recent years an advanced method of collecting seismic data known as three-dimensional, or "3-D," seismic surveying has been developed and has become a primary exploration and production tool. The 3-D technology uses denser shot point and geophone spacing. Coupled with advanced computer techniques, it provides much more detailed,

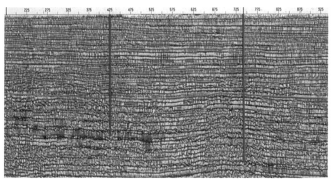

A computer-generated record of seismic data shows relative strength of reflected signals.

three-dimensional models of subsurface formations that can be viewed and printed as maps with virtually any geometric perspective. Saudi Aramco, keeping pace with rapidly expanding oil exploration technology, is using the new 3-D seismic methods to gain detailed insight into the geologically complex fields it discovered in the early 1990s. The key to Saudi Aramco's continuing success in the search for oil lies largely in the experience, development, and training of its geoscientists in these and similar new techniques.

Techniques such as these are obviously far advanced over the methods used in the early days of the industry, when instinct rather than science was often the guiding force, but even they cannot guarantee the presence of oil. The only sure way to find oil remains essentially the same as it was at Titusville, Pennsylvania, in 1859: Drill a well and see if it is there.

Whether their purpose is to find oil or gas or water, wells can be dug by three methods. They can be dug by hand with the old-fashioned pick and shovel; they can be hammered into the earth by percussion; or they can be drilled with what is basically the equivalent of the carpenter's or machinist's drill bit — and all three methods have been used in the search for oil.

The most primitive of these methods is obviously that used in digging a well by hand. Although the history of the

(Continued on page 180)

The Search For Oil

next step up the technological ladder, drilling by percussion, can be traced back for well over a thousand years, oil wells were being dug by hand in some parts of the world as late as the early twentieth century.

The drilling of wells by percussion or pounding consists basically of suspending a large sharpened "drilling

An early 1800s drilling machine used the "Teeter board" method of actuating the tools.

tool" at the end of a rope and repeatedly raising it and letting it drop into the hole being drilled. The rope is passed over a pulley and attached to a long board or pole which serves as a lever, and for many centuries it was simply the weight of the drilling crew at the end of the lever that raised the tool before letting it drop again. With the advance of mechanization in the nineteenth century, however, the efficiency of this very laborious method of drilling was greatly increased by using steam power rather than man power for the task of raising the drilling tool. This made it possible to employ a very much heavier drilling tool and to use a steel cable instead of a rope, and the new system came to be known as the cable-tool method. It was by an

Steam-powered engine and draw works used at Spindletop were the best to be had in their day.

early form of this cable-tool method that Colonel Drake's first well was drilled in 1859.

The next step in the improvement of drilling techniques came in the last decade of the nineteenth century with the introduction of rotary drilling. This makes use of a steel drill bit attached to a long string of steel pipe (the "drilling

string") and rotated at the bottom of the hole. One of the chief advantages of rotary drilling, as it is called, is that it enables a hole to be drilled much faster than with cable-tool drilling.

Rotary drilling in turn made possible the next advance in drilling methods — the introduction of the cone bit invented in 1909 by Howard Hughes, Sr. The cone bit, which led to a further significant increase in the speed of drilling, consists of an arrangement of three steel cones fitted with exceptionally hard, durable teeth that rotate as the bit is turned and grind their way through successive layers of rock. Rotary drilling did not supersede cable-tool drilling overnight. In fact, the cable-tool method continued to be used in some places for drilling oil wells until the 1950s, but after about 1930 rotary drilling became in most cases the method of choice.

No matter what the method used, the cutting tool (whether pick and shovel or high-speed rotary bit) is only one of the essential elements in drilling the well. Another is some means of shoring up the sides of the hole to prevent their caving in. As drilling techniques began to develop in the nineteenth century, iron pipe came to be used to line the well and eventually this was replaced by lengths of steel pipe which could be screwed together end to end and lowered into the hole as it progressed deeper into the earth. At the present time this "casing," as it is called, serves not only to line the well bore and prevent it from caving in but also as a channel through which the broken rock from around the rotating drill bit can be brought to the surface.

Yet another essential element in the drilling process is what is known as "mud" — which actually consists of water to which the driving engineer adds one or more types of special solids mixed with chemicals to produce the exact consistency and weight that is required. This fluid is pumped down into the hole through the drilling string which turns the bit, and it serves two purposes: It lubricates the teeth of the drill bit as they cut into the subterranean rock, and as it flows back to the surface through the space between the drill stem and the casing it carries with it the debris from the bottom of the hole. These pieces of broken rock or "cuttings" serve to show what type of rock the drill is cutting through. The drilling mud has one very important additional purpose: The weight of the column of mud in the hole acts as a plug or stopper and serves to restrain and in most cases to prevent any sudden eruption or "blowout" that might otherwise occur when the drilling tool penetrates an oil-bearing or gas-bearing formation.

The drilling string itself, the long hollow steel shaft which turns the bit, must meet exacting standards. Screwed together section by section on the surface, it must be strong enough to support its own considerable weight suspended from the rig and the additional weight of heavy steel "drill collars" at the bottom of the string that increase the rate of penetration of the bit. The pipe is also subjected to great strain each time the drillers, in order to change bits, hoist it out of the hole, dismantle it, screw the sections back together again, and lower the string back into the hole.

Drilling rigs have also changed considerably since the early days of the industry. The first rigs were no more than simple towers of cross-braced wood whose main purpose was to provide sufficient height so that the cutting tool at the end of the drilling rope or cable could be raised and allowed to drop again. When the well was completed it was usually easier to leave the tower in place than to dismantle it. This gave rise to the popular but mistaken image of an oil field as a forest of rigs or derricks. Today rigs have evolved into a variety of enormous, usually mobile steel structures able to cross deserts and oceans under their own power. The characteristic tower or derrick is no longer needed to provide height for the drop of the cable tool but instead serves as a means of drawing the drill stem from the hole and also as a place in which the drill pipe can be stacked vertically while the bit is being changed. Once the well is completed, the rig is removed, leaving only a few valves on the surface of the earth or, in the case of offshore drilling, a steel platform over the site of the well.

Modern rigs, far more powerful than their forerunners, are also safer and more reliable.

Drilling, of course, varies enormously from country to country and location to location in its difficulty, its cost, and the techniques required. Wells do not have to be vertical; modern drilling techniques can aim curved holes precisely to reach pockets of oil at considerable lateral distance from the rig on the surface. In the latest "horizontal drilling" techniques followed by Saudi Aramco, the lower bore can even be turned horizontally to tap longer exposures of oil-bearing strata and thus greatly increase oil production. Some wells are "wildcats," drilled in an effort to find oil in a new location. Others may be "delineation wells," drilled to find the limits of a known producing location. Still others are "development wells," drilled to bring additional production from an established field. But today, as always, the crucial well is still the "discovery well." In answering the key question — Is there oil there? — the discovery well is the bottom line in the search for oil.

would take Spindletop seriously. Higgins, insisting there was oil there, finally persuaded Lucas and his backers to drill, and on January 10, 1901, at a depth of about 350 meters, Lucas brought in the greatest "gusher" ever seen outside Russia: a plume of oil that came in "like a shot from a heavy cannon," as one witness described it, spouted 45 meters in the air, and set off the most spectacular oil boom in history. Within hours telegraph wires had carried the report across the country, and within days thousands of oilmen, merchants, and speculators were on their way for a share of the excitement and, they hoped, the profits. Some found both as, in the next three months, six other wells came in — one of them topped by a 66-meter geyser — and the tempo of exploration, as well as the price of land, increased frantically.

Patillo Higgins had spent eight years in unsuccessful oil ventures before Spindletop made him famous overnight.

Spindletop's gusher — a layman's term coined there to describe North America's first major free-flowing well — quickly became a symbol for the undisciplined oil exploration that characterized the industry before it developed more controlled methods of drilling. The discovery well at Spindletop, for example, disgorged about 800,000 barrels in the nine days required to cap it — and most of that went up in flames eight weeks later. In Mexico in 1910 a well called Potrero del Llano No. 4 spewed some 100,000 barrels a day into the jungle for two months before it could be capped. Another at Dos Bocas caught fire and burned for 40 days in a flaming column visible from ships far out in the Gulf of Mexico. Gushers are unquestionably spectacular, but they are also wasteful and dangerous and every effort is now made to prevent their occurrence.

Anthony Lucas, Higgins' partner, decided to use a steam-driven rotary rig at Spindletop in what proved to be the first large-scale success of rotary drilling equipment.

The impact of the Spindletop find was almost as spectacular as the wells themselves. In one stroke American oil production surpassed the rest of the world; according to one estimate six of the wells at Spindletop were producing more oil in one day than the rest of the oil fields in the world combined, and before they gave out the fields there yielded 142 million barrels. Spindletop also gave birth to other oil companies that would soon challenge Rockefeller's nearly impregnable position — companies such as today's Sun, Gulf, and Texaco. Because of the ingenuity of its drillers, Spindletop contributed to the development of the rotary drill and focused

400 A. D. 1848 A. D. 1849 400
DISCOVERED IN BORING FOR SALT WATER near
Wonderful MEDICAL VIRTUES DISCOVERED.
THE Bank of the Allegheny River, IN
ALLEGHENY COUNTY, PENNS'A.
about FOUR HUNDRED FEET below
the Earth's surface, is pumped up with the Salt Water, flows into
the Cystern, floats on top, when a quantity accumulates, is drawn off
into Barrels, is bottled in its natural state without any preparation
or admixture. For particulars, get a Circular.
Pittsburgh, 400 S. M. Kier,
Jan. 1st, 1852. Proprietor.
400 400

A scrapbook page of vintage photographs reconstructs the humble birth of an industry which, despite the cycles of success and failure that marked its adolescent years, grew up to alter the world. 1. Early pump station of the Ohio State Oil Company. 2. Pipeline construction in the pioneering days of oil in Persia. 3. Cement-coated pipeline in Louisiana, 1919. 4. Pseudo bank note issued in 1852 by Samuel Kier to advertise the medicinal virtues of petroleum. 5. Spindletop, the first gusher ever seen by American oilmen, in January 1901. 6. Primitive pumping apparatus in which a weight attached to the pumping level swung like the pendulum of a clock. 7. Pumping well in the Allegheny Mountain fields of Pennsylvania. 8. Traffic jam in a Humble Oil Company field in 1918. 9. "Eternal Fire," the flaming oil seep at Kirkuk in Iraq, known centuries ago and still burning. 10. Pioneer run at Oil Creek, Pennsylvania, 1865. 11. Drilling rig at Chiah Surkh on the border between Persia and Iraq, 1902.

attention on the importance of drilling "mud," a fluid that lubricates the bit and washes cuttings to the surface. Spindletop also showed that oil could be found around salt domes. The Spindletop discovery, furthermore, triggered another dynamic search for oil throughout the United States, a search that in the next 10 years would uncover great fields in California, Louisiana, Illinois, Kansas, and Oklahoma.

One of the earliest refineries in the United States was the Monitor Refining Company plant at Oil Creek, Pennsylvania.

THE INTERNAL-COMBUSTION ENGINE: Even more important than Spindletop was the development in the late nineteenth century of the internal-combustion engine and the realization, about 1900, that gasoline could fuel it better than kerosene.

Although a liquid-fuel engine had been developed as early as 1873, and Gottlieb Daimler and Karl Benz experimented with a gasoline engine in 1886, the internal-combustion engine was, until the turn of the century, impractical. For a time, in fact, electric automobiles were far more popular. But oil industry researchers were working hard to find a practical use for gasoline, a volatile by-product of kerosene refining. Although small quantities were used as solvents for paints and varnishes and for special lamps for street lighting, most of it was poured into rivers or burned. One effort to find a use for gasoline led to one of the first advertising campaigns in America. Launched in 1898, it urged housewives to try "cool, convenient, and economic gasoline stoves." The campaign was successful, but its impact was soon eclipsed by the growing awareness that gasoline was precisely the fuel that the internal-combustion engine needed.

Along with the automobile, the 1920s ushered in a new type of small business.

For both automobiles and petroleum these interlocking developments were turning points. Reinforcing each other they created an American automotive industry which after turning out only about 4,000 cars in 1900 produced 187,000 in 1910 — two years after Henry Ford came out with his famous Model T — and a petroleum industry that was thereafter closely geared to the need for gasoline. The leapfrogging demands of the two industries also stimulated such allied industries as steel, glass, and road construction. In addition they transformed petroleum refining into a highly sophisticated science that in the next 50 years improved shipping, revolutionized aviation, transformed agricultural techniques

Gasoline had little market until Henry Ford built a car that the public could afford.

and output, and created new chemical compounds on which numerous other industrial advances — and some entire industries — were based.

REFINING AND GEOLOGY: Before the automotive age, improvements in oil refining came mostly from trial-and-error experiments. Starting with simple distilleries — no more than large iron vats mounted on bricks over a wood fire — the early oilmen had to unlock the secrets of refining one step at a time, on the basis of observation and conjecture. As a result, the first refineries smelled, smoked, and sometimes exploded — as did the kerosene they produced.

Storage, shipping, and distribution methods were equally simple. In some places storage "tanks" were simply holes in the ground lined with planks, kerosene was shipped in oak barrels piled on oxcarts, and gasoline was carried to automobiles in open pails. Even when "gas stations" were introduced they consisted of no more than a metal tank and a garden hose.

The scientific approach, to be sure, was not unknown. As early as 1887 Herman Frasch, a brilliant industrial chemist in Canada, had devised a way of reducing the rank sulfur in oil by stirring metal oxides into raw kerosene during distillation. Frasch also improved several other processes and products. But it was not until the automotive age, with its need for more gasoline, that researchers began to zero in on the basic properties of petroleum, and not until 1912 that they perfected a process called "cracking" — perhaps the most important technological breakthrough in the history of the oil industry.

Cracking, essentially, is the use of intense heat to change the molecular structure of natural crude oil. Discovered accidentally about 1860, cracking was first used to increase the yield of kerosene from a barrel of oil. In 1912, however, as a result of attempts to adapt cracking to the production of gasoline, a process called "pressure cracking" was developed which not only doubled the volume of gasoline obtainable from a barrel of crude oil, but also opened the way to manipulation of petroleum molecules in other ways and to development of the processes on which the whole modern petrochemical industry is built.

From a military standpoint, the development of pressure cracking came none too soon. By then — on the eve of World War I — Western military leaders had begun to convert battleships from coal to oil, introduce gasoline-powered submarines, trade off mules for trucks, and try out the first clumsy tanks and flimsy warplanes. All of this stimulated the demand not only for greater volumes of gasoline, but also for more efficient gasoline.

The problem of efficiency was not satisfactorily solved until the 1920s, when the postwar growth of the automotive industry created additional demands for better gasolines. In trying to provide it, researchers noted that cracked gasoline seemed to "knock" less than "straight-run" or natural gasoline. This led to experiments with an additive called tetraethyl lead which reduced knocking. The process was used for almost 50 years until the passage of antipollution legislation led to its gradual abandonment. As automobile engines with greater compression ratios were introduced, however, refining experts had to delve still more deeply into gasoline's performance characteristics. Out of that research came, in the 1930s, what are called "thermal reforming" processes, which provided greater volumes of higher octane, better performance gasolines.

This period — the 1920s and 1930s — was what the oil historian G. A. Purdy calls "the golden years for petroleum engineers." In those days, he wrote, "almost every process could be improved by the application of engineering principles." But researchers made contributions too. Chemists, for example, eliminated the impurities in gasoline that clogged carburetors, left coatings of carbon in engines, and corroded processing equipment. Along the way they also developed new processes and stored up information that would lead to the 1939 breakthrough called "catalytic cracking" — the cracking of petroleum molecules by the use of chemical catalysts rather than by heat and pressure alone.

Oil geology also came into its own in this period. Although theories about anticlines, the arched rock strata in which petroleum is frequently found, had begun to provide guidelines as early as the 1880s, the discovery of oil remained for many years after that more a matter of observation and guesswork than of science. Even as late as 1930, when Columbus M. "Dad" Joiner brought in the giant East Texas field, intuition was still thought by many to work as well as geological study. In 1910, however, geologists C. Willard Hayes and Everette Lee DeGolyer led a British company to Mexico's Potrero del Llano No. 4, and geology began to gain ground. DeGolyer four years later began to determine the shape, location, and structure of underground strata by applying gravity survey information. In the next decade geologists greatly improved their ability to assess underground formations, adapted the seismograph to oil exploration, and, in applying their new techniques, led the industry to other major oil discoveries.

Dad Joiner proved he was right about the potential of what came to be America's top producing state when he brought in the East Texas field

Everette DeGolyer pioneered the use of geophysics in prospecting for oil in the U.S., with his greatest sucess the development of seismic techniques to detect fields associated with salt domes.

MORE DISCOVERIES: These advances were timely. World War I demands for oil had imposed alarming strains on American oil resources and after the war the growth in the number of tractors, buses, trucks, and cars — 23 million on the road by 1929 — required ever greater volumes of oil. Partly as a result of advances in the science of geology, however, individuals and companies went on to uncover still greater fields in Appalachia, Illinois, Indiana, Louisiana, and California. By the end of the 1920s some 87 percent of the world's production was coming from North America.

In the meantime, the first major discoveries had been made in what came to be called Mexico's "Golden Lane," and by 1918 Mexico was the second largest producer of oil in the world. In 1914 the first commercial field had been found on the eastern shores of Lake Maracaibo in Venezuela, and subsequent exploration by British, Dutch, and American companies established the Maracaibo Basin as one of the world's greatest oil reservoirs.

The discoveries in the United States, Mexico, and Venezuela, coupled with the drop in production during the post-revolutionary chaos in Russia, gave the Western Hemisphere a lead in oil production that would endure through the 1960s. They also gave the United States and its allies a crucial advantage as, in 1939, war again engulfed the world.

The Tia Juana field under Lake Maracaibo was a major source of oil for the Allied forces in World War II.

WORLD WAR II: Oil played an even more important role in World War II than it had in the First World War. Because Germany lacked sufficient sources of oil it was essential for her to capture the Baku fields of the Soviet Union, and failure to do so was a factor in Germany's defeat. Both Germany and Japan were forced to depend to a significant extent on synthetic oil made from coal which, in terms of both quality and quantity, was inadequate. Germany's aviation fuel, for example, could not compare with the high-octane fuel that the American oil industry began to produce after catalytic cracking was developed in 1939. On the Allied side, researchers turned to petroleum to develop a synthetic substitute for rubber after Japan's Far East conquests choked off supplies of the natural product.

The war's unprecedented demands for oil also stimulated construction of pipelines in the United States, increased the size of tankers, and improved exploration techniques. Production also increased; during World War II fully half of U.S. shipments overseas consisted of petroleum products.

In the immediate postwar period, the expansion of American industrial technology, the enormous scale of the physical reconstruction of war-devastated Western Europe and Japan, and the outbreak of hostilities in Korea all continued to stimulate demand. By 1948 the United States had become a net importer of petroleum rather than an exporter. This change, in absolute terms, was slight — on the order of 145,000 barrels a day. But it was a turning point nonetheless. From that day on the United States would be increasingly dependent on foreign oil and eventually on oil from the Middle East — where 40 years before another important chapter in the history of oil had begun.

OIL IN THE MIDDLE EAST: In one sense the history of oil in the Middle East is less colorful than its history in the United States. It is far more a story of prolonged and complicated national and corporate negotiations than of the boomtown individualism that characterized so much of American exploration. In the early years some individuals, however, do stand out: Baron Julius von Reuter, who founded the Reuters news agency; Calouste Gulbenkian, who became known to the public as "Mr. Five Percent"; William Knox D'Arcy, a British financier who fathered the first major oil discovery in the Middle East; George B. Reynolds, a tenacious field manager

who brought in the first well; and King 'Abd al-'Aziz of Saudi Arabia, who in 1923, even before he had unified his kingdom, granted a concession to Major Frank Holmes, a mining engineer from New Zealand. Holmes, despite many contrary opinions, insisted that there was oil in the Arabian Peninsula — and was proven spectacularly right.

Baron von Reuter was the first on the scene but his role was relatively small. In 1872 and again in 1889 he obtained from Shah Nasr-ed-Din concessions to search for and exploit minerals and oil in Persia. The company formed by von Reuter drilled three wells but found no oil and in 1899 the concession was canceled.

About the same time the young Armenian Gulbenkian, who had learned something about oil in Russia's Baku fields, submitted a report to the Ministry of Mines of the Ottoman Empire, which until the First World War had actual or theoretical sovereignty over all the Arab lands of the Middle East. This report, calling attention to the historic oil seepages of Mesopotamia, excited Sultan Abdul-Hamid and later led Germany to seek petroleum rights from the Ottoman Empire along the route of the proposed Berlin-to-Baghdad railroad. That move, in turn, lured other competitors into the Middle East, including Royal Dutch Shell and an American admiral representing the New York Chamber of Commerce.

The military coup d'état and Young Turk revolution of 1900 and the deposition of the sultan in the following year put a temporary stop to all such efforts to obtain concessions in the Ottoman Empire. In 1910, however, a National Bank of Turkey was founded at the urging of the British and with British capital, and with Gulbenkian as one of the directors. Gulbenkian persuaded the bank to negotiate an agreement by which in 1912 the Turkish Petroleum Company (TPC) was formed, with himself a shareholder. Through TPC Britain and Germany reached an agreement on Turkish oil development that played a considerable part in negotiations after World War I.

Meanwhile William D'Arcy, who had made a fortune in gold in Australia, had begun to investigate the possibility of finding oil in the Middle East. In 1900 D'Arcy sent representatives to Persia to negotiate a concession and scout the terrain. In the following years he

Baron Julius von Reuter was the earliest of several international financiers to win mineral and oil concessions in Persia.

Calouste Glbenkian, prime mover in negotiations involving territories of the old Ottoman Empire, was a founder of the Turkish Petroleum Company, forerunner of the Iraq Petroleum Company.

William D'Arcy lost over a million dollars in a fruitless oil search in Iraq and Persia before he succeeded.

obtained the concession, and after his geologists recommended "two unquestionably petroliferous territories" he picked a field manager and dispatched him to search for oil near Chiah Surkh, located in the border area between Persia and Iraq.

George Reynolds, the field manager, was a tenacious geologist who spent 16 months completing the first well, brought in one well that went dry five months later, and endured nearly five years of hardship — including an outbreak of plague — before his and D'Arcy's hopes were fulfilled.

They almost were not. By 1905 D'Arcy was running out of money — he had spent well over a million dollars and had been forced to seek loans. Two years later, as Reynolds began drilling in the Zagros Mountains of the Ahwaz region in western Persia, funds ran out again. D'Arcy tried to hold on but in May 1908 his syndicate decided to withdraw and sent a cable ordering Reynolds to stop work, dismiss his crew, and come home.

MASJID-I-SULAIMAN: At that point Reynolds had been drilling into a ledge in an area called Masjid-i-Sulaiman. As the ledge overlooked a stream where oil had been seeping out for years, Reynolds was certain he was on the verge of success and decided that he would not accept these instructions until he received a letter confirming them. Instead, he kept drilling and on May 26, 1908, at a depth of 360 meters, brought in the first commercial well in the Middle East — and uncovered the Masjid-i-Sulaiman field, which by the end of the 1970s had produced well over a billion barrels of oil.

The Masjid-i-Sulaiman field lies in a sedimentary basin rimmed by the mountains of Oman, Iran, Iraq, Turkey, Syria, Lebanon, the Hijaz, and Yemen — a basin that contains the prolific oil fields of the Tigris and Euphrates river regions, Saudi Arabia and the other Arab states lying in or on the Arabian Gulf, and the Iranian foothills and coast. No one knew that in 1908, however, and even experienced geologists refused concessions that later turned out to include exceptionally thick porous rocks and reservoir conditions ideal for the recovery of the oil. As a consequence, discoveries that were to alter world patterns of supply and demand were not made for another 25 to 30 years.

EARLY DEVELOPMENTS: Development in Persia started almost at once. In April 1909 D'Arcy's Anglo-Persian Oil Company was launched — with the encouragement of Admiral Fisher, First Sea Lord of the British Admiralty, who for many years had been pressing for conversion of the Royal Navy from coal to oil. Five years later, on the eve of the First World War, the British Government gave more concrete support. Under the prodding of Winston Churchill, Parliament voted to acquire a majority interest in the Anglo-Persian Oil Company, which later became the Anglo-Iranian Oil Company and — still later — British Petroleum.

In Persia, meanwhile, Anglo-Persian had laid out and supervised the construction of a 210-kilometer pipeline to the Arabian Gulf that was completed in 1911. The line ran between Masjid-i-Sulaiman and an uninhabited mudflat at the head of the Gulf called Abadan; there, in 1913, a refinery with an initial capacity of 2,400 barrels a day was completed. Shortly after, Churchill signed a contract under which Anglo-Persian would supply oil to the Royal Navy. As the Allies needed great quantities of oil in the war it proved to be a wise precaution, although in fact the new discoveries played no more than a secondary role in the eventual Allied victory over the Central Powers.

After the war Middle East oil again came to the fore as the Allies, in negotiating spheres of influence in the territories formerly ruled by the Ottoman Empire, discussed various agreements affecting future discoveries of oil in Mesopotamia. As the discussions were complicated and often involved controversy, some key matters were not finally settled for years. One was the objection of the United States to being excluded from Mesopotamian oil exploration. These objections led to the formation of the Near East Development Corporation, a consortium of five American companies, which spearheaded American participation in Middle East oil development.

In 1928 an agreement was reached under which shareholding in the Iraq Petroleum Company (IPC) was adjusted: Anglo-Persian, Royal Dutch Shell, the Compagnie Française des Pétroles, and the Near East Development Corporation each held a 23.75 percent share — with five percent for Gulbenkian, which

The perseverance of George Reynolds was rewarded in May 1908 when he brought in the first commercial oil well in the Middle East.

The Masjid-i-Sulaiman discovery opened up a field that had produced over a billion barrels by the end of the 1970s.

OIL AND GAS FIELDS OF THE MIDDLE EAST

The major oil and gas fields of the Middle East are located in a sedimentary basin lying between the Arabian Shield on the west and the mountains of Iran on the east and contain well over half of the world's total oil accumulations. Saudi Arabia has the largest proved and probable oil reserves in the world and two of the largest oil fields in the world are in the kingdom — the Ghawar field onshore and the Safaniya field offshore. The majority of the 71 oil and gas fields that Saudi Aramco had discovered by 1995 lie in the Eastern Province of the kingdom. By that year, however, fields in the central part of the kingdom were also contributing significantly to production and exports, and discoveries along the Red Sea coast had also helped increase reserves.

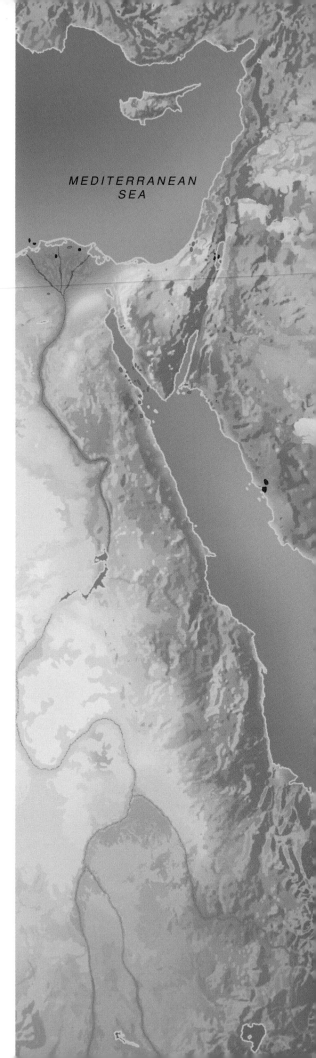

MEDITERRANEAN
SEA

Oil field   Gas field

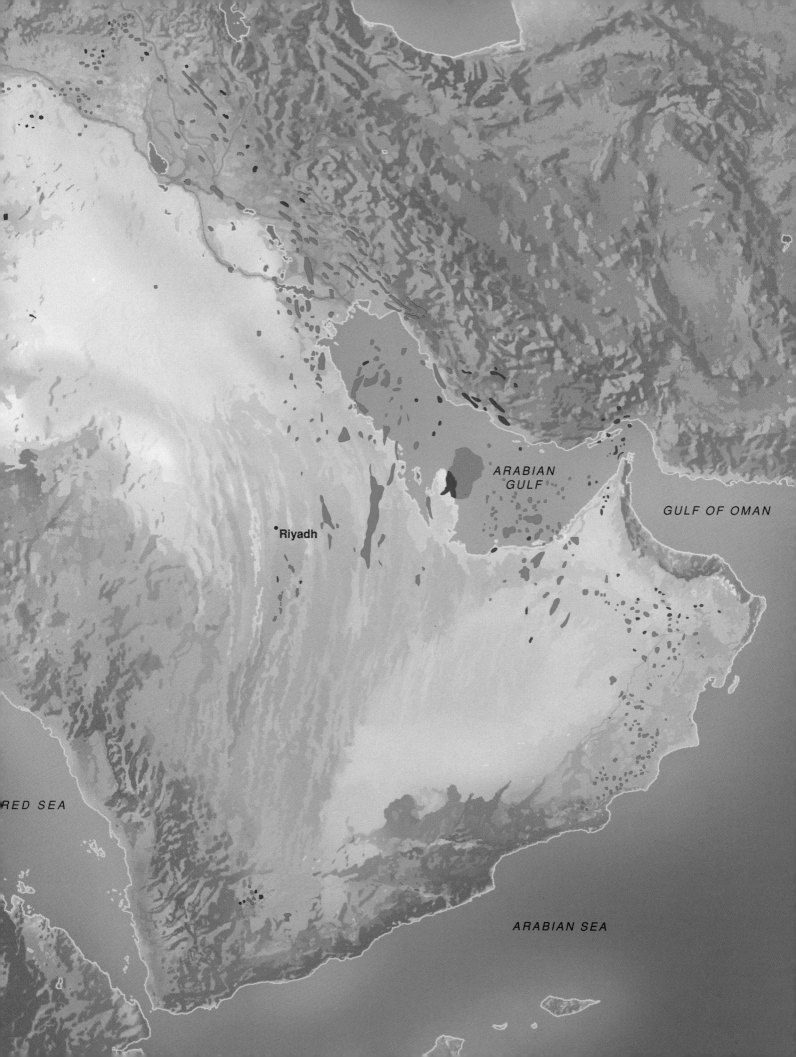

RED SEA

Riyadh

ARABIAN
GULF

GULF OF OMAN

ARABIAN SEA

earned him his nickname. Simultaneously, the companies concerned signed what was called the "Red Line Agreement," whereby each undertook not to seek individual concessions in a large area of the former Ottoman Empire without agreement by the other companies. The area included all of Arabia except Kuwait.

A year before IPC was formed, oil had been found at Baba Gurgur in Iraq, 225 kilometers north of Baghdad. Within 2,300 meters of Iraq's Eternal Fires — which had burned continuously at least since Nebuchadnezzar's time — oil burst to the surface on the night of October 14, 1927, and like the first Spindletop well in Texas spurted into the air in a gusher 43 meters high. It also took five lives and touched off a crisis that lasted 10 days while men in gas masks strove to master it. This well flowed at the rate of 95,000 barrels a day, marked the discovery of the Kirkuk oil field, one of the largest in the Middle East, and confirmed Gulbenkian's predictions made more than a quarter of a century earlier.

One of the largest fields in the Middle East was tapped with the discovery well at Baba Gurgur in Iraq.

In the 1930s exploration and development in the Middle East proceeded at a steady pace. In Persia, Anglo-Persian discovered and began to export oil from the Haft Kel field (1928), the Gach Saran field (1935), and Agha Jari (1944). In Iraq IPC discovered the Ain Zalah field and, through its subsidiary the Basrah Petroleum Company, brought in the Zubair field. In 1930 Standard Oil of California took over a concession originally negotiated by Frank Holmes on Bahrain and on May 31, 1932, found oil — a discovery that stimulated the search for oil in Saudi Arabia and Kuwait. Some two and a half years later the Kuwait Oil Company, formed in 1933 by Anglo-Persian and Gulf Oil Company, received a concession for Kuwait and in April 1938 discovered the giant Burgan field, one of the largest in the world. Exploration and development concessions for oil were granted in Qatar in 1935, in Oman in 1937, and in Abu Dhabi in 1939. And in 1938 oil in commercial quantities was discovered in Saudi Arabia.

In brief, as reported by a team of experts under the leadership of Everette DeGolyer which the U.S. Government sent to the Middle East in 1943, the center of gravity of world oil had begun to shift from the basins of Texas, Mexico, and Venezuela to the Middle East. Crude oil production from the Western

Major Frank Holmes won the first oil concession in what is now Saudi Arabia, but no exploratory work was undertaken and the agreement lapsed.

Hemisphere continued to be a significant part of the world total, but its proportionate share of the total declined steadily after the end of World War II; and while the United States continued to be the world's largest producer until 1974, its share of the total was likewise declining and in fact by 1948 it was already importing more oil than it exported. Certainly by the end of the 1930s it was becoming clear that the Middle East was where the world's biggest reserves were located and that therefore this was where the major share of world oil production would have to come from in the future. Eventually it became evident that the largest oil accumulations were located in the Arabian Peninsula — and especially in Saudi Arabia. The chief role in developing, operating, and managing the kingdom's petroleum enterprise was initially played by the Arabian American Oil Company (Aramco), a task which in 1988 passed into the hands of the Saudi Arabian Oil Company, better known as Saudi Aramco.

SAUDI ARAMCO FORERUNNERS: In one sense the story of the Saudi oil industry began on May 29, 1933, when the Saudi Arabian Minister of Finance and a lawyer from the Standard Oil Company of California (Socal) signed an agreement in Jiddah giving that firm the exclusive right to "explore, prospect, drill for, extract, treat, manufacture, carry away, and export" oil and oil products. In another sense the story began a few months later when two American geologists landed at the village of Jubail on the east coast of Saudi Arabia and launched what would be an often-frustrating five-year search for oil. In still another sense it began on March 3, 1938, the day Dammam Well No. 7 encountered oil in commercial quantities in what is now called the Arab Formation. But the real beginning might well be said to be the day in 1930 when King 'Abd al-'Aziz somewhat reluctantly decided to allow foreigners to investigate and evaluate the potential oil and other mineral resources of his country.

The King's first attempt to exploit these mineral resources had been in 1923, before he had fully unified what was to become Saudi Arabia. In that year he granted an oil concession to Major Frank Holmes, an energetic and amiable New Zealander. Holmes, however, represented a London financial syndicate, not an oil company, and the syndicate was unsuccessful in its attempts to persuade any oil company that the chances of discovering oil in Arabia were enough to risk the necessary capital expenditure. The Holmes concession, covering more than 77,000 square kilometers in eastern Saudi Arabia, was allowed to lapse through failure to make certain nominal annual payments.

In 1930, as an outgrowth of discussions with his English friend and unofficial adviser, H. St. John B. Philby, King 'Abd al-'Aziz set in motion a series of developments that would eventually bring the two American geologists to the eastern shore of his country. The first of these developments was an invitation to visit Saudi Arabia which 'Abd al-'Aziz sent to Charles R. Crane, a wealthy American businessman who had done philanthropic work in Yemen and who had been cochairman of the King-Crane Commission, the controversial fact-finding delegation which had been sent to the Middle East in 1919 to make recommendations about the future government of the Arab lands that had been part of the Ottoman Empire. Crane accepted the invitation, and during his visit to Saudi Arabia he promised to send the King a mining engineer to examine the country's water, agriculture, and mineral resources. The man Crane selected was Karl Twitchell, a Vermont Yankee who in 1931-32 crossed the peninsula looking for water sources, mineral outcrops, and oil seeps.

In the meantime, Socal had sent two geologists to Bahrain to make recommendations about oil prospects there. One of them was F. A. Davies, later to become chairman of the Board of Directors of Aramco. "From the day we first set foot on Bahrain," wrote Davies some years later, "we had a strong desire to examine the geology of the mainland in Saudi Arabia, not only because of its bearing on what might be expected below the surface of Bahrain, but also because it was thought there might well be oil possibilities in the vastly larger area of the mainland." When, as a result of Davies' recommendations, Socal drilled in Bahrain and discovered oil there in May 1932, it was only natural for Socal's executives to press with considerable vigor to explore Saudi Arabia.

Socal at first attempted to arrange a meeting with the King through Major Holmes (the same Major Holmes who had negotiated an agreement with the King in 1923 and with whom Socal had been associated from the start

in Bahrain), but then discovered that the King had already asked Twitchell to see if he could find an oil company interested in paying for a concession to explore for oil and produce it if found. Socal immediately retained Twitchell to assist in obtaining such a concession.

Socal also attempted to get in touch with the King through Philby. Philby was asked to propose to the King that Socal make a preliminary geological examination of eastern Saudi Arabia, with the idea that a concession would be negotiated if the preliminary investigation appeared to justify it. When Philby indicated that he was prepared to serve as an intermediary and reported that the King was willing to negotiate, Twitchell and Lloyd N. Hamilton, a Socal lawyer specializing in contracts and leases, traveled to Jiddah to negotiate a concession agreement, arriving in February 1933. There Philby joined the team.

NEGOTIATIONS: The negotiations took three and a half long months. Saudi Arabia's team of negotiators, led by 'Abd Allah as-Sulayman, the king's shrewd Minister of Finance, was capable, patient, and tenacious. The Saudi team knew the best terms already won by Persia and Iraq for oil concessions and was determined to obtain no less favorable conditions. There was also competition. The Iraq Petroleum Company (IPC) sent its distinguished negotiator Brigadier Stephen Longrigg to Jiddah at the same time. Although IPC seemed less interested in developing oil in Saudi Arabia than in keeping out other oil companies, it had the full backing and support of the British minister to Saudi Arabia. Even Holmes put in an appearance for a few days, but he was obviously not prepared to meet the minimum demands of the Saudi negotiators.

Hamilton, however, had several factors working in his favor. One was that IPC had at that time more oil available than it knew what to do with. Another was that Hamilton's company had already discovered oil in Bahrain and had demonstrated there that its interests were confined to business and involved no political entanglements or ambitions.

Socal was far from taking on a "sure thing" in moving to obtain a concession in Saudi Arabia. The negotiations were conducted shortly after the United States had gone off the gold standard and President Roosevelt had closed U.S. banks, and at a time

Philby served as go-between in the king's early talks with oil-company representatives.

Fred Davies, sent by Socal to explore for oil on Bahrain, was convinced the mainland held promise as well.

Negotiations over the Socal concession were headed on Saudi Arabia's behalf by the influential Minister of Finance, Shaykh 'Abd Allah as-Sulayman.

5

6

7

From Saudi Aramco's archives come photographs of the men who made the beachhead and views of a landscape they began to transform. Today, from the perspective of history, their accomplishment is formidable, but in the 1930s they were simply men who came to do a job. 1. View of Dhahran in 1937, as seen from the jabal south of Dammam Well No. 1. 2. Jerry Harriss, Fred Davies, Soak Hoover, and Max Steineke camped near Jabal Dam, 1936. 3. Hoover and Harriss at camp on the Abqaiq structure in 1936. 4. Tom Barger-Ernie Berg field party departing in the fall of 1937. 5. Krug Henry and Art Brown shooting the stars at Linah, 1934. 6. Joe Mountain, Dick Kerr, and Russ Gerow with the Fairchild used for aerial reconnaissance. 7. Felix Dreyfuss and Steineke at their Jubail office, 1935. 8. Dhahran in 1946, beginning to look like a town. 9. Ras Tanura tank farm and pier, 1947. 10. Early family housing in Dhahran, late 1937. 11. Abqaiq in 1947.

The first man in the vanguard of oil prospectors was Bert Miller.

In the pattern of most early explorers, Krug Henry wore a beard and Arab headgear.

Invaluable to the pioneering effort was a single-engine plane equipped for aerial photography.

The inventive Dick Kerr helped develop special methods for processing photographic film in desert conditions.

when crude oil was selling for less than 50 cents a barrel. The company was in fact taking a calculated business risk in a period when the economic outlook and business prospects were particularly gloomy.

The concession agreement was signed in Jiddah on May 29, 1933, by 'Abd Allah as-Sulayman and Hamilton. In the agreement Socal committed itself to loans totaling £50,000 gold (about $250,000 at that time), yearly rentals of £5,000 (about $25,000), and royalties of four shillings gold (about $1) per ton of oil produced. It also undertook to provide the government with an advance of £50,000 gold once oil was found in commercial quantities. From the government's standpoint, of equal importance were provisions calling for an immediate start of exploration, for drilling to begin as soon as a suitable structure was found (in any case no later than September 1936), and for construction of a refinery after oil was discovered.

Socal obtained the exclusive right to prospect for and produce oil in eastern Saudi Arabia and preferential rights (that is, the right to obtain additional areas by matching the terms of other offers) to most of the rest of the kingdom that had any chance of containing oil accumulations. The agreement was to run for a period of 60 years from 1933 and was later extended for an additional six years. It was ratified by the Government of Saudi Arabia by a Royal Decree issued on July 7, 1933, and became effective on July 14 when it was published in the official gazette.

BEACHHEAD: Less than three months later, on September 23, 1933, the first oil prospectors stepped ashore at Jubail. Like Twitchell, who had driven across Arabia to join them, Robert P. (Bert) Miller and Schuyler B. (Krug) Henry wore beards and were in Arab dress to avoid appearing unnecessarily conspicuous to the townspeople and Bedouins, few of whom had ever seen a Westerner. A house at Hofuf in the oasis of al-Hasa had been obtained for them, and later arrangements were made to rent a small compound at Jubail for use as headquarters.

On the afternoon of their arrival the geologists began to explore the coastal region south of Jubail, including a hill called al-Jubayl al-Barri (the Berri oil field was named after this many years later). On September 28 they

came to Jabal Dhahran, the distinctive feature they had observed from Bahrain. The Dhahran area is today the site of Saudi Aramco's headquarters, the King Fahd University of Petroleum and Minerals, and much else that has grown up in the intervening years. It is also a textbook example of a geological dome and it eventually proved to contain oil. They chose to give the name Dammam Dome to the structure underlying Dhahran. It was oil industry practice at the time to name newly discovered structures after a nearby town or village, and Dammam and the little fishing village of al-Khobar were the only settlements in the area. Dammam was chosen because it was larger and because it was felt it would be easier for Americans to pronounce correctly than al-Khobar.

This phase of the job — geological reconnaissance — was done using car odometer, alidade and plane table, geologist's pick and Brunton compass. Heat and sandstorms often confined the geologists to their tents or drove them back to their headquarters. In March 1934 their work was made easier when they acquired a proper tool for the exploration of an area bigger than Texas — an airplane.

To Socal, the use of airplanes was considered so important that permission to use them was sought during the original concession negotiations. By the time Henry and Miller came ashore at Jubail, Socal's chief geologist was already negotiating with a geologist, pilot, and aerial photographer named Richard Kerr to provide aerial reconnaissance, photographs, and support to the geologists on the ground. Socal bought a Fairchild 71 highwinged monoplane, equipped it with extra gas tanks and a big camera, arranged for special chemicals for developing film in Saudi Arabia's extreme heat, and shipped all these to Egypt on a freighter. From there Kerr and Charley Rocheville, as copilot and mechanic, flew the plane to Saudi Arabia via the Suez Canal, Gaza, and Baghdad, touching down at last on an airstrip made ready for them by the geologists at Jubail.

There were problems, even with the plane. The government at first insisted that it fly high, avoid the interior of the country, and use no radio. Blowing sand grounded it often. The large tires it needed to land on soft sand affected the plane's stability and reduced its

Geology Of Arabia

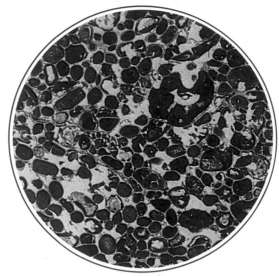

This photomicrograph shows the pore system of the oil-bearing Arab-D limestone.

Hundreds of millions of years ago vast seas washed up against the mountainous landmass of what is now the western part of Saudi Arabia. This ancient landmass is the Arabian Shield, the first of the peninsula's two distinct geological provinces. Sediments left behind by the primordial seas constitute the second and younger province, the Arabian Shelf.

Far more extensive than the present-day Arabian Gulf, or even the sea known to the peninsula's earliest man, the ancient waters spread north as well as west, intermittently covering the regions now occupied by Iraq, Jordan, Lebanon, and Syria. Through age upon age the sea alternately flooded and abated, exposing bare sediments to erosion and then covering them again. Tiny plants and animals living in the marine waters died and became part of the seabed sediments. Through the long, slow processes of nature the mineral sediments hardened into rock and organic matter was transformed into the petroleum now found so abundantly in the Gulf-Mesopotamia region.

The Arabian Shield is actually part of a greater Afro-Arabian Shield, split today by the Red Sea Rift. Bulging toward the east at its midsection and occupying about one-third of the Arabian Peninsula, the ancient landmass consists of rocks formed from molten magma and from folded, compressed sediments that were fused and altered by the enormous heat and pressure associated with mountain building. In some regions comparatively recent lava flows lie upon the shield. This igneous-metamorphic complex, Precambrian in age, slopes gently eastward beneath the sedimentary strata of the Arabian Shelf, forming the crystalline "basement" that surfaces again in parts of the Iranian mountains.

Laid down some 550 million years ago, the oldest of the Arabian Shelf sediments outcrops on the shield's eastern flank and, along with the progressively younger overlying beds, dips gently and thickens in the direction of the Rub' al-Khali and Arabian Gulf basins. Altogether the sediments are some 6,500 meters thick in the region below the present-day shoreline of the Gulf. Time has worn away the exposed softer beds at their thinner western edge, leaving those less subject to erosion standing high above the level of the general terrain in the shape of escarpments in which Paleozoic, Mesozoic, and Lower Tertiary beds outcrop. Most notable is the long, curved belt of westward-facing cliffs known as the Tuwayq Escarpment. Resistant limestones cap each scarp, and many of these rock units continue without significant interruption for hundreds of kilometers across the shelf. In eastern Saudi Arabia many of these are covered by essentially flat-lying Tertiary and younger deposits.

All the known accumulations of hydrocarbons in Saudi Arabia occur in the sedimentary shelf and most of them in rocks of the Late Jurassic. In that period in geologic time, roughly 150 million years ago, a progressive change in sedimentary environment — reflecting the withdrawal of the sea — began in what is now central Iraq and subsequently pushed south into the Arabian Peninsula as desiccation spread. The alternate transgressions and regressions of the generally receding sea set up a fourfold sedimentary cycle in which mainly shallow-water carbonate rocks were overlain by beds formed as seawater evaporated in shallow, enclosed depressions. These particular carbonates are the four members of the Arab Formation. They are of paramount importance (especially the porous and extremely prolific member labeled the Arab-D) for they contain the largest reserves of petroleum in the world.

Millions of years after deposition of the sediments, tectonic forces deformed the strata, and oil accumulated in structural traps where overlying beds of impermeable rock prevent its migration and escape. Dammam, the first oil field discovered in Saudi Arabia, is a domal uplift — probably the result of upward pressure from a deeply buried salt plug. The great majority of known accumulations in the kingdom, however, are in anticlines formed when earth movements forced the sediments to buckle. Major deformation occurred in Cretaceous and Eocene times and resulted in broad, shallow folds, unlike the high mountain ranges uplifted later in Iran. The axis of these gentle flexures in eastern Arabia trends generally north and south, and the longest — which includes Ghawar, the largest oil field in the world — extends for more than 400 kilometers.

RED

SEA

SURFACE GEOLOGICAL FEATURES OF ARABIA

The Arabian Peninsula can be divided into two main geological provinces: the shield and the shelf. The Arabian Shield, shown as "Precambrian basement" on the map, consists of igneous and metamorphic rocks and makes up the western one-third of the peninsula. Bordering the shield on the east is the Arabian Shelf, consisting of sedimentary strata. The shelf slopes gently away into the Rub' al-Khali and Arabian Gulf basins, interrupted by a series of escarpments in which Paleozoic, Mesozoic, and lower Tertiary beds crop out. The largest known accumulations of oil and gas in Saudi Arabia are in the eastern portion of the shelf.

ARABIAN GULF

ARABIAN SEA

Eolian sand and gravel, mainly quaternary

Volcanic rocks, mainly tertiary and quaternary

Oligocene

Miocene and Pliocene

Paleocene and Eocene

Cretaceous

Jurassic

Triassic

Paleozoic

Precambrian basement

Pilot-photographer Joe Mountain amassed an excellent pictorial record of the oil venture's early days.

Some Saudis called Max Steineke, the indefatigable geologist, "a big man with a big arm and a big voice."

Dammam No. 1 blew oil and gas during a flow test, but the oil accumulation proved to be too small for commercial production.

speed. The pace of the work, nevertheless, quickened and the reach of the exploration effort was extended. When permission was received to use the Fairchild's radio, the pilots were able to coordinate their aerial surveys with geologists working on the ground, and by the end of the first eight months in the field the reconnaissance teams had seen enough to justify a recommendation to drill the Dammam Dome and to study the general area with geophysical methods.

THE FIRST WELL: By October 1934, after a summer in the cool mountains of Lebanon, most of the original team, plus some new men, were back in Saudi Arabia working for Socal's new subsidiary, California Arabian Standard Oil Company (Casoc), to which it had assigned the concession.

Among the newcomers were two men who, in different ways, typified the pioneers of the reconnaissance period. One was Joe Mountain, a fine pilot and a professional photographer and darkroom technician who produced a fascinating pictorial record of those early days. The other was Max Steineke, whose contribution to the story of the oil industry in Saudi Arabia would be of inestimable value. Burly, enthusiastic, implacably methodical, Steineke was a field geologist who laid the foundations for the understanding of what lay beneath Saudi Arabia's deceptive sands. Aiding them were Saudi Arabs in a wide variety of jobs (115 were on the payroll by 1935) — the first of thousands who eventually formed the backbone of the company's multinational work force.

In most ways the 1934-35 season was like the first one, but there were differences too. For one thing, there were more jobs to do — the construction of a pier at al-Khobar, for example, and the construction of bunkhouses. The main difference, though, was that on April 30, 1935, they finally began to drill for oil.

But Dammam Well No. 1 was a disappointment. For one breathtaking day, the men at San Francisco headquarters thought they had struck oil on the first go. They received a cable reporting a flow of 6,537 barrels daily from the Bahrain zone, but that figure turned out to be garbled. The estimated flow was about 100 barrels a day. That would have been an oil well in Bakersfield, California, but it was hardly commercial in Saudi Arabia.

No. 1 was completed as a gas well, producing from a shallow horizon for domestic cooking and heating and for firing the boilers used to provide power for the drilling of other wells.

Drilling of Dammam No. 2 started in February 1936. By June 20, during a five-day test, it flowed an average of 335 barrels of crude a day. A week later, after acid treatment, it produced at the equivalent of 3,840 barrels a day. San Francisco was jubilant and approved four more wells in the Dammam Dome and a wildcat at Al Alat, some 32 kilometers northwest of Dhahran. In July plans were made for yet another well, Dammam No. 7. It was scheduled to be a deep test into a zone which had shown gas, but little oil, in Bahrain.

They drilled No. 1 deeper, down below 975 meters, without striking anything worthwhile. Then No. 2 went bad and started producing more water than oil. No. 3 and No. 4 were both drilled into the Bahrain zone in 1936. The first never produced more than 100 barrels of heavy road oil a day, and the second was a dry hole.

DAMMAM NO. 7: On December 7, 1936, the drillers spudded in Dammam No. 7. It was destined to become the most stubbornly defended route to oil yet encountered. There were accidents, cave-ins, lost circulation, and stuck drilling bits. Everything imaginable seemed to go wrong.

Meanwhile the geologists were beginning to probe still farther and deeper into Saudi Arabia's geological past. Aiding them was their first paleontologist, Richard A. Bramkamp. It was Bramkamp who confirmed and refined Steineke's intuitive grasp of the peninsula's geology and succeeded Steineke as Aramco's chief geologist.

In the spring of 1937 the first two wives of American employees came to join their husbands in the new houses being constructed not far from the drilling rigs, and in September four more wives and four children arrived. In that same spring Max Steineke crossed the Arabian Peninsula both ways. It was not the first trip (Miller and Hamilton had done it in 1934), but from the standpoint of an understanding of the geology of Saudi Arabia it was of great importance. By the time he returned, after having a comprehensive look at the kingdom's surface geology, Steineke had the data which enabled him to construct a skeleton of the general structure and stratigraphy of the peninsula on which all subsequent geological knowledge was built. It was a tremendous piece of reconnaissance geology.

When he got back to Dhahran, however, Steineke found the camp in gloom. Dammam No. 7, the deep test, was not going at all well. By October 1937, plagued with troubles, it was at a depth of 1,005 meters and still showing, in the terse words of a cable sent to San Francisco, "no oil, no water," and on the last day of the year, while drilling at only 58 meters above the depth at which oil was eventually found, it suffered a gas blowout. Drilling continued, but Casoc's worried officials contacted Steineke to ask for a crucial piece of advice: Shall we go on?

Keep drilling, he advised. As it turned out, Steineke's advice was sound, for on March 3, 1938, Dammam No. 7, now down to 1,441 meters in the Arab Formation, began to flow at a rate of 1,585 barrels of oil a day. By March 22, it was flowing 3,810 barrels a day. That fall, the government was formally told that Dammam was a commercially valuable oil field, and Casoc gladly paid the agreed upon advance of £50,000 gold. Saudi Arabia was in the oil business.

Dammam No. 7 had taken 15 months to complete. By the time it was finished, the wildcat at Al Alat had been abandoned as a dry hole at 1,381 meters. Dammam No. 2 and No. 4 were eventually deepened to the Arab Formation and became excellent producers. Lucky No. 7 was tough to drill but it was also a steady and reliable producer. Indeed the well was still capable of producing 1,800 barrels a day when it was taken out of production in 1982 because of slack demand, even *after* turning out a total of more than 32 million barrels in almost 45 years of service.

INTO PRODUCTION: The discovery of oil in commercial quantities in Dammam No. 7 was a turning point. Until then the Saudi Arabian venture had been a risky and costly proposition. Now, as a commercially sound operation, it needed large transfusions of money to expand some facilities and to build others. By 1935, in addition, Socal's production in the Middle East and the Netherlands East Indies (now Indonesia) called for additional marketing facilities. At the same time, the marketing facilities that had been developed by The Texas

Company (now Texaco) were in need of more accessible production. The two companies therefore combined their interests located between Egypt and the Hawaiian Islands into a new company called Caltex and in this way, in 1936, they became half-owners of Casoc. Additional funds thus became available for expansion of the operations in Saudi Arabia.

Casoc now had the go-ahead for more housing, for enlargement of the pier at al-Khobar, for pipelines and storage tanks, for a stabilizer plant to remove poisonous hydrogen sulfide gas from the oil before shipment, for a small refinery, and for a marine terminal for tankers. In September 1938 at Dahl Hith, a few kilometers southeast of Riyadh, Bramkamp and another Aramco pioneer, Tom Barger (who later became the company's chief executive officer), found an outcrop of the same thick anhydrite cap rock that had overlain the producing zone in Dammam No. 7. This confirmed a deduction that Steineke had made. By showing that conditions favorable for the accumulation and entrapment of oil probably existed not just in the Dammam Dome but over most of eastern Saudi Arabia, it justified a considerable expansion of the exploration program. At the end of 1938 the structure drilling program was extended, with a coordinated study of rock cores brought up to the surface when drilling in deep wells.

Dammam No. 7, the discovery well that put Saudi Arabia into the oil business, alone produced over 32 million barrels of oil between 1938 and 1982. It is still capable of producing 1,800 barrels a day without pumping.

In addition the men of Casoc had a special sort of deadline to look forward to in the winter and spring of 1939. They had invited King 'Abd al-'Aziz to come to the east coast to take his first look at the oil installations and, on May 1, watch the first tanker being loaded with Saudi Arabian oil.

The King, they knew, was well pleased with Casoc's success — pleased enough to reject some of the offers that in the wake of Dammam No. 7 had come in from Great Britain, Germany, and Japan, then on the verge of World War II. Still, Casoc's owners had decided it would be wise to be ready with matching offers in case they should be necessary. By January 1939 Casoc was again involved in complicated negotiations affecting the two Neutral Zones between Saudi Arabia and its neighbors to the north, Kuwait and Iraq, and parts of Saudi Arabia outside the exclusive area granted in the original concession agreement of 1933. The King's visit was indeed going to be important.

Floyd Ohliger led King 'Abd al-'Aziz on a tour of oil installations during His Majesty's first visit to the company.

In its continuing effort to gauge and quantify the kingdom's enormous oil reserves, Saudi Aramco exploits the full range of modern exploration techniques. Top, from left to right, advanced satellite imaging is analyzed and more traditional surface surveying is carried out. These help in determining seismic survey areas and the kind of energy to use to "see" underground. For dynamite surveys, used in thick sand areas, shot holes are drilled to receive the charges. In rock desert regions, specially equipped trucks use vibrating pads to generate seismic waves. At bottom, from left, recording cables are arranged and geophones placed in arrays over wide areas to record the seismic waves for storage, processing, and interpretation. In a similar procedure offshore, at right, a seismic vessel tows strings of recording hydrophones over an existing field. High density coverage produces the large volume of digital data required to create the computerized subsurface models, or 3-D "cubes" that translate into the precise simulation models Saudi Aramco requires for reservoir management. Three-dimensional surveys also provide the information required for the correct placement of development wells in structurally complex areas, far right.

The visit was a great success. In the spring of 1939 the King and his retinue moved east from Riyadh in a caravan of 2,000 people in 500 cars, and on April 28 they set up a city of white tents near Dhahran. After two days of banquets and formalities, the King inspected the installations, dined aboard the *D. G. Scofield* — the Socal tanker picked to load the first shipment — and, on May 1, 1939, opened the valve that let the first barrel of oil flow into the first tanker at Ras Tanura. During the festivities the King also made some mild suggestions with respect to the new round of negotiations and personally cleared the way for an amicable resolution of the differences between the negotiating teams. On May 31, 1939, Casoc and the government signed a supplemental agreement extending the northern and southern parts of the concession area westward and also granting the company rights with respect to Saudi Arabian interests in the two Neutral Zones.

The first cargo of Saudi Arabian crude oil was shipped aboard the tanker *D.G. Scofield* in May 1939.

WARTIME: Shortly after the King's visit, war engulfed Europe, and life in Dhahran began to change. At that time, of course, Saudi Arabia and the United States were still neutral. On a moonlit night in 1940, nevertheless, the war came unexpectedly to Casoc when four crews of Italian airmen flew from the island of Rhodes in the Mediterranean and bombs fell on Dhahran.

Years later Aramco learned that the raid was an attempt to knock out the Bahrain Petroleum Company (Bapco) refinery on Bahrain, which was a British protected state, and that the attack on Dhahran was a mistake. Approved by Mussolini personally and led by the secretary of the Fascist party, the raid involved four Savoia-Marchetti S-82s which took off from Rhodes, flew all the way to the Gulf, bombed the refinery in Bahrain, and recrossed the Arabian Peninsula to land in Ethiopia. It was a daring and dangerous flight. En route to the target one of the four pilots lost contact with the rest of the squadron, saw the lights of Dhahran, and, thinking it was Bahrain, dropped his bombs.

On the night of October 19, 1940, the Casoc crews and their families did not know it was a mistake. All they knew was that the quiet desert night was suddenly rocked with the thunder of explosions and that the explosions seemed distressingly close. Miraculously, no

Craters from small "50-pound" bombs remained as evidence of a misdirected air raid in the early days of World War II.

one was hurt and little was damaged, but it reminded Casoc that there was a war on. Blackout restrictions were imposed, air-raid shelters were excavated, first-aid instructions were given to employees, and shortly afterwards American women and children were evacuated — the first group by tanker to Bombay.

Oil had been struck at 3,084 meters at Abu Hadriya in March 1940 and, soon after the bombing of Dhahran, Abqaiq No. 1 came in. At 9,720 barrels of oil per day, it was a whopper. The war at sea was severely limiting shipping, however, and this and wartime shortages slowly strangled oil development. The two new fields were shut in. The 3,000-barrel-a-day "teapot" refinery at Ras Tanura that had been started up in January 1941 was shut down in June. The tank farm and marine terminal lay idle. The most important activities remaining were the production and shipment by barge of from 12,000 to 15,000 barrels of oil per day from al-Khobar to the Bapco refinery in Bahrain.

The skeleton group that remained in Saudi Arabia found plenty to do. In May 1940 the first school for Aramco's Saudi employees was started in the sitting room of Hijji ibn Jasim's home in al-Khobar. Within two months there was a second school and in six months there was a third. Some of the geologists and engineers became involved in the development of irrigation facilities in the area of al-Kharj southeast of Riyadh. At the request of the government a food hauling service to Riyadh was instituted with trucks that became increasingly difficult to keep rolling as the war went on. On one occasion three Hurricane fighters were hauled out of a salt flat where their RAF pilots had made emergency landings. Most of the fuel, cement, and other material used in the drilling of two wells at Abqaiq and a wildcat at Jauf in 1943 was transported by camel caravan. The Jauf well proved dry, unfortunately, and the movement of materials by camel proved to be neither cheap nor efficient.

To ensure their supplies in a food-short Saudi Arabia, the oilmen began growing their own vegetables and raising poultry, rabbits, cows, and sheep. They introduced incubators and experimented with crossbreeding. At one time they had a flock of some 5,000 sheep, and, thanks to an extraordinary Bedouin named Mutlaq, they raised a lot of cattle. On three separate occasions Mutlaq and his son, alone and on foot, rounded up cattle in Yemen and drove them 1,600 kilometers to Dhahran. By breeding them with stock obtained from Bahrain the men in Dhahran built up respectable herds of dairy and meat cattle.

As the war progressed the United States Government became concerned about the rate of oil consumption, and the importance of Saudi Arabia as a source of additional oil became increasingly evident. The United States gave consideration to plans ranging from the purchase of all or part of Casoc to the construction of a pipeline to the Mediterranean in return for guaranteed access to Saudi crude. None of these proposals was carried out, but late in 1943 scarce steel and other materials were officially allocated for the construction at Ras Tanura of a 50,000-barrel-a-day refinery, to be financed by the American owner companies. Trucks and construction equipment were made available, and high priorities were given to the movement of men and supplies.

The project included not only the large refinery, tank farm, and marine terminal at Ras Tanura but also a submarine pipeline to Bahrain, where the Bapco refinery was to be enlarged. While gearing up for all of this, the California Arabian Standard Oil Company on January 31, 1944, changed its name to the more appropriate Arabian American Oil Company, soon shortened to Aramco.

The quiet of Aramco's restricted operations turned into roaring activity. The tumult and confusion of any large construction project far from its base of supplies were intensified by wartime bottlenecks and shortages. A ship carrying urgently needed supplies was torpedoed and sunk. Men arrived before they could be housed. Men and materials arrived in the wrong order.

Construction of the refinery and related facilities such as tanks and marine installations was put in the hands of a number of American construction companies. In December 1945 these firms alone had some 600 Americans working in Ras Tanura, and workmen were recruited also from Iraq, India, and Eritrea.

As difficult as it was to obtain drilling equipment, more oil was needed. Drilling therefore resumed at Abqaiq in 1944, and what turned out to be a successful wildcat was started in Qatif. There were times in 1944 when drilling crews, however, were needed even more desperately elsewhere. For a couple of weeks some of the drillers helped build bunkhouses.

The submarine pipeline to Bahrain went into service in March 1945. Two weeks after the formal terms of Japanese surrender had been signed on September 2 in Tokyo Bay, the first crude still of the new Ras Tanura refinery was fired. With the firing of the second crude still in December, Aramco brought to a conclusion, on schedule, an extraordinary wartime project.

POSTWAR EXPANSION: In September 1945 Aramco posted and celebrated its first 2-million-barrel month, and by the end of 1945 it had loaded 38 tankers at the new terminal and had produced over 21 million barrels of oil. Nearly three times the total production in the previous year, this figure marked the first giant step toward the record achieved in 1976 when Aramco for the first time produced more than 3 *billion* barrels of oil in a single year.

The American wives who left in 1940 were back again. In October 1945 ten children were enrolled in a one-room, American-curriculum school which had been closed for the duration of the war. At the same time, some 170 Arab young men and boys were enrolled in training classes in what was known as "The Jabal School." It was no Phillips Exeter or Eton, but it was the prep school for a number of the company's future Saudi Arab executives. Charley Rocheville, who had accompanied Casoc's Fairchild aircraft to Saudi Arabia, came with another single-engine airplane, a Norseman, to provide support for exploration parties and for pipeline patrols. Aramco had chartered a few C-47s from the U.S. Air Transport Command in 1945, and work was under way to construct a first-class airport on the site of the old Dhahran airstrip.

There was at this time no road across the sands to Abqaiq. One proceeded there across the desert by driving from target to target placed on the highest dunes. To get to Bahrain it took two hours by fast launch and three hours by regular launch. American employees and their families going on leave could catch a flying boat in Bahrain and fly to Cairo via Basra, Lake Habbaniya, and the Dead Sea. Or

The building of a refinery and small tank farm at Ras Tanura was a major undertaking amid wartime restrictions.

Schoolmaster Sam Whipple had 10 pioneer offspring in his one-room school in 1945.

Many of the company's earliest Saudi employees took instruction at what came to be called "The Jabal School."

they could take a ship to Abadan in Iran and then make a memorable three-day flight to New York via Egypt, France, Ireland, and Newfoundland. In January 1946 TWA made the first commercial passenger flight out of Dhahran.

King 'Abd al-'Aziz made his second and last visit to Aramco installations in February 1947. This time he and his entourage traveled in a fleet of the kingdom's own aircraft. He spent the better part of a week entertaining and being entertained. He was obviously pleased by what he saw, and he provided visiting journalists with fascinating stories and attractive photographic opportunities.

The year 1948 was an especially important one in the postwar history of the company because in November of that year arrangements were completed for Standard Oil Company (New Jersey) and Socony-Vacuum Oil Company (subsequently renamed Exxon and Mobil respectively) to acquire shares in the ownership of Aramco. The period of phenomenal growth and expansion that followed in the 1950s and later was thus in part a reflection of the cooperation among four of the major oil companies of the world.

Production was up to 500,000 barrels a day by the end of 1949. This was achieved only with an extraordinary expansion of pumping, pipeline, and treating facilities, not to mention warehouses, shops, offices, and housing. Meanwhile the search for new reserves of oil went on. Drilling in the late 1940s was concentrated in the Abqaiq field, but wildcats also discovered oil at 'Ain Dar and at Haradh. Although the geologists recognized that these two places were in one geological structure, it was several years before they could prove that they were not separate fields but all part of the world's largest oil accumulation, the Ghawar field, nearly 260 kilometers long and 32 kilometers wide at its greatest width.

Since a pipeline shortcut to the Mediterranean would eliminate the long haul by tanker around the Arabian Peninsula and through the Suez Canal, various plans for such a project had been considered during World War II. In July 1945 Aramco's owner companies incorporated the Trans-Arabian Pipe Line Company, a name easily shortened to Tapline, for the purpose of building such a pipeline, the world's largest privately financed

Abqaiq — one day to become the largest crude oil processing center in the world — was little more than an outpost in 1946.

Tapline, crossing some 1,200 kilometers of desert and mountains, cut a much needed route to the Mediterranean coast.

Exploring The Aramco Concession

A 1934 field party — Tom Koch, Max Steineke and their Saudi guide — takes time out for coffee.

The concession agreement signed at the end of May 1933 by the Government of Saudi Arabia and Socal required the company to "commence plans and preparations for geological work" at once and to begin actual field work within four months. The systematic investigation of the sedimentary geology of Saudi Arabia accordingly began in September 1933 with the mapping of the dome in the Dammam area that oilmen on Bahrain had seen and speculated about. S. B. (Krug) Henry and J. W. (Soak) Hoover did the initial work and they were later joined by others, including Max Steineke, who arrived in 1934 and two years later was named the company's chief geologist. As more men and equipment arrived, the pace of geological reconnaissance increased, expanding to nine field parties by 1937 and eventually radiating out across the desert until by 1959 roughly 1.3 million square kilometers of sedimentary outcrop had been surveyed.

Some of the early findings, however, seemed to show little or no correlation between surface features and what lay beneath them, and geologists were perplexed until Steineke identified the problem. Because percolating groundwater had dissolved and washed away a calcium sulfate (anhydrite) member of one of the formations underlying much of the original exploration area, the rocks above it had collapsed in some places and the consequent slumping of surface topography gave a false picture of the subsurface configurations. To get around, or below, this problem Steineke late in 1936 instituted a program of structure drilling — that is, drilling not for oil but for information that would help locate structures containing oil. Careful study of the rock samples taken from these structure holes and their correlation with the results of surface geological work eventually gave a more positive idea of the continuity, depth, and flexure of the strata.

Observation and mapping had shown that the main structural features of eastern Arabia trended north and south. Structure holes, spaced usually half a degree of latitude apart, were therefore drilled in lines running east and west. It was reasoned that in this way the greatest number of structures and the biggest structures (oil-bearing, it was hoped) would be defined most quickly and most economically. In the beginning the holes were shallow, just deep enough to penetrate below where the dissolved anhydrite had been washed away, but as the drills moved westward to central Saudi Arabia and southward to the Rub' al-Khali it became necessary to drill to other deeper horizons (eventually reaching 915 meters) to obtain a true structural picture. Today, looking back, it can be said that it was a combination of structure drilling and surface mapping that has identified the structures containing the bulk of the oil found to date in Saudi Arabia.

In 1953 the company moved on to stratigraphic drilling (probing for information on the composition and sequence of rock beds). In the stratigraphic program fewer wells were drilled and they were drilled deeper — to depths of from 1,525 to 3,050 meters — to include the deep-lying and possibly oil-bearing formations. At first spaced mainly around the periphery of the concession area, they were later spotted at selected locations within the company's operating area. The lithologic information obtained from these holes was basic in making concession-wide correlations of strata, in recognizing changes in rock composition over broad areas, and in pinpointing places where strata either had been removed or were never deposited.

Along with the direct means of obtaining information in all types of exploratory drilling, that is, by examining and testing rock samples brought to the surface, geologists also obtain information indirectly, by running wire-line logs. With testing devices lowered into the hole, this procedure measures and "logs" such data as the electric current within the earth, the earth's magnetic field, or the naturally occurring radioactivity inherent in all sediments. Logs make it possible to infer the presence of particular rocks and to calculate such information as porosity and the nature of reservoir fluids.

To help them reconstruct the history of the earth, geologists — and more especially paleontologists — identify the fossilized remains of tiny animals (often consisting of only a single cell) that yield valuable clues to the environment in which rock beds were deposited, and hence provide a clue to their age. In the early 1960s, Aramco paleontological studies were expanded to include the then new science of palynology, the study of spores and pollen of the earliest known plants. All these investigations served to reinforce the analyses of data obtained by other means, which all together helped determine which areas warranted concentrated oil exploration.

Two additional techniques that the early explorers put to use were employed in a survey that began along the coast in 1939 and 20 years later had covered most of the concession area. One was the gravity meter survey, which measures from point to point the minute variations in gravitational pull caused by differences in either the depths of rocks or their relative densities. The other was the magnetometer survey, which detects variations in the earth's magnetic field. Relatively inexpensive, the gravity-magnetic approach provides limited information that must be supplemented by other geophysical means, and today it is used primarily in attempts to define and refine the deep-seated aspects of certain structural anomalies.

Other geophysical techniques with an even longer history have in recent years come to account for a major share of the company's exploration efforts. Since 1937 Aramco has used seismic exploration, which measures the speed of artificially induced shock waves in the earth and by using these measurements infers the presence of subsurface structures. Programs have alternated between the reflection method, which is more precise, and the refraction method, which provides rapid coverage and can accurately detect any large-scale structural features.

An especially ambitious part of the seismic program got under way in 1958 when Aramco embarked on a refraction survey that eventually resulted in a network blanketing the entire concession area. By integrating this work with results achieved by both drilling and surface reconnaissance, the company was able to outline the structural framework and establish the stratigraphic succession of the entire area in which it operated.

By the time this refraction survey was completed, the largest surface structural features had been drilled. Of necessity the oil search moved on to the smaller, deeper, and more subtle expressions of structure, and to investigate these hard-to-find flexures Aramco turned to high-resolution reflection seismic prospecting. Using conventional energy sources as well as weight-dropping techniques, field parties surveyed tens of thousands of line kilometers, recording the data on multichannel computer tapes. Such information is processed by computer to produce subsurface contour maps that help mark the structural highs that are the object of the search — maps which would require innumerable manhours to produce by conventional means.

Because of the vast size of the area to be covered and because there were so few commanding elevations from which to get a broad view, an airplane equipped with an aerial camera was essential to even the earliest exploration of eastern Saudi Arabia. Beginning with the arrival of Dick Kerr and his Fairchild 71 in March 1934, photogeology has been used extensively, for from the superior vantage point of the cockpit explorers could pinpoint rock exposures for ground investigation, photograph surface structures, and from the overlapping photos piece together bigger and more accurate maps. Between 1949 and 1958 most of the concession area was mapped on a 1:60,000 scale. Using these photos to tie together and reconcile various ground surveys, Aramco then produced a series of geologic and geographic quadrangle maps of the sedimentary rocks of Saudi Arabia in a project which culminated in 1963 with the publication by the United States Geological Survey of English and Arabic editions of a 1:2,000,000 map of the peninsula with supporting publications.

This mapping project represented a cooperation among government, oil company, and private individuals that was unprecedented and it turned the Arabian Peninsula, virtually unknown to the explorers who had arrived 30 years before, into one of the better-mapped areas of the world.

Working closely with specialists in exploration and petroleum engineering, Saudi Aramco drilling teams prospect for oil in an area covering more than 1.5 million square kilometers of land and sea. On the facing page, Saudi Aramco's mammoth Drilling Platform 3 at work in the Arabian Gulf. Clockwise from top left, this page: Mobile satellite communications technology is used to transmit data or receive feedback; a drilling rig probes the earth at Farwan in the northwestern region of the kingdom following a hydrocarbon discovery nearby in 1991; core analyses are routinely conducted by electron microscope at advanced laboratories in Dhahran; drilling is accomplished in both traditional vertical wells or horizontal wells that make possible impressive gains in reservoir recovery factors under certain conditions; drillers work year-round, through harsh summertime temperatures; and valves and piping at Saudi Aramco's first horizontal well, completed in the offshore Berri field in 1991. New fields continued to be found in the 1980s and 1990s, with 15 fields, many south of the capital city of Riyadh, discovered between 1989 and 1994 alone.

construction project up to that time. As originally conceived and surveyed, the line would have followed a great circle route from Abqaiq to Haifa in the British mandate of Palestine.

Political tension and later outright war between the Arab states and Israel caused rerouting of the line and delayed the construction program. Though it was first planned to begin construction at both ends at about the same time, the war delayed work on the western end until the autumn of 1949. In early 1948 construction was begun of an Aramco pipeline from Abqaiq to Qatif Junction and thence to Qaisumah, which would be the actual starting point of Tapline, and work proceeded westward from there, operating from a port and camp built for the purpose at Ras al-Mish'ab, some 260 kilometers north of Dhahran. Later work began from the western end as well, and the final weld connecting the western and eastern sections was made on September 2, 1950. A few months later, the first tanker was loaded with Saudi Arabian crude at the four-berth western terminus of the pipeline a few kilometers south of the ancient city of Sidon in Lebanon.

Nearly half of Tapline was constructed above ground, reducing its cost and facilitating the job of inspection crews.

Towns grew up around the four main pump stations in Saudi Arabia, and the road that parallels the pipeline became an important Middle Eastern transportation artery. The main Tapline pump stations located at Qaisumah, Rafha, Badanah, and Turaif, all in Saudi Arabia, provided an initial capacity of 320,000 barrels daily. In 1957 the capacity was increased to 450,000 barrels daily by installation of auxiliary pumping units midway between the main stations.

Tapline continued to be a major outlet for Aramco's west-bound crude exports until the early 1970s, when the development of ultra large tankers began to give an economic advantage, even for many European destinations, to shipments by sea. Finally, after major civil disturbances in Lebanon and after the pipeline was damaged during the Israeli invasion of Lebanon in 1982, the parts of the pipeline and Tapline's other immovable assets in Syria and Lebanon were abandoned effective December 31, 1983. Tapline in 1984 continued oil deliveries at greatly reduced rates as far as Jordan, supplying the Jordanian refinery at Zerqa. In 1990 Tapline's assets in the kingdom were transferred to Saudi Aramco, of which it became a division. Deliveries to

A self-propelled Budd car traveled the new railroad link transporting passengers between Riyadh and Dammam.

Jordan were suspended that same year.

Even before Tapline was completed Aramco had become responsible for construction of another big project, the 580-kilometer Saudi Government Railroad between Dammam on the Arabian Gulf and the capital city of Riyadh. Part of this project consisted of building a deep-water port at Dammam. For this an 11-kilometer causeway and trestle was pushed out into the Gulf, ending in a two-berth wharf. The government has enlarged the port several times over the years to provide berthing facilities for 40 ships.

Aramco did not actually build the railroad and port, which were paid for by deductions from royalties due the government. The company contracted the job to a consortium of American engineering firms, but supervised the project and saw that it was finished almost on schedule. In October 1951 King 'Abd al-'Aziz presided over ceremonies in Riyadh featuring the driving of the traditional golden spike into the final tie, and a new diesel locomotive pulled into the Riyadh station to inaugurate the new service. In 1985 the General Saudi Railways Organization, a government-owned body, opened a new dual line between Dammam and Riyadh (via Hofuf), shortening the distance between the two cities by 20 percent by avoiding the loop to Haradh to the south.

CONSOLIDATION: Once the boom period of postwar construction and expansion was ended, Aramco settled down to a relatively quiet period of consolidation and reorganization. The company's manpower — swollen by the temporary forces needed to construct new oil installations as well as new offices, houses, schools, roads, hospitals, and other support facilities — reached a peak of 24,000 in 1952 and then began to decline.

This is not to imply that Aramco lapsed once more into the inactivity of the wartime years. On the contrary the company's production of crude oil and refined products, once the producing and shipping facilities were in place, climbed steadily, and during each year from 1945 through 1974 the amount of oil produced by Aramco increased at an average rate of about 19 percent. Average daily crude production, which was running at about 500,000 barrels per day (BPD) at the start of 1950, doubled by 1958, doubled again seven years later, doubled once more to 4,000,000 BPD by the end of 1970, and by 1974 it had doubled again. Production peaked at 9,631,366 BPD in 1980.

The company's cumulative total production had reached 10 billion barrels by 1968; during 1974 it passed 20 billion; and in 1978, 40 years after oil was discovered in commercial quantities, the total exceeded 30 billion barrels. Ten years later cumulative production had nearly doubled again, reaching 56 billion barrels. Production of refined products also climbed steadily. In spite of this enormous production, discoveries continued to result in annual increases in the kingdom's crude oil reserves. Refinery runs in 1947 (the first full year of operation after the expansion of the refinery begun during the last years of the war) were slightly over 100,000 BPD and reached a peak of over 500,000 BPD in the early 1970s.

Nor did the company continue to limit its production to crude oil and a basic slate of refined products such as fuel oil, diesel oil, kerosene, and gasoline. Beginning in the early 1960s it became increasingly involved in the production, processing, and export of natural gas liquids (NGL). NGL includes liquefied petroleum gas (LPG), composed of butane and propane, and natural gasoline.

Aramco had been selling relatively small quantities of LPG in Saudi Arabia for cooking and heating since 1950. This LPG was extracted from refinery process gases. In 1960, however, the company took a step that would eventually bring it into the LPG business in a big way. A plant was constructed in Abqaiq to separate the liquid component of what is called "associated gas," the gas produced along with crude oil. The original idea was to inject these liquids (which were not otherwise economically usable at the time) into the 'Ain Dar field as a means of maintaining the field's reservoir pressure and as a simple conservation measure. Additional studies and cost analysis indicated that while this was a good method of maintaining reservoir pressure, it was economically even more desirable to sell this product than to inject it. After construction of suitable plants this raw LPG was removed at Abqaiq and then sent by pipeline to Ras Tanura for further processing, storage in refrigerated tanks, and sale. In December 1961 the first shipment of LPG was loaded into a specially designed tanker. It amounted to 50,000 barrels, the

equivalent of 2.1 million cubic meters of gas.

Demand for LPG climbed so rapidly that Aramco's facilities, designed initially to process 3,400 barrels of LPG a day, were expanded again and again. Starting in September 1967 the natural gasoline portion (that is, the pentane and heavier hydrocarbon fractions) of the associated gas was removed at new facilities in Abqaiq and shipped to Ras Tanura for blending and sale with LPG. Production of NGL rose from 2,900 BPD in 1962 (the first full year of processing) to over 300,000 BPD in 1979, thus making Aramco the world's largest producer of NGL.

In 1960 most of the tankers leaving Ras Tanura and Sidon with Aramco crude and products were sailing for Europe, which in that year took 42 percent of the company's total exports. Asia and Australia together took the next largest share, or 37 percent. These proportions remained roughly the same into the mid-1970s when North America began to take a growing export share. Loadings for North America accounted for more than 19 percent of the total in 1979, reflecting the increased dependence of the United States on foreign oil. In 1993 North America was taking 19 percent and Europe 24 percent while the share of Asia, with its burgeoning Pacific Rim economy, had reached 53 percent.

Maintaining the enormous flow of oil which made Aramco the largest producing company in the world required a constant search for new oil, and the company kept up an active exploration and development program. Even during the quiet days of World War II, when production slowed to a trickle and the work force was reduced to a low of 1,825, exploration continued. Both the very large Abqaiq oil field and the smaller Abu Hadriya field were discovered in 1940, and the Qatif field was found in 1945. The search for new sources of oil continued at an accelerated pace after the war. 'Ain Dar, even bigger than Abqaiq and eventually proved to be part of the mammoth Ghawar field, was discovered in 1948; and Safaniya, the largest offshore oil field in the world, was found in 1951. Saudi Aramco has enjoyed consistent success in exploration, as did its predecessor, Aramco.

As production continued to grow after World War II, new and enlarged facilities for storing and shipping oil were required. Facilities serving Aramco's three outlets — the pipeline to Bahrain, Tapline, and tanker-

The sub-zero temperatures of refrigerated LPG frost loading lines at Ras Tanura.

The first Sea Island berths, built to accommodate supertankers, went into service at Ras Tanura in 1966.

The discovery of Safaniya, the world's largest offshore oil field, added tremendously to Saudi Arabia's reserves.

Saudi Arabia is the world's largest exporter of crude oil and natural gas liquids, and Saudi Aramco's marine terminals are among the busiest anywhere. Mammoth crude carriers and many smaller NGL and other tankers — some from the fleet of Vela International Marine Limited, a Saudi Aramco subsidiary — call at Ju'aymah, Ras Tanura, and Yanbu' every day, following weeks of planning by Saudi Aramco offices at home and abroad to coordinate arrivals. Clockwise, from right: Chiksan arms designed to load different products simultaneously help accelerate loading times at all terminals; Saudi Aramco's newest loading terminal, at Yanbu' on the Red Sea; the 10-kilometer trestle at Ju'aymah on the Arabian Gulf coast carries butane and propane to a two-berth terminal; an operator helps coordinate an NGL loading operation; a full complement of tankers takes on crude at Ras Tanura's eight-berth Sea Island; a harbor pilot brings in a tanker to load; an NGL tanker takes on a cargo of liquefied petroleum gas; and an aerial view of Ras Tanura's giant tank farm, where crude, refined products, and NGL are stored for export. There are several 1.25 million-barrel floating-roof crude oil tanks at Ras Tanura, while nearby Ju'aymah has tanks holding up to 1.5 million barrels. The dome-roofed tanks hold NGL.

loading terminals at Ras Tanura on the Arabian Gulf — were successively enlarged and improved. The pipeline to Bahrain (the destination for Aramco's first oil shipments back in the 1930s) was "looped" or doubled for part of its distance in 1948 and again in 1952.

After the end of WWII, however, the major outlet for the oil produced by Aramco was shipment by tanker in the Gulf, and facilities for tanker loading were repeatedly enlarged and added to in order to keep pace with rising production. The marine terminal at Ras Tanura started with a single pier capable of docking two small tankers. It grew to two piers with a total of 10 berths and then to a major oil port able to load giant tankers. To provide berths for the bigger ships Aramco built structures known as Sea Islands. Four Sea Islands have been built and provide berths and loading facilities for eight tankers ranging in size from 60,000 to 500,000 deadweight tons.

As tankers continued to grow in size, and as oil production continued to increase, even the Sea Island facilities were not enough to handle Aramco's export shipments, and other solutions were soon found to be needed. One such solution, a departure from any oil-handling procedure Aramco had previously followed, was implemented in connection with the production of the offshore Zuluf field.

In 1969 Aramco had built a gas-oil separator plant (GOSP) offshore to separate gas from crude oil produced from the Safaniya field. This plant, however, merely supplemented the principal gas-oil separation facilities for that field, which were on land. Now, for the Zuluf field, Aramco carried out the entire gas-oil separation process offshore. Crude oil from Zuluf wells flowed into a plant mounted on pilings in the Gulf for removal of the gas, and was then sent on through underwater lines to a 1.8 million-barrel capacity floating storage vessel.

This vessel, actually an oil tanker named the *F. A. Davies* after the former chairman of Aramco's Board of Directors, received the oil and then pumped it to tankers being loaded 1½ kilometers away. Both tankers were meanwhile free to swing about into the wind while receiving and discharging crude through single-buoy moorings. With the *F. A. Davies* Aramco had two tanker outlets in the Gulf. But with production soaring and both outlets working to

capacity, it realized that Saudi Arabia would have to have another new terminal of considerable size.

About 25 kilometers up the coast from Aramco's original oil terminal the company therefore built a new terminal, Ju'aymah, to rival Ras Tanura in size and importance. The most conspicuous installations on land at Ju'aymah are fourteen 1.25 million-barrel storage tanks and five 1.5 million-barrel tanks, each of them big enough to accommodate a 30-story building on its side. Other facilities include 142-centimeter loading lines reaching 11 kilometers out into the water to a platform mounted on piles where meters gauge the crude before it is loaded aboard tankers.

The Ju'aymah terminal went into operation in November 1974, loading crude aboard a tanker tied up at a single-buoy mooring of the same type introduced at Zuluf. Ju'aymah terminal was built to load the largest tankers afloat. It thus complements Ras Tanura's Sea Island facilities where tanker loading is limited due to channel depth. Giant tankers may take on partial loads at Ras Tanura and then "top off" their cargo tanks at Ju'aymah, which can load oil at six moorings located at least 1½ kilometers from each other. The terminal facilities at Ras Tanura and Ju'aymah combined can load more than 10 million barrels of crude oil per day. The *F. A. Davies* was taken out of service at the beginning of 1976.

CONSERVATION MEASURES: Oil production from most of the company's fields makes use of the natural pressure of oil and gas within the oil-bearing reservoirs to flow the oil and gas through the well bores to the surface and to move it through flowline gathering systems to plants where the oil and gas are separated. As production continues, pressure within the reservoir naturally declines, and unless something is done to maintain the pressure, flowing production could cease before even a small percentage of the oil in the ground has been removed.

In many fields, good oil field practice, which aims at producing in the long run as much oil from the reservoirs as economically possible, requires that this pressure be maintained. This is usually achieved by injecting either gas or water into the oil reservoir. Aramco's injection programs date back to the early 1950s. In 1954 the company's first gas

The tanker *F.A. Davies* served as a vessel for offshore storage until 1976.

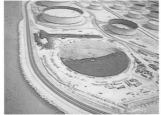

Tanks holding up to 1.5 million barrels each were built to support shipping facilities at Ju'aymah.

Production from the Zuluf field called for an offshore GOSP and later a second which, at the time it was built, ranked among the largest of its type anywhere.

injection plant was completed as part of a pressure maintenance program for the Abqaiq field. Dissolved gas produced with oil was separated from the oil, compressed to a pressure of about 172 bars (2,500 pounds per square inch) and returned to the oil-bearing formations through injection wells in the higher part of the structure. A second gas injection plant was installed at 'Ain Dar in 1959.

During the development of Abqaiq and Ghawar, Aramco decided that the most efficient means of maintaining reservoir pressure in these fields would be through injection of water into the flanks of the producing reservoirs. Water injection can help sustain higher producing rates as well as increase ultimate oil recovery. Water pumped into subsurface producing reservoirs maintains pressure by moving in behind the oil and gas that are produced. At the same time water acts to sweep the oil ahead of it toward the centrally located producing wells.

To supplement gas injection at Abqaiq, the injection of water in the northern end of the field was begun in 1956. By the end of the year about 40,000 barrels a day were being injected through three water wells and company engineers were designing additional facilities to bring the daily rate up to 300,000 barrels. Since 1956 the company's water injection programs have expanded considerably. The quantities of water injected increased dramatically during the 1960s and early 1970s, eventually amounting to some 1.1 million barrels a day. The sources of injection water were the aquifers of the Wasia and Biyadh formations, which overlie the oil fields of the Eastern Province and in this area are too saline for domestic or agricultural use.

To lessen the dependence on aquifer water, plans were made to utilize seawater in its place, and construction began in 1976 of a giant seawater intake and treatment plant on the shore near Qurayyah, east of Abqaiq. This installation initially was designed to supply 4.2 million barrels a day of treated seawater for injection into the central portion of the Ghawar oil field. Its capacity was increased to 5.1 million barrels a day in 1994. Seawater taken from the Gulf is treated and pumped 98 kilometers through three large-diameter pipelines to water injection systems replacing and augmenting the one previously used to inject acquifer water.

GROWTH IN THE 1970s: As Aramco moved into the 1970s it soon found itself faced with a demand for oil which called for expansion that would dwarf the efforts of its earlier years. In some ways the period was reminiscent of the years following World War II when, with both men and materials once again available, Aramco was able to get down to the serious business of defining and developing Saudi Arabia's vast oil resources. In 1947 the company produced less than 250,000 barrels of crude per day, drawing from just three of the four fields it had discovered, yet it stood on the threshold of a period of growth involving projects of unforeseen size and complexity that ultimately made it the number one oil-producing company in the world. Thirty years later, with the oil fields in its concession area — each one major to giant by world standards — numbering over three dozen, Aramco's annual production of crude for the first time exceeded an average of 9 million barrels a day; moreover, it had been transformed from a company whose primary interest was oil to one engaged in a variety of other industrial activities.

The postwar development of the 1940s and 1950s, if exciting and challenging to those involved, was steady, predictable, and by and large undramatic, and it went forward with the expectation that huge reserves meant steady, sure growth. Going into the 1970s Aramco was producing less than 3 million barrels a day, not even one-third of its crude production eight years later. World events altered that situation drastically, however. In quick succession the Western world enacted strict conservation laws, encountered the energy crisis, and felt the first pangs of a fuel shortage; and other top producing countries imposed production cutbacks just at the moment when demands from Western Europe, Japan, and the United States were skyrocketing. It was time for a sweeping reassessment of the world petroleum picture, and for Aramco the results indicated that it must expand fast.

With an enthusiasm not seen since its pioneer days, Aramco went to work hiring specialists in fields ranging from reservoir engineering to computer programming, making and revising plans, estimating and reestimating budgets as expansion became the byword of every department. Activity accelerated dramatically as the decade got under way: between 1972 and the end of 1975 Aramco built more than 1,250 kilometers of major pipelines, drilled some 1,000 deep wells

Installations at 'Ain Dar compress gas to be injected back into the reservoir for pressure maintenance.

Seawater treated at the Qurayyah plant replaces a significant amount of non-potable saline aquifer water formerly used for injection into areas of the Ghawar field.

With the consolidation of all facets of exploration, production, and reservoir management into the Exploration and Petroleum Engineering Center (EXPEC) in 1983, Saudi Aramco became a world leader in advanced earth sciences. Today, advanced technology enhances every facet of Saudi Aramco operations. This page, clockwise from top left: a specialist in corrosion/electrochemistry prepares a sample for analysis; facilities planning engineers review a pipeline simulation model; scientists use an isotope ratio mass spectrometer to distinguish different oil sources; a technologist evaluates corrosion inhibitors; and a petroleum engineer manipulates a deviated well plot. Facing page, clockwise from top left: a petroleum engineering systems analyst views a 3-D image of an oil reservoir; a technologist checks the chemical properties of reservoir injection water; a lab scientist prepares a "biocide loop" to study oil-field water injection operations; and scientists prepare a catalyst sample prior to viewing on the scanning electron microscope.

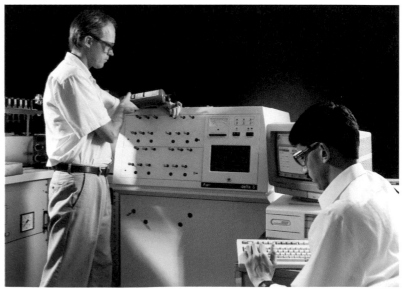

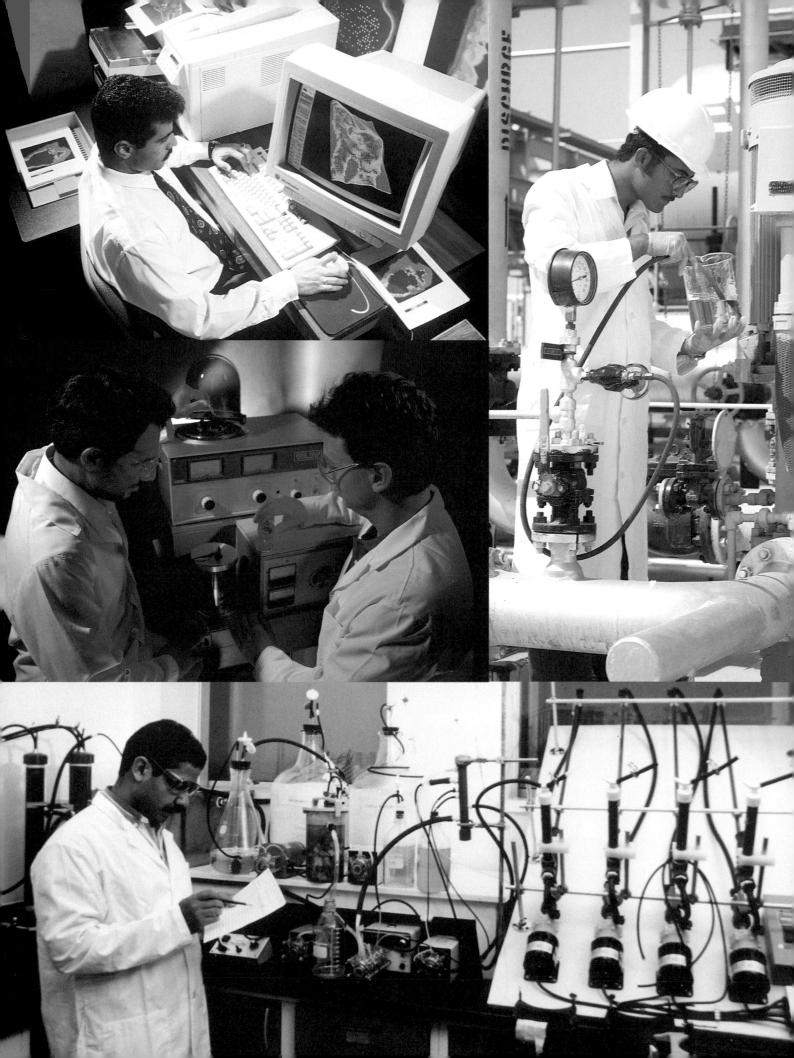

(four times the number drilled in the previous four years), built 24 GOSPs, and in increment after increment added more than a billion dollars worth of new turbines, generators, and stabilizing columns to its physical plant. At Ras Tanura the first-anywhere 1.25 million-barrel storage tanks went up, the fourth Sea Island went into operation, and in 1974 alone, the year the new Ju'aymah terminal opened, 4,470 tankers were loaded with the aid of new crude-loading systems, new and bigger loading arms, and record-sized pumps.

Data-processing systems were expanded to make them faster and more accessible to users.

Production jumped by roughly 25 percent in 1971, again in 1972, and again in 1973. To keep pace with this growth and the continuing rise in world demand for oil, the company enlarged and strengthened its exploration program and carried on an aggressive search for new oil reserves. Two seismograph crews were operating as the decade began; five years later there were eight, including one using Hovercraft for survey work in shallow marine areas. Counting the personnel involved in technical support and computer processing and interpreting data from the field, each of these crews added up to about 500 men.

As new manpower began pouring in, the demand for both living and office space soon outstripped what was available. A work force that totaled roughly 10,000 in 1970 was essentially double that by the end of 1975 and was six times as big by the end of the decade. By that time, furthermore, the company's own work force was augmented by the numbers working for scores of local contractors (many of them partners of giant international engineering and construction firms) who were working on Aramco's own projects as well as on projects that Aramco was supervising and carrying out on behalf of the Government of Saudi Arabia. With about 60,000 Aramco employees supplemented by another 20,000 contractor employees, it was as if the entire working population of a good-sized city had gone to work on a single huge project.

Storage tanks holding more than a million barrels each and dome-roofed LPG tanks boosted capacity at the Ras Tanura tank farm during the growth period of the 1970s.

Retractable and highly maneuverable Chiksan arms streamline loading procedures at Saudi Aramco shipping terminals.

To accommodate these masses of people, Aramco found it necessary to build and operate construction-worker camps at eight sites in the Eastern Province, and the overflow was housed in three- and five-story barges anchored offshore. In Aramco's own residential communities of Abqaiq, Dhahran, and Ras Tanura, two-man trailers, prefabri-

Oil Operations

The oil produced by Saudi Aramco is a complex mixture of organic compounds known as hydrocarbons, most of which are normally liquid but which are often accompanied by some gaseous fractions that remain dissolved in the liquids as long as they are kept under pressure in the ground. The mixture is different in each field and each producing reservoir — not all crude oil is the same. Some is "light," with a low specific gravity; some is thick and viscous. Some is "sour," with a high content of dissolved hydrogen sulfide, and some is almost hydrogen sulfide-free or "sweet." The different types of crude require different kinds of handling and processing, and Saudi Aramco groups them into five standard "grades" known as Arabian Heavy, Arabian Medium, Arabian Light, Arabian Extra Light, and Arabian Super Light. Market demand, capital and operating costs, conservation considerations, and availability of manpower and facilities determine how much of any particular grade is produced in any given period. To achieve a proper balance of all these factors calls for a high level of skill in both long-term planning and short-term scheduling.

Saudi Aramco operates today in an area totaling over 1.5 million square kilometers, exceeding the combined areas of Texas, California, Oklahoma, and Utah, or of France, Spain, and Germany. From its northernmost tip to its farthest southward extension in the Rub' al-Khali is a distance of 2,100 kilometers. This exploration and production area includes all the sedimentary terrain of Saudi Arabia as well as the kingdom's offshore areas in the Gulf and the Red Sea. The bulk of the company's production comes from fields in the coastal plain of the Eastern Province in an area extending 300 kilometers north and south of Dammam, a region that includes both onshore fields up to some 160 kilometers inland and offshore fields in the Gulf. Important new fields, however, came on stream in 1994 in central Arabia, south of Riyadh. This region of newer discoveries produces the company's most valuable crude grade — Arabian Super Light.

The Saudi Aramco operating area at the beginning of 1995 was estimated to hold some 259 billion barrels of recoverable crude oil reserves. Saudi Arabia thus has the largest reserves in the world. The kingdom is also the world's largest exporter of oil and Saudi Aramco is responsible for 97 percent of its production. In 1994 the company produced nearly 3 billion barrels, at rates of about 8 million BPD. This was accomplished with fewer than 1,500 producing wells, with the average well flowing over 5,300 BPD. In contrast, the United States required some 590,000 wells to produce 6.6 million BPD, with the average U.S. well producing only about 11 BPD.

In nearly all Saudi Aramco fields, the wells flow freely to the surface because of the expansion of the dissolved gas which is trapped under pressure with the crude oil in

216

the reservoir. When the reservoir is produced, the oil flows to the surface together with this dissolved or "associated" gas, while water underlying the oil displaces it. As the oil rises up the well bore, pressure decreases and gas starts to separate out of the oil, much as gas bubbles out of a bottle of carbonated beverage when it is opened.

As oil and the gas associated with it are produced, the pressure in the reservoir naturally declines. To maintain this pressure Saudi Aramco injects large quantities of non-potable water at the edges of the reservoir to supplement the influx of water surrounding the oil. A large part of the water used for this purpose is drawn from the Gulf, while the rest comes from saline underground aquifers. The company injects some 10 million barrels of water daily into selected oil fields. The seawater intake and treatment plant on the coast south of Dhahran at Qurayyah is by far the largest such plant in the world.

Production wells are drilled at carefully planned intervals, so that oil and its associated gas from one field may reach the surface at wellheads spaced over an area of many square kilometers. A cluster of such wellheads is connected to a central point by a spoke-like "gathering system" of relatively small-diameter pipelines called flowlines. The mixture of oil and gas travels to the central point where the two are physically separated in a series of vessels or traps at a gas-oil separator plant or GOSP. Two or more flowlines are sometimes joined into a

More than 5 million barrels of seawater can be treated daily at Qurayyah.

single "trunk line" before reaching the GOSP. This is especially true in offshore fields. Saudi Aramco has built more than 8,000 kilometers of oil flowlines and trunk lines.

GOSPs consist primarily of a series of large steel vessels designed to release the dissolved gas from crude oil in gradual stages at successively lower pressures. This is basically a physical process in which the gas, which is lighter, collects at the top of the vessel, still carrying a very small amount of oil, while gravity causes most of the heavier oil to collect at the bottom. Depending on the initial pressure — which as the oil comes from the well may be as high as 35 bars (500 pounds per square inch) — pressure is reduced in two or three stages to about 3.5 bars (50 pounds per square inch). At this pressure the crude still contains some dissolved gas when it leaves the GOSP to be pumped to downstream processing plants. The gas which has been separated out is also collected at the GOSP.

Pipelines stretch across the sands moving oil and gas to and from Abqaiq.

Saudi Aramco pipes the gas to downstream centers for further processing and eventual sale. The Master Gas System provided the major installations needed to collect and process most of the gas from onshore fields. The system was later extended with links to virtually all onshore and offshore fields with significant gas production. As part of the gas program, gas compression facilities were built at many GOSPs. In addition, wet crude-handling facilities — not a part of the gas system — were included at a number of GOSPs to ensure that the crude oil meets required specifications for water and salt content. With such additions, many GOSPs became sizable plants in themselves. At the beginning of 1995 Saudi Aramco had 55 GOSPs, both onshore and offshore.

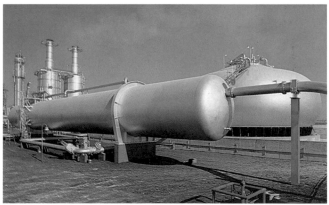
Gas-oil separator plants numbered 55 by 1995.

Saudi Aramco's oil operations are divided geographically into two areas. The southern area, which includes the world's largest oil field and the company's two largest gas plants, is administered from Abqaiq, a major processing center for crude oil and NGL located about 65 kilometers southwest of Dhahran. The company's

(Continued on page 218)

Oil Operations

fields in the central Arabia region, south of Riyadh, are also attached administratively to the southern area. The northern area is administered from Ras Tanura, about 65 kilometers north of Dhahran, the site of an oil refinery, liquefied petroleum gas (LPG) plants, and marine terminals handling both crude oil and oil and gas products. Crude oil from the largest fields in the southern area is piped from GOSPs in the area to Abqaiq for further processing, while most of that from GOSPs in the northern fields proceeds directly to Ras Tanura. Arabian Super Light production from the central Arabia fields is linked to the East-West Crude Oil Pipeline system and pumped west in batches for export at Yanbu' on the Red Sea. Where compression facilities have been added to a GOSP, "raw" gas condensate — composed of all the various readily condensable hydrocarbons present in the separated gases and some impurities such as hydrogen sulfide — is also piped downstream to processing plants.

The pipelines which carry the oil, gas, or gas condensate from the GOSPs are large compared with the flowlines of the gathering systems and range in diameter from 15 to 142 centimeters. Overall, Saudi Aramco operates over 20,000 kilometers of hydrocarbon pipelines.

Crude oil leaving Saudi Aramco GOSPs, with the exception of that from the offshore Safaniya and Zuluf fields in the north and from the central Arabia fields, is sour crude. That is, it contains toxic hydrogen sulfide, which in gaseous form is poisonous and when dissolved in water forms a highly corrosive acid. Sour crude can be shipped by pipeline. It can also be used as refinery feedstock, because the hydrogen sulfide can be removed as part of the refinery operation. It is obviously undesirable, however, to transport

The NGL stream reaching Ras Tanura for fractionation becomes LPG and naphtha for export.

sour crude on board tankers, and before it can be safely shipped it must be "sweetened." This is done by a process called stabilization, which consists of partial distillation in the course of which the hydrogen sulfide gas is boiled off.

Sour crude from the southern area is stabilized at Abqaiq, which handles nearly 70 percent of all crude produced by Saudi Aramco. That portion of the northern area crude that requires stabilization is processed at Ras Tanura. First, however, a final stage of separation brings the crude from the pressure of 3.5 bars (50 pounds per square

inch) at which it leaves most GOSPs down almost to atmospheric pressure. This is accomplished in large spherical vessels called spheroids, operating at 0.2-0.3 bars (three to four pounds per square inch), which release most of the remaining dissolved gases.

In the stabilizing process, sour crude containing 350-600 parts per million of poisonous hydrogen sulfide is sweetened to 10 parts per million or less by heating it and passing it downward over a series of trays in the stabilizer

Nearly 70 percent of all crude oil produced by the company is gathered and sweetened at Abqaiq.

column. After cooling, the sweetened crude is pumped to storage tank farms at the Ju'aymah and Ras Tanura marine terminals, to refineries in central and western Saudi Arabia, or to the Red Sea terminal at Yanbu' for export. The Ras Tanura Refinery and the Bahrain Petroleum Company refinery on Bahrain Island, which processes some of Saudi Aramco's oil, also use some of the crude as feedstock.

Spheroid gas, gas separated from crude during the stabilization process, and gas piped from the GOSPs can be compressed to a pressure of about 28 bars (400 pounds per square inch) then cooled and liquefied to recover additional natural gas condensate. The condensate is then stabilized by feeding it into "stripper" columns to remove the light gases such as ethane and methane as well as hydrogen sulfide. This stripping is done at a plant at Berri, near the industrial complex at Jubail in the northern producing area, at gas plants in the southern area, and at Abqaiq. The natural gas condensate, now in the form of stabilized NGL, is mixed with the NGL recovered from the gas in the gas plants and moved on by pipeline to fractionation plants for final processing.

Saudi Aramco's refineries each consist of a number of plants that break oil down into such products as jet fuel, gasoline, kerosene, naphtha, and diesel oil. Other plants remove the last traces of sulfur compounds from stabilized NGL and fractionate it into ethane gas and three main liquid components. The two lightest liquids, propane and butane, are commonly referred to as liquefied petroleum gas (LPG). To be kept as liquids at low pressure, they must be refrigerated and stored and shipped in specially equipped tanks. The third NGL product is natural gasoline. Saudi Aramco's four domestic refineries in 1994 produced almost 670,000 BPD of products, while NGL recovery companywide totaled some 688,000 BPD. 1994 work was started at the Ras Tanura Refinery on an addition of new processing units giving greater yields of high-value gasoline, kerosene, and diesel fuel.

Both crude oil and products can be shipped from the marine terminal at Ras Tanura, with its two loading piers and the four offshore Sea Islands, which provide berths for vessels of up to 500,000 deadweight tons. A second terminal at Ju'aymah, located northwest of Ras Tanura, can load the largest crude carriers now in service from moorings in deep water 11 kilometers out in the Gulf. Large storage tanks at both terminals provide loading flexibility. The crude tank farm at Ju'aymah holds 25 million barrels, while the tank farm at Ras Tanura stores over 30 million barrels of crude and products, nearly five times the daily oil production of the United States. Yet another crude oil export terminal at Yanbu' on the Red Sea provides alternative outlets to Western markets. The terminal has 12.5 million barrels of crude storage and the capacity to ship 4.2 million BPD of oil from the East-West pipeline system. Smaller terminals meet the coastal shipping needs of the Rabigh and Jiddah refineries.

All of the oil operations, from wellhead to terminal, are links in a long and intricate chain, with each operation dependent on many others. A change in reservoir pressure, the shutdown of a processing unit for maintenance or modification, an emergency of some kind, fluctuating requirements for various grades of crude or different products, even a few days of unusual weather — any one of these can ultimately have an effect on production planning and ship scheduling. These factors and more are constantly balanced by men and computers in a Dhahran nerve center that monitors virtually every aspect of Saudi Aramco's oil, gas, NGL, and electric power operations. Because oil operations are complex and potentially dangerous, intensive training and safety measures require more attention than in many other industries.

To support the many specialists who are directly engaged in all these operations there are hundreds of other

Oil piped from central Arabian fields enters the East-West crude line at Pump Station 3.

skilled employees running water, steam, and cooling plants and still others driving trucks, flying airplanes, or piloting boats. In this oil company there are employees who manage warehouses, operate laboratories, keep accounts, care for the sick, teach classes, and build houses. But all these jobs are ancillary to Saudi Aramco's primary task: to find, produce, process, market, and transport hydrocarbons in the form of crude oil, liquid products, sulfur, and gases.

cated efficiency apartments, and new family and bachelor houses were added by the hundreds, along with new schoolrooms and new recreation, shopping, and medical facilities. A three-story addition that added over 40 percent more space to Aramco's main administration building in Dhahran in the mid-1970s soon proved inadequate and a new 10-story administrative headquarters had to be built. The camp at 'Udhailiyah, mothballed years before, was reopened and expanded to provide working and living accommodations for some 1,500 employees involved in oil development, water injection, and gas facilities in Aramco's southern producing area.

Aramco's payments to contracting firms working for it went up about 800 percent between 1970 and 1973. The company's annual outlay for materials amounted to $65 million in 1970, reached the neighborhood of $450 million in 1974, and climbed to almost $1.8 billion in 1978. Of these amounts the purchases through Saudi suppliers of locally manufactured and imported materials rose from $36 million to $737 million in the 1970-1978 period. In 1977, a year of major expansion, Aramco added some $1.7 billion worth of properties, plants, and equipment for its own operations and another $2 billion in connection with government programs in which the company was involved.

During the construction boom of the 1970s, contract workers lived aboard "floating hotels" such as this one anchored offshore at Ju'aymah.

THE MASTER GAS SYSTEM: In the midst of this all-out effort to increase its oil producing and handling capacity and build the necessary supporting facilities, Aramco took on a new role as a working partner in Saudi Arabia's industrialization program, one of the most ambitious any country has ever undertaken. During the 1970s the world pricing structure for hydrocarbons, as well as a growing concern with environmental factors, made the conservation and utilization of the high-pressure gas produced in association with crude oil economically feasible for the first time. Accordingly, early in 1975 the Saudi Government asked Aramco to design, build, and operate on its behalf a nationwide system to gather and process this gas to supply fuel and raw material for the kingdom's multibillion-dollar industrialization program (see page 163) and for export. The task was an enormous undertaking, involving at its height tens of thousands of workers both on the ground in the kingdom

Expansion during the 1970s brought new life — and housing for hundreds of new employees — to 'Udhailiyah in the southern area of operations.

Oil Operations

The crude oil produced by Saudi Aramco from both onshore and offshore fields goes first to gas-oil separator plants for removal of gases, after which it is sent for further processing at stabilizers or refineries. Most of the crude oil is delivered to tankers at Ras Tanura, Ju'aymah, or Yanbu', and smaller quantities are exported by pipeline from the Eastern Province to Bahrain. Saudi Aramco also delivers crude oil within Saudi Arabia for use as fuel and to feed refineries turning out products for domestic use and export. See map of Saudi Aramco operations on pages 240-241.

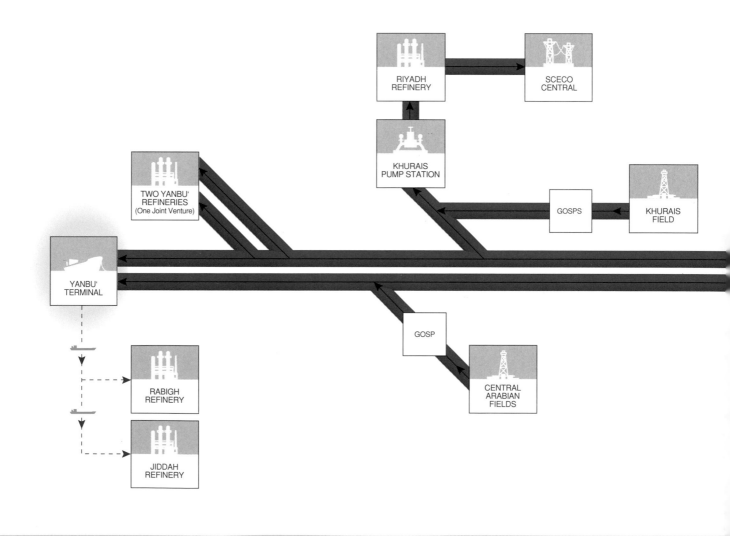

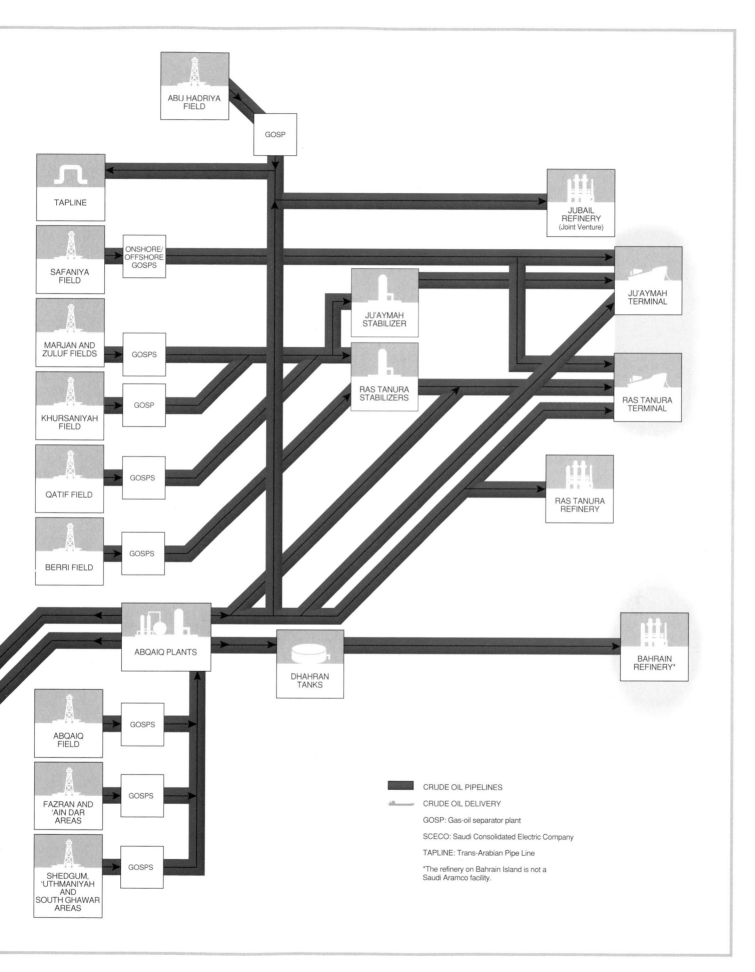

ABU HADRIYA FIELD

GOSP

TAPLINE

SAFANIYA FIELD

ONSHORE/ OFFSHORE GOSPS

MARJAN AND ZULUF FIELDS

GOSPS

KHURSANIYAH FIELD

GOSP

QATIF FIELD

GOSPS

BERRI FIELD

GOSPS

JUBAIL REFINERY (Joint Venture)

JU'AYMAH STABILIZER

RAS TANURA STABILIZERS

JU'AYMAH TERMINAL

RAS TANURA TERMINAL

RAS TANURA REFINERY

ABQAIQ PLANTS

DHAHRAN TANKS

BAHRAIN REFINERY*

ABQAIQ FIELD

GOSPS

FAZRAN AND 'AIN DAR AREAS

GOSPS

SHEDGUM, 'UTHMANIYAH AND SOUTH GHAWAR AREAS

GOSPS

CRUDE OIL PIPELINES

CRUDE OIL DELIVERY

GOSP: Gas-oil separator plant

SCECO: Saudi Consolidated Electric Company

TAPLINE: Trans-Arabian Pipe Line

*The refinery on Bahrain Island is not a Saudi Aramco facility.

Three heavy-duty tractors were needed to offload the first of three surge bullets — each 750 tons — when it arrived for the Ju'aymah fractionation plant.

The Berri plant, forerunner in design and principle of the entire gas program, is located not far from the Jubail industrial complex.

At the gas treating area of the Shedgum plant, gas is sweetened by the removal of carbon dioxide, hydrogen sulfide, and other sulfur compounds.

and in offices and fabrication plants on several continents. Indeed, it was the largest industrial project ever undertaken by a single firm.

When the Master Gas System, or MGS, started up in the early 1980s it enabled Saudi Arabia to use for the first time nearly all the gas produced onshore along with crude oil. At that time the MGS could handle some 85 million cubic meters (3 billion cubic feet) of associated gas daily. Desulfurized and treated gas began fueling and feeding industrial complexes built from scratch at the historic site of Jubail on the Gulf and across the Arabian Peninsula at Yanbu' some 350 kilometers north of Jiddah on the Red Sea. It was also used to generate thousands of megawatts of power for community and industrial needs, to produce large amounts of desalinated fresh water, and to provide NGL for export.

The basic components of the MGS remain the same today as in the early 1980s, although the gas input system has been substantially enlarged. Gas is collected at GOSPs and sent to one of three processing centers in the Eastern Province. The Berri plant, the first to be completed, was inaugurated in 1977 and stands only a few kilometers from Jubail Industrial City. The two others, at Shedgum and 'Uthmaniyah, lie to the south at the heart of the kingdom's most prolific oil field; roughly equal in size, they became fully operational in 1981 and 1982 respectively.

In 1984 associated gas from three offshore fields was brought into the MGS and the input system was further enlarged in an accelerated program to gather high-pressure, nonassociated gas produced independently of crude oil, from the deep Khuff reservoir in the company's southern area. At the same time the MGS was expanded to receive, as an emergency backup supply, gas which had been injected into the Abqaiq field for storage and conservation purposes.

The addition of Khuff gas to the MGS has been a boon to the kingdom since it is an important source of fuel gas whose supply does not vary with fluctuating crude oil production. This was particularly important in the mid-1980s, when crude output fell in response to low international demand and remained well below the record levels of the previous decade. However, Khuff gas is produced from reservoirs lying between three and four kilometers underground and comes to the surface at higher pressures than associated gas, which flows from reservoirs at shallower depths. Like

associated gas, Khuff gas also has a high content of toxic and corrosive hydrogen sulfide gas. High-pressure piping and valves and equipment modifications were therefore required at the Shedgum and 'Uthmaniyah gas plants before they could receive it. Even as crude oil and associated gas production rebounded in the early 1990s, Khuff gas continued to figure importantly in the kingdom's domestic energy picture.

By 1994 the peak processing capacity of the smaller Berri Gas Plant was 21 million cubic meters (750 millions standard cubic feet) of gas per day, while the combined peak capacity of the Shedgum and 'Uthmaniyah plants was about 108 million cubic meters (3.8 billion standard cubic feet) per day. The plants operate at peak capacity during the summer months when fuel gas consumption in the Eastern Province is greatest, due largely to heavy electrical power demand for air conditioning. The daily MGS collection and processing capacity — some 130 million cubic meters (4.5 billion standard cubic feet) — adds the equivalent of more than a million barrels of oil per day to the world's energy supply.

Gas is first treated at the gas plants to remove hydrogen sulfide. Further processing produces a sweet, dry gas — primarily methane — and a liquid stream of NGL and ethane. The dry gas is compressed and distributed to seawater desalination plants operated by the Saline Water Conversion Corporation, to the Eastern Province power grid, and to Jubail Industrial City, where it is used as fuel or petrochemical feedstock. Elemental sulfur is produced at a rate of several thousand tons per day as a byproduct of the gas-sweetening process; the bulk is pelletized at a company facility at Jubail's King Fahd Industrial Port and exported. The liquid gas is piped to fractionation plants at Ju'aymah and Yanbu' where it is split into ethane for industrial plant fuel and feedstock, and propane, butane, and natural gasoline.

Propane and butane, or LPG, is refrigerated and exported from Ju'aymah and Yanbu', while natural gasoline goes out in conventional tankers from Yanbu' and Ras Tanura. By the end of the 1990s, however, petrochemical plants at Jubail and Yanbu' are expected to consume most of the butane and natural gasoline. Because of the extended shallowness of the coastal waters at Ju'aymah, the company built a 10-kilometer-long prestressed reinforced concrete trestle to serve tankers loading LPG produced at the fractionation plant there. Both the fractionation plant and two-berth export terminal were inaugurated by King Khalid ibn 'Abd al-'Aziz in November 1980. At Yanbu', NGL products are exported from a two-berth offshore terminal which opened in 1982. The terminals at Yanbu' and Ju'aymah can handle the largest LPG tankers afloat or planned.

Because it crosses some extremely harsh terrain, the pipeline that feeds the gas fractionation plant at Yanbu' was a major engineering feat in itself. The longest and most advanced computer-controlled NGL pipeline ever built, the 1,170-kilometer high-pressure line can deliver up to 290,000 barrels per day of ethane and NGL to Yanbu'. Its diameter varies from 66 to 76 centimeters. Special design features include a supervisory control and data-acquisition system located in the Operations Coordination Center (OCC) in Dhahran, where the dispatcher receives information from 38 locations via microwave channels. The OCC issues operating orders to Shedgum, where the pipeline originates, to Yanbu', and to pipeline operators over the same microwave network.

The NGL pipeline parallels the five million-BPD East-West Crude Oil Pipeline system, which consists of two pipelines whose common pumping stations are powered by NGL from the gas pipeline. The crude lines run 1,200 kilometers from Abqaiq to Yanbu'. They supply two refineries and the Yanbu' Crude Oil Export Terminal which, like the NGL terminal, provides alternative outlets for the kingdom's hydrocarbon exports to international markets. A third pipeline, built to carry crude oil for export from the new terminal of al-Mu'ajjiz south of Yanbu' under an agreement with the South Oil Company of Iraq, runs south from the Iraqi border to a point east of Riyadh where it joins the East-West pipeline corridor. This pipeline and al-Mu'ajjiz Terminal opened in 1990 but operations were shut down in August that year and remained suspended in 1995.

GROWTH AT JUBAIL AND YANBU': The crude oil and NGL pipelines that link Yanbu' to the oil fields and gas facilities to the east are the base for industries similar to those at Jubail. Together the two complexes form part of an industrial mosaic which has diversified Saudi Arabia's economy and provided a lasting foundation for future growth. In January 1983 King Fahd ibn 'Abd al-'Aziz visited Yanbu' to inaugurate facilities there. Plants at Yanbu' include an NGL fractionation plant, two refineries, and a large petrochemical facility, while Jubail boasts a refinery, a steel mill, and petrochemical and fertilizer plants among the industries at its heart. By 1994 Jubail and Yanbu' together had 22 of these "basic" industries, and Saudi Arabia had become an industrial force outside the traditional oil and gas sectors.

Saudi Aramco and government agencies including the Royal Commission for Jubail and Yanbu', the General Petroleum and Minerals Organization (Petromin), and the Saudi Arabian Basic Industries Corporation (SABIC) have worked together to lay the foundations for the industrial cities, widening the country's industrial base, creating business opportunities for enterprising Saudis, and generating thousands of jobs. Although the benefits of this alliance can best be seen in Jubail and Yanbu', which by 1990 had a total population of more than 200,000 people, they also extend across the land.

Constructed in the early 1980s, the East-West NGL Pipeline originates at Shedgum and carries natural gas liquids across the peninsula to Yanbu'.

While this fruitful partnership continues at Yanbu', Jubail, and elsewhere, the company's responsibilities have grown significantly over the years as the Government of Saudi Arabia has entrusted it with progressively more duties in the oil and gas sector. The direction and pattern of the company's development were set as the government acquired an increasingly larger stake in Aramco and in 1988 established the wholly government-owned Saudi Arabian Oil Company (Saudi Aramco); through training programs that sharpened the skills of talented young Saudis in order to place more and more operational and managerial tasks in their hands; and through a drive to acquire and develop advanced petroleum technologies.

Towers of the fractionation plant at Yanbu' rise in silhouette against the sunset.

ESTABLISHMENT OF SAUDI ARAMCO: Saudi Aramco was created by a Royal Decree on November 8, 1988, to take over the operation and management of the kingdom's oil and gas fields from Aramco. In fact the Saudi Government had begun to purchase Aramco's assets from its shareholders — Standard Oil of California (later Chevron), Texaco, Exxon, and Mobil — in 1973. The process was completed in 1980, retroactive to 1976, although Aramco continued to operate and manage the kingdom's oil fields for the government.

The foundations for the new enterprise

Gas Operations

Gas produced with crude oil is collected at gas-oil separator plants. There, impurities are removed, hydrogen sulfide is recovered for conversion to elemental sulfur and sweet, dry gas is extracted for use as an industrial fuel or feedstock. From gas processing centers located at Shedgum and 'Uthmaniyah, natural gas liquids (NGL) and ethane are piped to plants at Yanbu' and Ju'aymah for fractionation into their separate components. After removal of the ethane, the NGL is further fractionated into LPG (propane and butane) and natural gasoline. NGL production from Berri goes to Ju'aymah or Ras Tanura for fractionation. LPG is exported from Yanbu', Ju'aymah, or Ras Tanura. In the fractionation plants, ethane is produced in gaseous form for use as a petrochemical feedstock in the industrial complexes at Yanbu' and Jubail. Some butane is used as a feedstock at the Jubail Industrial Complex.

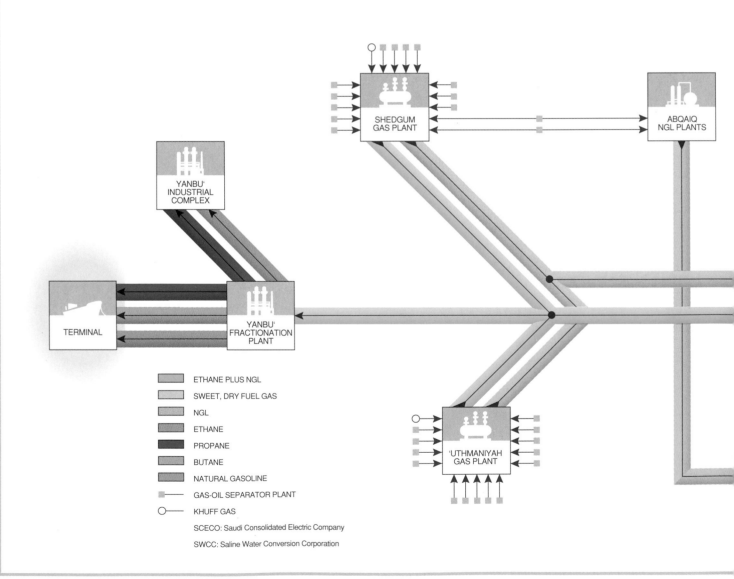

ETHANE PLUS NGL

SWEET, DRY FUEL GAS

NGL

ETHANE

PROPANE

BUTANE

NATURAL GASOLINE

GAS-OIL SEPARATOR PLANT

KHUFF GAS

SCECO: Saudi Consolidated Electric Company

SWCC: Saline Water Conversion Corporation

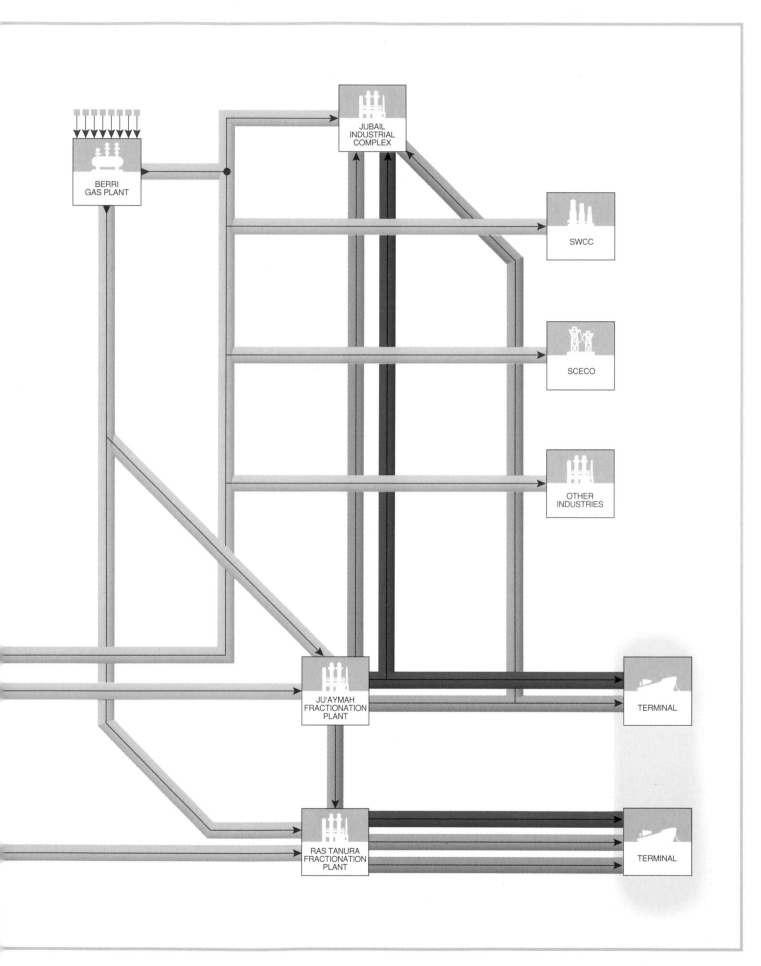

One of the most ambitious construction programs ever undertaken by any country, the Master Gas System enables Saudi Arabia to use nearly all the gas produced along with crude oil, adding the equivalent of more than a million barrels of oil a day to the world's energy supply. Gas is collected at gas-oil separator plants and sent to processing centers at Berri, Shedgum, or 'Uthmaniyah, where it is split into dry gas (methane) and natural gas liquids (NGL). The dry gas is used as a fuel and feedstock by local industries; the NGL is sent to fractionation plants at Ju'aymah or, via the East-West NGL Pipeline, to Yanbu' where it is separated into ethane for industrial fuel and feedstock, and propane, butane, and natural gasoline for export. Gas is also produced from deep natural reservoirs known as Khuff. Top from left: a control room at Berri; the East-West NGL Pipeline and the East-West Crude Oil Pipeline system share a cross-country corridor that cuts through a mountain range in the west; a gas rig drilling into the Khuff zone. Below from left: views of the 'Uthmaniyah and Shedgum gas plants, the Yanbu' NGL plant, and spheroids at Ju'aymah.

Ali I. Naimi was appointed Saudi Arabia's Minister of Petroleum and Mineral Resources in 1995 after serving as Saudi Aramco President and CEO.

Minister of Petroleum and Mineral Resources Ali I. Naimi, right, named Abdallah S. Jum'ah, left, to act as President and CEO of Saudi Aramco in 1995.

Bulldozers cleared the way for seismic exploration in the volcanic plains of northwestern Saudi Arabia in the early 1990s.

were well prepared: At the time of Saudi Aramco's establishment Saudis held almost all the company's management positions, while all plant operators' jobs were in Saudi hands. In November 1983 Ali I. Naimi — who had joined Aramco as a youth in 1947 and climbed the ladder of professional development through the company's training system — was appointed president of Aramco, becoming the first Saudi to hold the post. In early 1988, at the same time Minister of Petroleum and Mineral Resources Hisham M. Nazer was named Aramco Board Chairman, Naimi became Chief Executive Officer. When Saudi Aramco was established later that year Nazer became its first Chairman of the Board and Naimi was named its first President and CEO. In 1995 Naimi succeeded Nazer as Minister of Petroleum and Mineral Resources and Saudi Aramco Board Chairman. From that position he named Abdallah S. Jum'ah to act as Saudi Aramco's President and CEO; Jum'ah had most recently served as the company's Executive Vice President of International Operations.

Saudi Aramco reports to its owner, the government, through the Supreme Council which is chaired by King Fahd ibn 'Abd al-'Aziz. The Supreme Council sets the company's broadest policy and objectives. Saudi Aramco's Board of Directors makes key planning, budgeting, project, and operating decisions. The company has five business areas: Exploration and Producing, Manufacturing, Engineering and Operations Services, International Operations, and Finance and Relations.

Saudi Aramco's charter defines it as "an integrated international oil company which is to engage in all activities related to the oil industry, on a commercial basis and for the purpose of profit." The company's predecessor, Aramco, had already begun the metamorphosis into an integrated global oil enterprise, expanding from the traditional "upstream" activities of exploration and production into the "downstream" areas of crude oil transportation and marketing. In 1984, for example, it had formed subsidiaries to provide crude-oil tanker shipping services and acquire oil-storage facilities abroad and in 1985 it had begun marketing directly to refining companies rather than solely to Petromin and the four Aramco partners.

Saudi Aramco stepped up the pace of both upstream and downstream activities and broke new ground overseas, expanding marketing activities and setting up business ventures with regional refining and marketing companies. Furthermore, in 1993 the company assumed responsibility for nearly all facets of the kingdom's oil industry as a result of a Royal Decree. Saudi Aramco had developed into a world enterprise with operations reaching downstream from wellhead to corner service station.

EXPANDED EXPLORATION: Between 1989 and 1994 Saudi Aramco discovered 15 oil fields and gas fields in an exploration campaign over a prospecting area enlarged sevenfold to more than 1.5 million square kilometers — a region larger than Germany, France, and Spain combined. The discoveries, in central and northwestern parts of the kingdom and on the Red Sea coast, raised to 71 the total number of company-discovered commercial oil and gas fields. The central Arabian strikes included the discovery of super-light, sweet (low-sulfur) crude oil equal to the world's finest grades.

The campaign began in 1986 when the government asked Aramco to expand exploration outside the 220,000-square-kilometer "retained areas" on the eastern side of the kingdom, where the company had long focused its activities, to the limits of the original concession (see map, pages 230-31), an area covering some two-thirds of the kingdom. In 1990 exploration was further expanded to include the Red Sea coastal plain and the adjacent offshore area. In the first strike in the effort Saudi Aramco discovered sweet, super-light crude oil in June 1989 at al-Hawtah in central Arabia about 190 kilometers south of Riyadh. Further discoveries in the region revealed several hydrocarbon-bearing geological trends.

The company discovered gas at Kahf in the northwest of the kingdom in 1991 and in 1992 it achieved success on the Red Sea coast, proving the presence of oil and gas near Jaizan in the south, and finding gas and condensate in the Midyan region in the north. These discoveries were followed by new strikes, including oil at Midyan and gas at two fields lying about 150 kilometers north of Yanbu'.

Drilling also continued in the Eastern Province to explore for new gas reserves near existing production facilities. In mid-1994 Saudi Aramco discovered sweet gas and condensate in deep "pre-Khuff" formations more than four kilometers underground on the flank of the giant Ghawar field.

This pace of discoveries means that Saudi Aramco has continued to post annual increases in crude reserves even while maintaining its standing as the world's largest oil-producing company. From 1988 through 1994 crude oil reserves rose by 6.6 billion barrels to 259 billion barrels in spite of production totaling nearly 18 billion barrels during that time. Meanwhile, gas reserves grew by 250 billion cubic meters (8.8 trillion standard cubic feet) to 5.27 trillion cubic meters (186.1 trillion cubic feet) despite production of 284 billion cubic meters (10 trillion cubic feet) during the same period.

CRUDE PRODUCTION PROGRAM: In 1989 Saudi Aramco embarked on a major program to reach a maximum sustained crude production capacity of 10 million BPD. The aim was to meet customers' requirements for a mix of crudes in grades according to their seasonal needs, from terminals on either side of the kingdom, into the next century. The program, which was completed in 1994, included a series of oil and gas projects onshore and offshore in the Eastern Province and central Arabia, as well as expansion of the strategic East-West Crude Oil Pipeline system and the Yanbu' Crude Oil Export Terminal. The project boosted production capacity to a level last seen in 1980 and 1981. Output then had reached a record 9.6 million BPD before dropping in the face of declining international demand. Production declined to only a little more than 3 million BPD in 1985 before beginning a strong rally later in the decade.

As part of the production capacity expansion program, Saudi Aramco recommissioned many oil wells, GOSPs, and associated gas-compression and wet crude-handling facilities which had been taken out of service earlier because of low demand. The result was work marking the greatest challenge to the enterprise and its employees since the huge oil, gas, and electrical projects of the late 1970s.

In August 1990, in response to the Gulf Crisis, the company accelerated the program and increased its scope. During the last five months of 1990 Saudi Aramco raced to return to operation 17 previously shut-in GOSPs and raised crude production by more than 60 percent to 8.5 million BPD. That extraordinary effort made up for about 75 percent of the 4.6 million BPD of crude oil exports lost in the international embargo on crude sales from Iraq and occupied Kuwait. Oil prices, which had soared to more than $40 a barrel on fears of tight supplies, quickly retreated and an energy shortfall that could have seriously harmed an already depressed world economy was avoided.

The centerpiece of the offshore expansion was the huge Marjan project in the Gulf, completed in 1993, followed closely by projects in the Zuluf field — efforts which substantially raised Arabian Medium crude oil production capacity. The Marjan project included completion of two 250,000-BPD GOSPs and an offshore gas-compression plant with a capacity of 17 million cubic meters daily (600 million standard cubic feet) — one of the largest such facilities in the world. Pipelines deliver crude oil and compressed gas from the field's three GOSPs to a processing complex on the coast at Tanajib which was completed in 1994. Wet crude-handling facilities, used to remove water that is produced with oil, were completed at the Safaniya GOSP complex onshore to provide the capacity to process 1.2 million BPD of Arabian Medium crude from the Zuluf field.

In the early 1990s Saudi Aramco completed several major offshore construction projects, including Marjan GOSP 2.

The onshore expansion program focused on the 'Uthmaniyah and Hawiyah areas of the giant Ghawar field, which is the company's primary source of Arabian Light crude. In projects completed in 1993, the company added gas-gathering and wet crude-handling facilities to three GOSPs and built two new GOSPs with 300,000-BPD capacities. A new central water injection plant capable of handling 1.5 million barrels of treated seawater daily was built in South Ghawar to maintain reservoir pressure to achieve production targets. To meet increased requirements for injection water to maintain field production pressure three new seawater treatment modules were added to the Qurayyah Seawater Treatment Plant. This increased the capacity of the plant — already the largest such facility in the world — by 20 percent to 5.1 million barrels daily.

Hawiyah GOSP 4 was one of the projects completed in the giant Ghawar field in 1993.

In central Arabia the company completed a program in 1994 to produce up to 200,000 barrels per day of Arabian Super Light crude oil from 85 wells in the Hawtah, Ghinah, Hazmiyah, and Umm Jurf fields. The oil is

Newly discovered oil fields in central Arabia were connected to the East-West Crude Oil Pipeline system in 1994.

HISTORY OF THE ARAMCO
CONCESSION AREA TO 1986

The concession area granted to Socal by the Saudi Arab Government in 1933 consisted of an exclusive area covering all of eastern Saudi Arabia as far as the westerly edge of the Dahna and a further area of preferential rights extending west of that as far as "the contact between the sedimentary and igneous formations." In 1939 the concession was extended to include additional areas in the far northwest and southwest of the kingdom as well as Saudi Arabia's interests in the two Neutral Zones to the northeast.

The concession area was reduced by subsequent relinquishments until, in 1973, it consisted of six separate pieces covering a total of 220,000 square kilometers. In 1986 Aramco, as part of its changing relationship with the Saudi Arab Government, was reassigned oil exploration rights to all of the territory of the original concession and its supplemental areas within the kingdom's modern boundaries.

—— - - - 1933 Exclusive and Prefential Areas
—— - - - 1939 Additional Areas
Relinquishments

1947	1948	1949	1952	1955
1960	1963	1968	1973	

Remaining Concession Areas
1973 -1986

SAUDI ARAMCO PROSPECTIVE AREA
1986 TO THE PRESENT

In 1990 the Red Sea coastal strip with adjacent waters was added to the prospective area. This gave Saudi Aramco exploration responsibility for all parts of the kingdom which have sedimentary rock strata and which thus could potentially hold hydrocarbons. The company's exploration, refining, product distribution, pipelining, and support operations now extend kingdomwide. See Saudi Aramco operations map on pages 240-241.

—————— Prospective area since 1986

—————— Additions in 1990

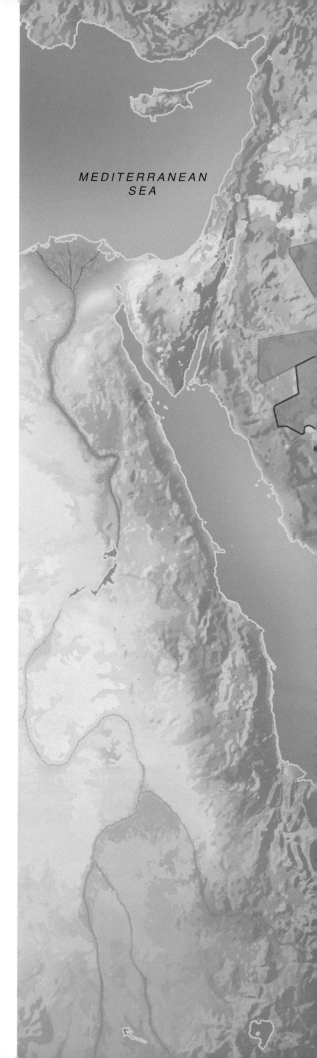

MEDITERRANEAN SEA

King Fahd and Crown Prince 'Abd Allah visited Aramco in May 1983 to officially inaugurate the company's new Exploration and Petroleum Engineering Center (EXPEC).

EXPEC is a focal point for sophisticated earth-science and engineering technology.

Sophisticated computer-generated data enables EXPEC geoscientists to select promising drilling sites.

piped 330 kilometers north to the East-West Crude Oil Pipeline system for shipment to Yanbu'.

The capacity of the East-West Crude Oil Pipeline system was raised from 3.2 million BPD to a peak of 5 million barrels, increasing the flexibility of supply to customers. The project, completed in 1993, was the second expansion since the government gave Aramco responsibility for the pipeline's operation in 1984. In 1987 capacity was boosted from 1.85 million BPD to 3.2 million barrels by "looping" the original pipeline, 122 centimeters in diameter, with a parallel 142-centimeter pipeline using the original line's pump stations. The latest expansion involved the addition of two 1.6 million-BPD "super pumps," each driven by a combustion gas turbine as powerful as that on a Boeing 747, to the system's 11 pump stations. Modifications were also made to plant facilities in the Eastern Province to allow the system to transport Arabian Heavy as well as Arabian Light crude.

At Yanbu' the capacity of the crude oil export terminal was raised by 60 percent to 4.2 million BPD. This project included construction of a fourth supertanker berth capable of handling vessels of up to 400,000 deadweight tons and enlargement of an existing berth to load ships up to 500,000 deadweight tons. The company also built a new control center to allow the terminal to manage the flow of crude through the pipeline system and erected a 1.5 million-barrel crude storage tank — the largest-diameter tank in the world, with a circumference of 125.5 meters — increasing the terminal tank farm capacity to 12.5 million barrels.

EXPEC: King Fahd ibn 'Abd al-'Aziz inaugurated the Exploration and Petroleum Engineering Center, or EXPEC, in Dhahran in May 1983, signaling the kingdom's intent to acquire, use, and develop independently the most advanced computer-centered technology available in the industry. The facility links top-of-the-line computer, exploration, petroleum-engineering, and laboratory facilities in one of the largest earth-science and engineering centers in the international oil industry. Through EXPEC, Saudi Aramco has essentially eliminated its dependence on upstream technological support from other oil companies and provides technical expertise and special services in-

house in all facets of engineering and producing operations. Data processing and analysis, and the mapping of production strategies, take place in close proximity, enhancing coordination, improving efficiency, and speeding the pace of work.

Not only has EXPEC enabled the company to keep pace with new developments in the industry, but Saudi and expatriate specialists there carry out independent research and development to answer questions of singular importance to the kingdom. The Laboratory Research and Development Center, an important part of the EXPEC complex, provides services to enhance technologies relating to exploration, production, processing, and transportation.

Commensurate with the huge size and complexity of the kingdom's hydrocarbon reservoirs, Saudi Aramco employs some of the world's largest reservoir simulation models including several with more than 240,000 three-dimensional subdivisions or "cells." By introducing varying factors into these models, for instance, petroleum engineers can calculate whether, where, and how much pressure support will be required to produce oil from a reservoir at a given rate on a certain date — determinations that are vital to long-term planning. New reservoir models under consideration are expected to contain as many as 5 million cells.

The EXPEC Computer Center (ECC) employs the latest-technology supercomputers, data communications networks, workstations, and applications software to support exploration, petroleum engineering, and drilling. Beginning operations with a single IBM 370/168 computer, by the mid-1990s the ECC's computing and on-line data-storage capacity had undergone a quantum increase. The facility's inventory includes two supercomputers and numerous advanced workstations linked through an ultra high-speed communications network. This powerful computing environment enables the processing and interpretation of all seismic and well bore data; it also enables simulations of all reservoirs using sophisticated 3-D modeling techniques and assists in designing production facilities. Another powerful computer provides general-purpose scientific capabilities for specialized petrochemical studies and is a focus for communications with remote field operations.

The impact of EXPEC has been substantial. It played an important role in six years of intensive studies completed in 1988 that increased the known crude oil reserves in company-managed fields by more than 50 percent to more than 250 billion barrels. At the same time, studies increased known gas reserves by 25 percent. EXPEC also provides an unparalleled training ground for young Saudi geoscientists and petroleum engineers who work alongside specialists from the kingdom and abroad.

INTERNATIONAL OPERATIONS: Saudi Aramco has forged ties with oil refining and marketing companies around the world as part of its drive to gain a higher value from each barrel of oil it produces. By 1995 the company had acquired long-term crude oil sales commitments totaling as much as 1.1 million BPD through business ventures in the United States, the Republic of Korea, and the Philippines while continuing to search for additional profitable refining and marketing investments in the United States, Europe, and the Far East.

The first downstream partnership, a joint venture between subsidiaries of Saudi Aramco and Texaco Inc., created Star Enterprise, which became the sixth largest marketer of gasoline in the United States when it opened for business on January 1, 1989. The joint venture gave Saudi Aramco a 50-percent share in three major refineries and a petroleum marketing network covering all or part of 26 eastern and southeastern states and the District of Columbia, providing an outlet for up to 600,000 barrels of oil a day. In 1992 facilities were completed at Star's 250,000-BPD refinery at Port Arthur, Texas, to allow it to extract more valuable light products from heavier grades of crude oil.

In 1991 a Saudi Aramco subsidiary bought a 35 percent interest in the SsangYong Oil Refining Company, Ltd., the Republic of Korea's third largest refiner and leading lubricant manufacturer. SsangYong, with a 525,000-BPD refinery at Onsan, has a strong marketing position in the fast-growing Pacific Rim region and Saudi Aramco is committed to supplying at least 70 percent of the refinery's crude requirements. A major refinery upgrading and expansion program to produce high-quality gasoline and other products from atmospheric residue was to be completed in 1995.

In early 1994 Saudi Aramco and an affiliate signed an agreement with the Philippine National Oil Company (PNOC) to buy 40 percent of the outstanding shares in Petron Corporation, PNOC's refining and marketing affiliate. Petron has a 155,000-BPD refinery and Saudi Aramco is committed to supply at least 90 percent of its crude requirements. Petron markets products through some 1,000 retail outlets throughout the Philippines and commands 45 percent of the country's petroleum product market.

Earlier, as a result of a Royal Decree issued on July 1, 1993, Saudi Aramco took over Petromin's 50 percent interests in three joint-venture refineries in the kingdom, at Jubail on the Gulf and at Yanbu' and Rabigh on the Red Sea. In 1995, the company became sole owner of the 325,000-BPD Rabigh Refinery. The two remaining joint venture refineries for which Saudi Aramco assumed Petromin's ownership share have a combined processing capacity of more than 600,000 BPD. They are operated by the Saudi Aramco Mobil Refinery Company at Yanbu' (320,000 BPD) and the Saudi Aramco Shell Refinery Company at Jubail (300,000 BPD). Saudi Aramco supplies the refineries with 100 percent of their crude oil requirements, and the bulk of the company's 50 percent share of lighter products (gasoline, diesel, jet fuel, and kerosene) is consumed domestically.

SHIPPING AND MARKETING: Saudi Aramco has established international marketing support services affiliates in New York, London, Tokyo, and Singapore to provide crude oil and product sales assistance. They give a variety of services including weekly crude analyses to assess worldwide demand and help the company schedule voyages.

The company's shipping subsidiary, Vela International Marine Limited, was established in 1984, when rights to Vela's name and four existing tankers were acquired. By 1995 the fleet had expanded to 23 large crude oil tankers and four refined product carriers. The tanker fleet consists of four ultra large crude carriers (ULCCs) and 19 very large crude carriers (VLCCs) including 15 new, technologically advanced VLCCs. Ranging from 290,000 to more than 300,000 deadweight tons, these vessels were built at shipyards in Japan, the

Saudi Aramco acquired a 35 percent interest in the SsangYong Oil Refining Company in the Republic of Korea in 1991.

In 1992 Saudi Aramco and Texaco officials inaugurated a delayed coker unit at the Star Enterprise joint-venture refinery in Port Arthur, Texas.

The *Phoenix Star* and *Libra Star* were two of 15 very large crude carriers constructed for Vela between 1993 and 1995.

Aramco Services Company, a subsidiary in Houston, Texas, provides a wide range of services to Saudi Aramco.

SAUDI ARAMCO SUBSIDIARIES, JOINT VENTURES, AND PRINCIPAL EXPORT ROUTES

Saudi Aramco is an integrated international oil company with interests, operations and services extending across the kingdom and around the world. Since 1988 the company has formed partnerships with major refiners and marketers in the United States, the Republic of Korea, and the Philippines, firmly establishing itself in those important petroleum-consuming areas. Star Enterprise, a joint venture between affiliates of Saudi Aramco and Texaco in the United States, owns three refineries and sells products in a region covering parts of 26 states and the District of Columbia. Partnerships with SsangYong Oil Refining Company in the Republic of Korea, and Petron Corporation in the Philippines have given the company a strong presence in the Pacific Rim region, an area of rapidly expanding economic activity. Within Saudi Arabia the company holds a 50-percent share in two joint-venture refineries at Yanbu' and Jubail, operated by the Saudi Aramco Mobil Refinery Company (SAMREF) and the Saudi Aramco Shell Refinery Company (SASREF) respectively.

Saudi Aramco has also established marketing support subsidiaries in New York, London, Singapore, and Tokyo to serve customers in those key petroleum-consuming regions. Owned and leased crude oil storage facilities in the Netherlands and the Caribbean provide the strategic capacity to meet customer requirements in Europe and the United States.

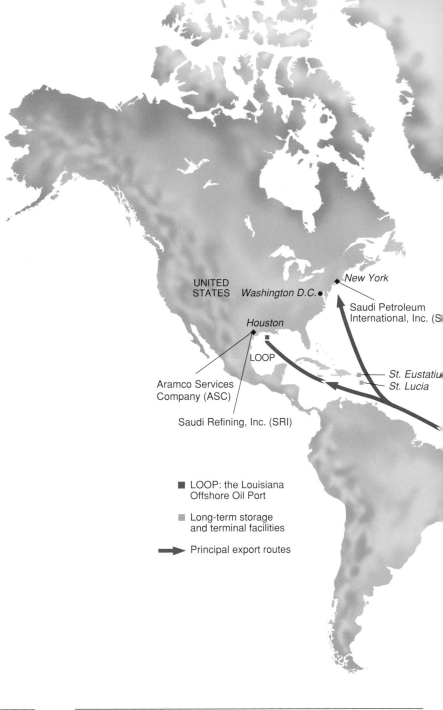

UNITED STATES
Washington D.C.
New York
Saudi Petroleum International, Inc. (S
Houston
LOOP
Aramco Services Company (ASC)
St. Eustatiu
St. Lucia
Saudi Refining, Inc. (SRI)

■ LOOP: the Louisiana Offshore Oil Port

■ Long-term storage and terminal facilities

➡ Principal export routes

SAUDI ARABIA

Saudi Aramco Shell Refinery Company — Jubail
Arabian Gulf — Gulf of Oman
Ras Tanura/ Ju'aymah ● Dhahran
Saudi Aramco Mobil Refinery Company
Yanbu'
★ Riyadh
SAUDI ARABIA
Saudi Arabian Oil Company (Saudi Aramco)
Vela International Marine Limited
● Jiddah
Red Sea
Arabian Sea
Gulf of Aden

UNITED STATES

UNITED STATES
Star Enterprise Delaware City Refinery
Star Enterprise Headquarters Houston
Star Enterprise marketing area covers all or parts of 26 states and the District of Columbia
Star Enterprise Port Arthur, Texas Refinery
Star Enterprise Convent, Louisiana Refinery

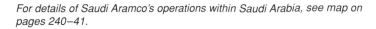

For details of Saudi Aramco's operations within Saudi Arabia, see map on pages 240–41.

234

Aramco Overseas
Company B.V.(AOC)

Saudi Petroleum
Overseas, Ltd.
(SPOL)

London
Leiden
Rotterdam

Sumed Pipeline

Sidi Kerir
Ain Sukhna
Ras Tanura/
Ju'aymah
Yanbu'
SAUDI
ARABIA

Saudi Petroleum,
Ltd. (SPL)

KOREA
Tokyo

PHILIPPINES

Singapore

Saudi Petroleum,
Ltd. (SPL)

REPUBLIC OF KOREA

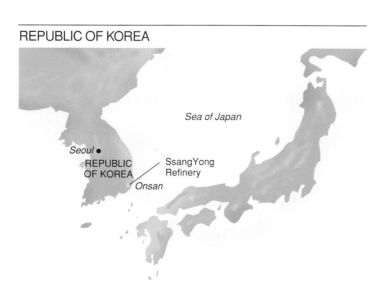

Sea of Japan

Seoul •
REPUBLIC
OF KOREA
SsangYong
Refinery
Onsan

PHILIPPINES

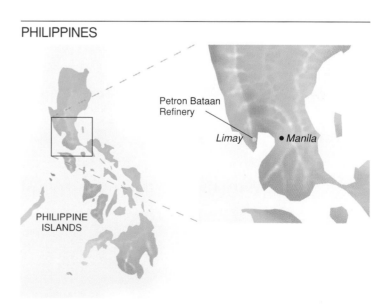

PHILIPPINE
ISLANDS

Petron Bataan
Refinery

Limay
• *Manila*

235

Once confined to the Eastern Province, Saudi Aramco operations today span the globe, with refining and marketing interests worldwide. In 1988 affiliates of Saudi Aramco and Texaco entered into an agreement establishing Star Enterprise, a joint venture to refine and market petroleum products under the Texaco brand name in the eastern and Gulf Coast areas of the United States. Other agreements brought Saudi Aramco interests in the SsangYong Refining Company in the Republic of Korea and Petron in the Philippines. Clockwise from right: The Star Enterprise refinery in Delaware City, Delaware; a Texaco Star Mart filling station; a Petron tank farm and service station in Manila; and the SsangYong refinery and terminal at Onsan. Marketing support services offices in London, Tokyo, Singapore, above left, and New York, above right, facilitate crude oil and refined products sales. Vela International Marine Limited, a Saudi Aramco subsidiary, provides crude and refined products transportation through owned and chartered vessels. At left, Vela's *Libra Star,* commissioned in 1993 as part of an order for 15 new very large crude carriers, takes on crude at Ju'aymah.

The integration of the kingdom's oil industry placed three additional domestic refineries under the Saudi Aramco umbrella, including a 140,000-BPD refinery at Riyadh.

Rabigh Industrial City is the support facility for the Rabigh refinery on the Red Sea.

The Jiddah refinery provides refined products principally for domestic consumption.

Republic of Korea, and Denmark and delivered between 1993 and 1995. Each new ship was built to high safety standards and is capable of carrying at least 2 million barrels of crude. The ships are expected to remain in service for 20 to 25 years.

Vela delivers crude oil to customers in Europe and the United States on its owned-fleet vessels and on time-chartered and spot-chartered ships. Indeed, Vela is the world's largest charterer of VLCC tonnage. Historically customers in the Far East have handled their own crude oil transportation with the assistance of company marketing support subsidiaries in the region. Vela also carries refined products between ports on both the east and west coasts of Saudi Arabia and delivers LPG to customers abroad.

To facilitate the speedy turnaround of crude deliveries, Saudi Aramco subsidiaries also lease or own capacity in several strategically located storage tank farms in both hemispheres. Located on the islands of St. Eustatius and St. Lucia in the Caribbean, and in Rotterdam, they have the capacity to store over 16 million barrels of crude oil.

INTEGRATION OF THE OIL INDUSTRY: The Royal Decree issued on July 1, 1993, also merged into the company all of Saudi Arabia's state-owned oil refineries and the product-distribution and marketing operations that had been operated by the former Petromin marketing and refining project known as Samarec. Under the same decree, Saudi Aramco assumed Petromin's 50-percent interests in the three joint-venture refineries in the kingdom. The corporation's new refining interests in the kingdom, combined with its Ras Tanura Refinery and its joint-venture and shareholding interests in five other refineries in the United States and the Far East, place the company in the top rank of world refiners.

As part of the reorganization carried out under the Royal Decree the company integrated into its ranks approximately 9,700 Saudi employees and some 1,500 expatriate contract staff who had been responsible for the assets merged into the company. This increased the Saudi Aramco work force by about 25 percent to approximately 57,500 employees.

Principal assets transferred to Saudi Aramco by the Royal Decree included three refineries — at Yanbu' (190,000 BPD), Riyadh (140,000 BPD), and Jiddah (90,000 BPD). These facilities and the 300,000-BPD Ras Tanura Refinery operate principally to meet domestic demand for gasoline, diesel fuel, kerosene, and asphalt. Saudi Aramco fully owns the domestic refineries in Riyadh and Yanbu', while 25 percent of the Jiddah plant is held by private investors. Furthermore the decree increased the number of Saudi Aramco's Red Sea terminals used to ship or receive crude oil, NGL, or refined products.

Saudi Aramco also assumed responsibility for operating the kingdom's vital petroleum-product distribution system. The company delivers gasoline, diesel fuel, jet fuel, kerosene, LPG, fuel oil, and asphalt to some 5,000 bulk customers and through them meets the needs of millions of consumers in the industrial, agricultural, and private sectors in three districts that span the country. Approximately 700,000 barrels of refined products are shipped daily to virtually every city, town, and village in the kingdom by truck, rail, pipeline, and ship. Moreover more than 300,000 barrels of refined products are transferred between bulk plants every day to meet regional needs. System facilities include 18 bulk plants and 14 air-fueling units as well as several temporary bulk plant/air fueling units and short-haul pipelines. The company also supplies crude oil to domestic industrial customers.

Studies began immediately to assess the kingdom's long-term refining and product-distribution requirements. In 1994 the company began to upgrade the Ras Tanura Refinery to produce more high-value light products, primarily gasoline, kerosene, and diesel fuel.

ENVIRONMENTAL PROTECTION: It is Saudi Aramco's policy to ensure that operations do not create undue risks to the environment or public health and are carried out with full concern for protecting the land, air and water from harmful pollution. The company's environmental protection activities have increased markedly in recent years in line with its growing responsibilities in the national and international oil industry. Important initiatives have included the development of low-sulfur, light crude reserves to meet rising demand for grades with favorable environmental characteristics. Environmental concerns are also reflected in plant modifications, materials orders, chemical usage, and oil spill protection.

The company's first line of defense against oil spills lies in their prevention. Strict operations guidelines and training underpin this effort. However Saudi Aramco emphasizes preparedness in case a spill occurs and has skilled manpower and a range of equipment, including pollution-control vessels and aircraft, ready to combat spills. Saudi Aramco played a major role in helping the Government of Saudi Arabia combat the 1991 Gulf oil spill, the world's largest. During the spill, the company successfully acted with government agencies to protect vital facilities along the coast. Saudi Aramco recovered more than 1 million barrels of oil, the most ever collected from a spill.

Saudi Aramco has mapped out a Global Oil Spill Contingency Plan to meet needs that might arise in any region in which it operates. The regions covered are the Arabian Gulf and Red Sea, the U.S. Gulf Coast and East Coast, the Caribbean Sea, Europe, the Mediterranean Sea, South Africa, and the Far East. To ensure the capability to combat oil spills, the enterprise carries out regular training exercises in the kingdom and abroad. The company is a charter member of several regional and international oil spill protection agencies. These include the Gulf Area Oil Companies Mutual Aid Organization, a group of oil firms whose common objective is to protect the Arabian Gulf from oil pollution through joint action, and the Oil Spill Service Center in England, an organization providing a worldwide spill-response capability.

Saudi Aramco conducts periodic environmental performance surveys of its facilities under a corporate-wide program established in 1989. More than 30 surveys, including assessments of the Ras Tanura, Riyadh, and Jiddah refineries, had been carried out by the beginning of 1995. The company continuously monitors its major facilities with respect to air emissions, wastewater discharges, and potential groundwater contamination. Waste minimization, solid and hazardous waste management, and wastewater management programs are also in place. Marine scientists carry out studies in another part of Saudi Aramco's environmental protection effort. Designed to assess and alleviate industrial impact on the marine environment, these include research in biological, chemical, and physical oceanography. Furthermore the company has important, long-term scientific links with universities, such as the Research Institute at Dhahran's King Fahd University of Petroleum and Minerals, in marine environmental studies and other fields.

Saudi Aramco is active in several international industrial environmental organizations and takes part in international environmental conferences, frequently as a member of the delegation of the Government of Saudi Arabia. The company was part of the government body that participated in the United Nations Conference on Environment and Development in Rio de Janeiro, Brazil, in 1992 and in follow-up meetings of the Intergovernmental Panel on Climate Change and the Convention on Biological Diversity.

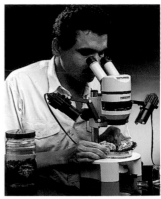
A marine biologist inspects a sea urchin for signs of toxins.

SAUDI ARAMCO AND WORLD ENERGY: Saudi Arabia's economic well-being and future growth are heavily dependent on oil. Although the government has taken major strides to diversify its economy, the sale of oil still provides much of the income for its day-to-day operation, the kingdom's social services, and development plans. The kingdom's oil — its production capacity and its massive reserves — also supports the economic and political role that Saudi Arabia plays on the world stage.

In fact about 75 percent of the Saudi Government's income is generated by the three companies that now produce its oil. Two foreign concessionaires produce about three percent of this oil from the kingdom's half share of production in the Partitioned Neutral Zone, which Saudi Arabia shares with Kuwait. The American company Saudi Arabian Texaco operates the onshore oil fields in the Partitioned Neutral Zone, while the Japanese-owned Arabian Oil Company operates the zone's offshore fields. By far the largest share of Saudi Arabia's crude production, however, comes from Saudi Aramco. The company's 1994 output of some 8 million BPD made up the remaining 97 percent of Saudi Arabia's crude oil production and an even higher proportion of the government's oil revenues.

The MISFAH 9, one of a number of specialized vessels designed to combat oil spills, is fitted with advanced fire-fighting equipment.

Saudi Arabia depends on other nations of the world to buy its oil. But this is far from being a one-sided relationship. These nations — especially the industrialized countries of the West, Japan, and other rapidly-growing economies of Asia — depend on the other oil-producing countries to supply a significant portion of their petroleum needs. Thus consumers depend on producers and producers depend on the consuming countries. Ever since it moved into the

A Houston-based oil-spill response team for the East and Gulf coast regions of North America and the Carribbean Sea holds regular hands-on drills.

SAUDI ARAMCO OPERATIONS

The company's petroleum activities, once limited to producing and refining oil in the Eastern Province, have expanded enormously since the mid-1970s and now cover virtually all the kingdom. Saudi Aramco facilities include five large gas-processing plants and the world's biggest seawater treatment plant, which provides millions of barrels of water daily to maintain reservoir pressure in the world's largest oil field — Ghawar. The pipeline network (whose key elements are shown on the oil and gas operations charts on pages 220–21 and 224–25) totals some 20,000 kilometers in length, including links with new oil fields in central Saudi Arabia. Major pipelines are the Trans-Arabian Pipe Line (Tapline), the Saudi Arabia-Bahrain Pipeline, and the East-West Crude Oil and Natural Gas Liquids (NGL) pipelines to Yanbu' on the West Coast.

Refineries located on both coasts and in the capital, Riyadh, process crude oil into products for customers in the kingdom and around the world. The company's major crude oil and NGL export terminals are at Ras Tanura and Ju'aymah on the Gulf and at Yanbu' on the Red Sea, while a number of smaller facilities handle refined products for export or domestic shipment. Saudi Aramco has responsibility for the distribution of refined products kingdomwide, and for operating bulk storage plants and aircraft-fueling facilities throughout the country.

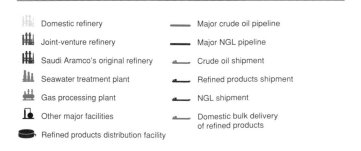

⊞ Domestic refinery		━━ Major crude oil pipeline	
⊞ Joint-venture refinery		━━ Major NGL pipeline	
⊞ Saudi Aramco's original refinery		◄━ Crude oil shipment	
⊞ Seawater treatment plant		◄━ Refined products shipment	
⊞ Gas processing plant		◄━ NGL shipment	
⊞ Other major facilities		◄━ Domestic bulk delivery of refined products	
⬤ Refined products distribution facility			

MEDITERRANEAN SEA

JORDAN

• Tabuk

Duba •

Yanbu'

Rabi

Jidd.

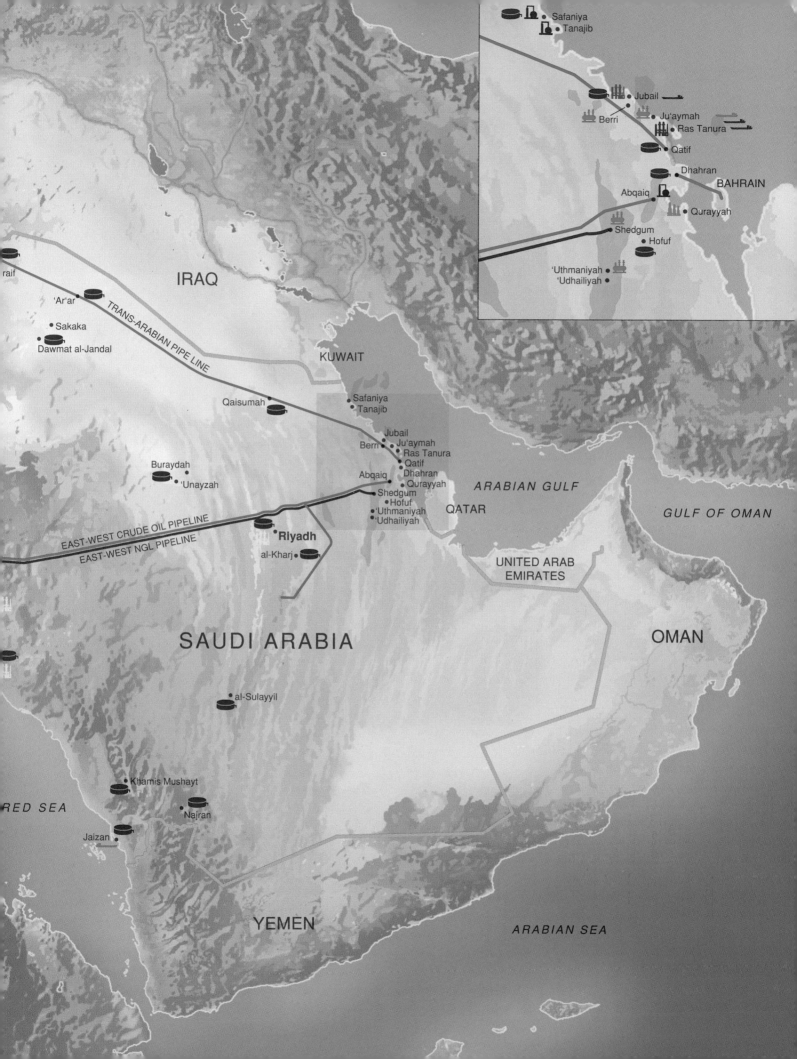

IRAQ

raif

'Ar'ar

TRANS-ARABIAN PIPE LINE

Sakaka

Dawmat al-Jandal

KUWAIT

Qaisumah

Safaniya
Tanajib

Jubail
Berri   Ju'aymah
Ras Tanura
Qatif
Abqaiq   Dhahran
Qurayyah

ARABIAN GULF

GULF OF OMAN

Buraydah

'Unayzah

Shedgum
Hofuf
'Uthmaniyah
'Udhailiyah

QATAR

EAST-WEST CRUDE OIL PIPELINE

EAST-WEST NGL PIPELINE

Rlyadh

al-Kharj

UNITED ARAB
EMIRATES

SAUDI ARABIA

OMAN

al-Sulayyil

RED SEA

Khamis Mushayt

Najran

Jaizan

YEMEN

ARABIAN SEA

Safaniya
Tanajib

Jubail
Berri
Ju'aymah
Ras Tanura
Qatif

Dhahran
Abqaiq
BAHRAIN
Qurayyah

Shedgum
Hofuf

'Uthmaniyah
'Udhailiyah

A 1993 Royal Decree entrusted Saudi Aramco with responsibility for Saudi Arabia's domestic petroleum product distribution facilities, its domestic refineries, and its joint-venture refinery interests. The company today manages refineries in Ras Tanura, Riyadh, Rabigh, Yanbu', and Jiddah. It has offices across the kingdom, including, at right, Jiddah. Its joint-venture refineries are the Saudi Aramco-Shell refinery at Jubail, bottom, this page, and top left, facing page, and the Saudi Aramco-Mobil refinery at Yanbu', bottom left, facing page. Facilities enlarged or constructed as part of the crude capacity expansion program included the East-West Crude Oil Pipeline system where, bottom right, facing page, a new stabilizer column was added at Pump Station 3 to bring Arabian Super Light crude oil on stream; Marjan GOSP 2 and Zuluf GOSP 4, this page, left and right center; and Hawiyah GOSP 4, top right, facing page.

forefront of the international petroleum picture the Government of Saudi Arabia has shown its awareness of this interrelationship, and has taken this interdependence as a major factor in shaping the kingdom's oil policies.

Stated in simplest terms, the United States, Western Europe, Japan, and the rest of Asia all consume more energy than they produce, and by far the greatest part of what they consume is in the form of oil and gas. The growing deficit in the energy balance of these regions is being made up by oil and gas from other areas of the world — primarily the Middle East.

In looking at the energy balance of a country or region, oil production and oil consumption are the two main factors to be taken into account. In forecasting what the future position of a country is likely to be, however, and in determining whether any country or region can make a significant long-term contribution to the needs of the energy-short nations, a third factor is equally important: It is necessary to consider its reserves — that is, how much oil and gas it has in the ground. Large-scale production over long periods obviously must be backed by large reserves.

The three charts on this page show these three factors for the major industrialized areas of the world in 1994. For example, the United States consumed 26 percent of the total oil produced in the world, but was producing only 11 percent and had only 2.3 percent of the total oil reserves. Western Europe, with less than two percent of world reserves, produced over nine percent of the world's oil but consumed 20 percent. The three elements are even farther out of balance for Japan, which has practically no production or oil reserves but consumes eight percent of the world's oil production. The rest of the Asia and Pacific region consumed 17 percent of the world's petroleum against production of 11 percent, with reserves limited to less than five percent. Therefore the United States, Western Europe, and Asia are all clearly in a deficit position with respect to petroleum energy.

These areas' petroleum import needs must be supplied by countries that produce more oil than they use at home, allowing them to become net exporters. Saudi Arabia is the world's leading oil exporter: Production in 1994 exceeded domestic requirements by more than 7 million barrels daily, thereby providing about a quarter of the world's import requirements. The kingdom's neighbors in the Arabian Gulf are also net oil exporters. When Saudi Arabia's exports are combined with those of its neighboring member countries in the Arabian Gulf Cooperation Council, or GCC, they together account for more than 40 percent of world oil exports. If the export contributions of other Middle Eastern countries are also added, the total reaches more than 50 percent. The Middle East is thus clearly the world's center of supply for oil, with Saudi Arabia supplying fully half of this region's exports.

FUTURE DEMAND; SOURCES OF OIL: What will demand for oil imports be in the future, and where will the required oil come from? Demand from the net importing areas including the United States, the countries of Western Europe, and Asia is expected to rise over time rather than to decline. This rise is expected because economic growth will result in higher energy requirements and because the major importing countries have limited potential for production increases of their own. In the United States oil production has been on a gradual decline since the 1970s and the mature oil fields there cannot maintain even present production levels. In Western Europe oil production is expected to peak in the late 1990s as the major fields discovered in the North Sea pass full development. The Asian region similarly has a relatively low level of reserves to production; it will not be able to support increases in production unless gigantic new oil fields are discovered, which is not likely. Moreover, oil consumption in Asia is expected to continue its recent trend of rapid expansion as its economies grow.

The reserves chart makes clear that the world will be looking to Saudi Arabia, its GCC neighbors, and other countries in the Middle East for an increasing share of its future oil needs. Only these Middle Eastern countries have both large reserves and domestic consumption rates significantly below their production levels. The chart shows that Saudi Arabia, with over 260 billion barrels of oil still in the ground, alone accounts for 26 percent of the world's proven reserves. It is followed by Iraq, the United Arab Emirates, Kuwait, and Iran, each of which has considerably less than half of Saudi Arabia's oil resources. The combined reserves of Saudi Arabia and its sister GCC states — Kuwait, Bahrain, Qatar, the U.A.E., and Oman — contain over 46 percent

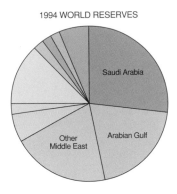

1994 WORLD RESERVES

1994 WORLD PRODUCTION

1994 WORLD CONSUMPTION

Saudi Arabia

Arabian Gulf: Kuwait, Bahrain, Qatar, U.A.E. and Oman

Other Middle East, excluding North Africa

Africa

United States

Other Western Hemisphere

China

Japan

Other Asia/Pacific

Western Europe

Eastern Europe and Former Soviet Union

of world reserves. Adding the other states bordering the Arabian Gulf, the total area contains about two-thirds of the world's oil reserves. It is clear that the Gulf region, and Saudi Arabia in particular, will be called on to supply a greater part of the world's supplies in the future. Over the years, as a result of its active programs for oil exploration, Saudi Aramco has been able to discover sufficient new reserves to more than make up for or equal the oil that it produces. Thus despite producing huge quantities of oil, its reserves keep rising. By 1995 Saudi Aramco had brought into production only about a third of the more than 70 oil fields it had discovered.

The Saudi Arabian Government, though mindful of the great importance of its oil reserves, is also keenly aware of the fact that oil is a depletable resource. Reserves that took millions of years to form may be used up in a few generations — and, unlike a forest, a river, or a wheat field, an oil field does not renew itself. Saudi Arabia therefore cannot afford to squander the wealth it now has, but must use it carefully to build alternative sources of income for the future. This is one of the primary objectives of the kingdom's five-year Development Plans.

Saudi Arabia's known oil reserves are expected to last for more than a hundred years, and the kingdom takes a long-term view when it shapes its national oil policies. In this respect it has consistently taken a moderating position within the Organization of Petroleum Exporting Countries (OPEC) on oil pricing issues, striving whenever possible to prevent dramatic spikes in oil prices that would provide short-term revenue gains but would reduce petroleum's energy market share in the long run. When Saudi Aramco completed a re-expansion of its production capacity to 10 million BPD in 1994, this was some 2 million BPD above demand. This extra capacity provides a "cushion" against any major disruption to world supply. Saudi Arabia has for many years maintained such a cushion, which helps to keep prices moderate and helps avoid problems of rationing and other hardships that may be associated with sudden supply shortages. Saudi Arabia's policy of maintaining surplus production capacity is not without cost, however; on-going expenditures must be made to maintain this capability.

If demand for Saudi Aramco's oil rises in the long run, its production capacity can be raised by further developing oil fields already in production or by developing its many other oil fields that have not yet been tapped. Owing to the ample reserves and favorable reservoir conditions in Saudi Aramco's operating area, Saudi Arabia is one of the most cost-effective areas of the world in which to develop additional production capacity.

THE PEOPLE OF SAUDI ARAMCO: Saudi Aramco's work force in 1995 totaled some 57,500 people, nearly 80 percent of them Saudis, with individuals from over 50 countries making up the remainder. Saudi Aramco has historically provided health care and other benefits to its employees. Saudi employees also benefit from the company's long-running Home Ownership and School Construction programs and from advanced training offered to them. This training — which in some cases extends to studies abroad in the sciences or in medicine — enables Saudis to provide for themselves and their families, while at the same time contributing to the growth of the company and the kingdom.

TRAINING AND DEVELOPMENT: The Saudi work force has changed markedly over the years. Sixty years ago educational opportunities in Saudi Arabia were extremely limited. Fishing, pearl diving, and small-scale agriculture in the oases were the mainstays of the local economy in the eastern part of the kingdom. When the oil pioneers arrived to start work in 1933 almost none of the people of the region had any acquaintance with modern technology or any experience in working in modern industry. Schools were scarce, literacy was rare, and there was almost a complete absence of technical and clerical skills needed by an oil company. Training began immediately.

The concession agreement of 1933 provided that the company would employ Saudi nationals as far as practicable, a requirement that historically determined both the composition of the enterprise's work force and to a large extent the direction in which its employee relations and training policies developed.

The first training was informal: Drillers, craftsmen, and office workers taught their specialties on the job. A more intensive effort was required, however, and 22 Saudi employees were receiving special training in Dhahran by 1940, while another 26 had been sent to Bahrain for training. These activities expanded

To meet growing demand for lighter grades of crude oil, Saudi Aramco converted Zuluf GOSPs 3 and 4 from Arabian Heavy to Arabian Medium crude production in the early 1990s.

A fourth supertanker loading berth was completed at Yanbu' in 1992, increasing shipping capacity to more than 4 million barrels a day.

Most of Saudi Aramco's professional positions are now held by Saudis, who make up nearly 80 percent of all employees.

Science, engineering, finance, management, and medicine — Saudi Aramco needs a work force with training in these fields and others, and through programs in Saudi Arabia and abroad provides its employees with the skills they will require to lead the company into the 21st century. Above, young Saudi geologists receive training in the field. Clockwise from immediate right: a materials engineer conducts an experiment as part of his doctoral studies at MIT in the United States; lab technology is taught on the job; and a construction engineer (left) learns the intricacies of concrete specification from a company specialist in the Specialist Development Program. Facing page, clockwise from top left, Saudi Aramco provides qualified physicians with training in specialties at Harvard, Georgetown, Baylor, and more than 20 other medical schools; drafting, computer, laboratory, and heavy equipment operator skills are honed through a variety of in-house programs.

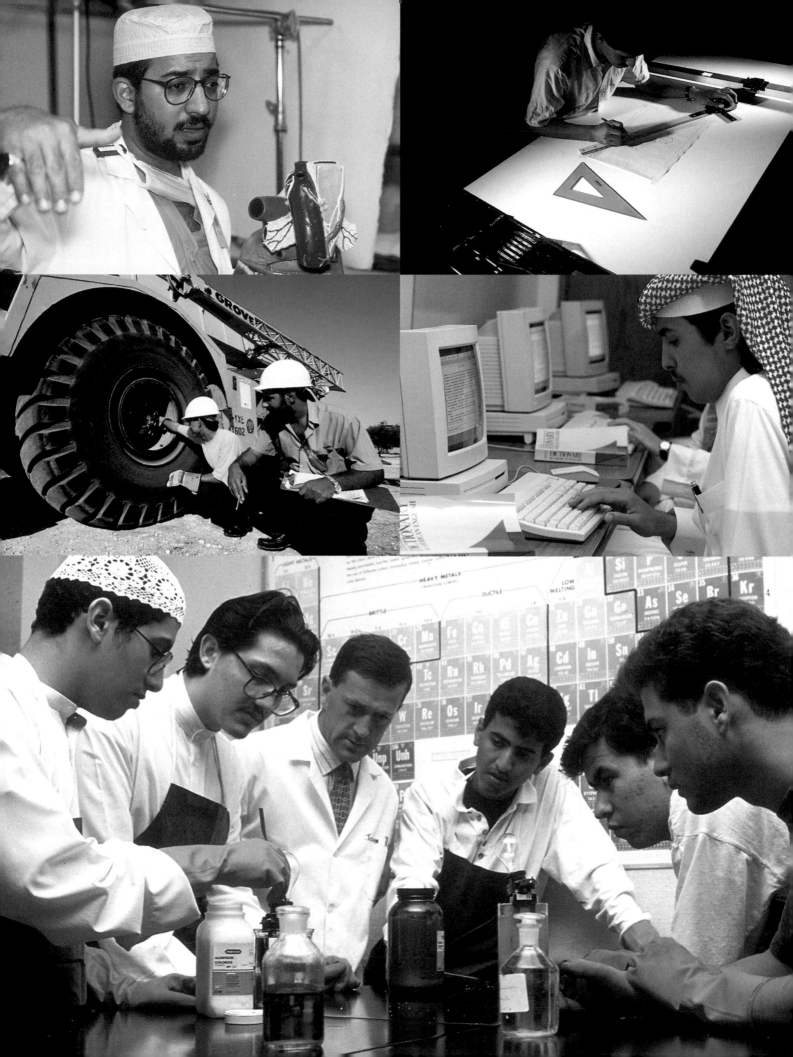

greatly after World War II and by 1950 more than 4,000 Saudi employees (about 40 percent of the Saudi work force) were learning some 144 different crafts, trades, and skills from a staff of about 250 full-time and part-time teachers.

In the 1950s the company's training was centralized in three industrial training centers (ITCs) in the three main operating areas of Dhahran, Abqaiq, and Ras Tanura and three industrial training shop (ITS) facilities. The ITCs concentrated on academic subjects, while the ITS curriculum focused on such skills as blueprint-reading, house-wiring, and plumbing. The 1960s saw the expansion of these training and development programs.

In 1974 Aramco opened a new ITC and workshop in Dammam and in al-Mubarraz in the oasis of al-Hasa. With these additions and with an enlargement of the Dhahran center in 1977, the combined capacity of the system rose to more than 5,000 trainees. ITS facilities, meantime, were enlarged and the curriculum broadened to include subjects ranging from basic industrial principles to advanced training in such areas as electronic instruments and industrial control devices. Additional training facilities were leased in Dammam and al-Mubarraz and new job skills training centers were added in Dhahran, Abqaiq, and Ras Tanura.

The Career Development Department was formed in the early 1980s to better manage career tracking and development activities for Saudi employees. It also provides coordinated development programs for Saudi college graduates and serves companywide supervisory, technical, professional, and management training needs. Furthermore, new training facilities were established in Ras Tanura, Abqaiq, and al-Hasa to house the expanded Industrial Maintenance Training Program. In 1986 King Fahd visited Ras Tanura to inaugurate the maintenance training complex and the refinery modernization project there.

The 1980s also witnessed the introduction of the two-year Apprenticeship Program and the four-year College Degree Program for non-employees, in which recruits may join Saudi Aramco after successfully completing the programs, depending on company needs. In 1993 Saudi Aramco established a new training department with divisions in Riyadh, Jiddah, and Yanbuʻ to meet the needs of up to 2,500 of the Saudi employees acquired under the Royal Decree that integrated large new refining and

A Saudi on a company scholarship in the U.S. participates in advanced laboratory studies.

Language studies play an important role in the company's industrial training centers.

At company industrial training shops, hands-on experience helps students learn such valuable skills as electronics repair.

Growth

The evolution of what was a small, foreign-owned exploration organization into Saudi Aramco — an integrated international oil giant owned by Saudi Arabia — is a fascinating story in itself. Some key developments were the result of business moves aimed at improving the bottom line others were related to the exceptionally open relationship of Saudi Aramco's predecessor, Aramco, with the Saudi Government, as the government moved to take full formal control of its oil wealth. The most momentous change, the establishment of a wholly Saudi-owned company, formally took place on November 8, 1988. In fact, the turnover had progressed steadily through talks over a number of years beginning in the 1970s, thus maintaining the continuity and stability of the kingdom's vital oil industry.

That is not to say the relationship between Aramco and the government was always smooth. There were differences — and sometimes major differences. These, however, were normally resolved through negotiation and agreement rather than through confrontation and crisis. Floyd Ohliger, an Aramco vice president who was frequently involved in talks with King ʻAbd al-ʻAziz in the 1930s, ʻ40s and ʻ50s, summed up the special relationship of those years like this: "King ʻAbd al-ʻAziz was very fair in the consideration of whatever the topic of discussion might be. There were times when we might have some real arguments about things. But we managed to work out the answers that were satisfactory to

King ʻAbd al-ʻAziz and then-Prince Faysal toured Aramco's new refinery at Ras Tanura in January 1947. Aramco Vice President James MacPherson, the company's highest-ranking resident official is to the King's left; T.V. Stapleton is to his right.

both parties." That positive working rapport with Aramco is reflected today by the government's decision to refer to the Saudi Arabian Oil Company as "Saudi Aramco."

The structure of the company evolved through the years in response to changing business circumstances. Shortly after the concession agreement was signed in May 1933, Standard Oil Company of California (Socal) assigned its concession rights to a wholly-owned subsidiary, California Arabian

And Transformation

Standard Oil Company (Casoc), with headquarters in San Francisco. This name was retained and the company's offices remained in San Francisco even after The Texas Company (forerunner of today's Texaco) acquired a 50 percent interest in Casoc in 1936. But on January 31, 1944, the enterprise became a Delaware-registered corporation called the Arabian American Oil Company and from that time on was generally known by its acronym, Aramco.

Late in 1946, after it became apparent that Saudi Arabia's petroleum reserves were extremely large and would call for large market outlets and enormous capital investments for their full development, negotiations were begun with Standard Oil Company of New Jersey (now Exxon) and Socony-Vacuum Oil Company (now Mobil Oil Corporation) to become part of Aramco. The negotiations were complicated by the fact that both these companies were members of the Iraq Petroleum Company group and thus subject to the Red Line Agreement, which prevented them from acting individually in a large area of the Middle East. Final arrangements were completed in November 1948, and from that time until 1975, when Mobil increased its ownership by five percent, shareholdings in Aramco were: Standard Oil of California 30 percent, Texaco 30 percent, Exxon 30 percent, and Mobil 10 percent. Saudi Arabian oil as a result gained extensive new marketing outlets, particularly in Europe.

In 1948, in anticipation of the change in ownership, Aramco moved its offices to New York. At that time offices also were maintained in four other cities in the United States, including Washington, D.C., to carry out various functions in support of operations in Saudi Arabia. In 1952 corporate headquarters were moved from New York to Dhahran, which became the residence of the chairman of the Board of Directors, the president, and most of the principal officers of the company. In 1959 two representatives of the Saudi Government — Hafiz Wahbah, a former Ambassador of Saudi Arabia in London, and Abdullah H. Tariki, the Director General of Petroleum and Mineral Affairs — were named to Aramco's Board of Directors.

In 1974 almost all the operations of the New York office were transferred to Houston, where they became the responsibility of Aramco Services Company (ASC), a subsidiary whose employees have important support responsibilities in the fields of engineering, computer and communications services, purchasing, shipping, and the U.S. training of Saudi employees, as well as various financial and recruiting functions. ASC also maintains an office in Washington, D.C. A second Houston-based subsidiary, Saudi Refining Inc., is Saudi Aramco's partner with a Texaco Inc. affiliate in Star Enterprise.

Another support subsidiary, Aramco Overseas Company B.V. (AOC) in Leiden, the Netherlands, has roots going back to the late 1940s when, following the end of

World War II, many nations had neither dollars nor other convertible currencies to pay for their oil imports. Aramco and its

King Khalid cuts the ribbon to officially open the Berri Gas Plant in October 1977. Accompanying him were then-Petroleum and Mineral Resources Minister Ahmed Zaki Yamani, center, and then-Aramco Board Chairman Frank Jungers, right.

owner companies therefore arranged to sell oil for non-dollar currencies and use these currencies to buy supplies and services. To carry out these arrangements, Aramco in 1948 formed a subsidiary company, Aramco Overseas Purchasing Company, later called Aramco Overseas Company. Located first in Rome, it moved to The Hague in Holland in 1954, and to Leiden in 1984. Until 1986 it played an important role in the design and engineering of many facilities, but in 1988 it once again became primarily a purchasing office. In 1989 the original Aramco Overseas Company was replaced by a new firm, Aramco Overseas Company B.V., a limited liability company formed in the Netherlands.

One reason for Aramco's good relations with the Saudi Government was that the company adhered strictly to Article 36 of the 1933 concession agreement, which stated that "it is distinctly understood that the company or anyone connected with it shall have no right to interfere with the administrative, political or religious affairs within Saudi Arabia." Another reason was that the company listened to the government's point of view and was willing to make changes in its arrangements with Riyadh when these seemed reasonable and necessary.

The original concession agreement was revised several times over the years. The first modification took place in 1939, just seven months after the company had discovered oil in commercial quantities. The revision dealt with various additional payments to the government and a number of other matters and also extended the concession area to the northwest and southwest as well as over Saudi Arabia's undivided half interest "of a maritime or of a territorial nature" in the two Neutral Zones — one shared with Iraq and the other with *(Continued on page 250)*

Growth

Kuwait. With the signing of this agreement the area of the company's concession rights, comprising an Exclusive Area and a Preferential Area, reached its maximum extent until the government increased the company's prospecting mandate between 1986 and 1990.

In their next major agreement, in October 1948, the government and Aramco defined the offshore portion of the company's concession area. The agreement stipulated that the government would receive an additional five cents per barrel in royalty on each barrel of oil produced from the offshore area and guaranteed that royalties paid on offshore oil would amount to at least $2 million a year.

The concession agreement of 1933 specified that "from time to time" the company would relinquish parts of its concession area which it decided not to explore further or make other use of, but for many years the government did not wish the company to exercise that option. Indeed the supplemental agreement of 1939 specifically provided that for the next 10 years the company would be under no obligation to relinquish any part of its Exclusive Area. Under the 1948 agreement the company gave up all its concession rights in the Neutral Zone between Saudi Arabia and Kuwait and all of its Preferential Area west of longitude 46 degrees E and undertook to relinquish an additional 85,500 square kilometers on six specified dates from 1949 to 1970. This accord was superseded by an agreement in March 1963 under which Aramco relinquished all the rest of its Exclusive Area except for approximately 324,000 square kilometers in five separate blocks or "retained areas" and undertook to make six further relinquishments at five-year intervals. But only two more relinquishments occurred, in 1968 and 1973, before the relinquishment program was dropped as part of a fundamental shift in the company's relationship to its government sponsor.

What was probably the most important change in the concession relationship took place in December 1950 with the signing of the so-called 50-50 agreement. Under this agreement, which followed by some months a precedent set earlier in Venezuela and was the first of its kind in the Middle East, Aramco submitted to an income tax (the concession agreement of 1933 had exempted it from all taxes) and undertook to pay the government 50 percent of its gross income less operating expenses. An additional tax of five percent on export sales of crude oil and refined products was imposed by a Royal Decree issued in December 1970.

By this agreement the Saudi Government's income from Aramco's operations came to be linked primarily not to the number of barrels of oil produced and sold, but rather to how much profit the company made. After 1950, therefore, the government showed increasing interest in the prices charged for oil, the cost of operations, and the accounting methods used in determining these things, since all of them affected the size of the kingdom's revenues. Many subsequent modifications of the Aramco concession dealt with matters such as these. At the same time, as the government was increasingly successful in developing a group of technically trained oil experts in its Ministry of Petroleum. It also became more and more interested in and involved with the actual operations of the company — such things as exploration programs, drilling practices, and pressure maintenance.

The theory of "participation," first put forward during the reign of King Faysal by then-Minister of Petroleum and Mineral Resources Ahmed Zaki Yamani in June 1968, was the logical culmination of these trends. Under participation the government would not only share in the income of the company but also in the ownership of its assets and in the top-level direction of its operations. On December 20, 1972, after long negotiations, participation became a reality when the major oil concessionaire companies operating in the Arabian Gulf signed a so-called "general agreement" with their host governments. As a signatory of this agreement, the Government of Saudi Arabia acquired, as of the beginning of 1973, a 25 percent interest in Aramco's crude oil concession rights, the crude oil it produced, and its crude production facilities.

King Fahd visited Aramco in 1986 to inaugurate a new training center in Ras Tanura. H.R.H. Prince Sultan, Minister of Defense and Aviation, and H.R.H. Prince Muhammad, Governor of the Eastern Province, are behind the King. Minister of Petroleum and Mineral Resources Hisham Nazer is to the King's left, and then-Aramco Board Chairman John Kelberer and Aramco President Ali Naimi are to his right.

Discussions continued during 1973 and led to an agreement increasing the Saudi Government's participation interest to 60 percent at the beginning of 1974. Negotiations leading toward complete government ownership resulted in the completion of a draft agreement in 1977, and final agreements were signed in 1980. Under that accord, retroactive to 1976, the participation interest increased to 100 percent when the government paid the shareholders for substantially all of their assets. Aramco, however, continued to operate and manage the country's oil fields on the kingdom's behalf until 1988, when Saudi Aramco was created to carry out those tasks.

In many respects the participation agreements that helped pave the way to the establishment of Saudi Aramco confirmed and reinforced the amicable and continually evolving relationship between Aramco and Saudi Arabia as a sovereign state exercising its rights over its oil resources. The new agreements did not shift the focus of policy- and decision-making: The crucial aspects of the company's oil operations, such as those of production and pricing, were already prerogatives of the government long before it had a share in ownership. Today, Saudi Aramco continues to have good relations with the former U.S. partners, participating with them in such important activities as technology transfer and training. The former U.S. partners, in turn, retain close marketing ties with the country that has consistently stood, within OPEC, for moderation in price and stability of supply of the crude oil that the energy-hungry world relies on.

product distribution assets into the company.

Saudi Aramco's training program is today one of the largest of its kind in the world, with a staff of more than 1,800 teachers, trainers, and support personnel. In 1994 approximately 10,000 Saudis participated either full- or part-time.

Saudi Aramco sponsors a selected number of Saudi employees annually to pursue bachelor's and associate degrees in required fields. The company also seeks out Saudi college graduates with degrees in disciplines such as science, engineering, or business administration. Meanwhile Saudi employees are sponsored on a highly selective basis to pursue advanced degrees up to the Ph.D. level in specialties of particular importance to the company, such as the earth sciences and engineering, while Saudi physicians and dentists may also be sponsored for residency and advanced training. In the Specialist Development Program, Saudi Aramco prepares engineers and computer professionals for careers as company technical experts in their fields through concentrated, hands-on field experience and, if required, advanced degrees. Experienced Saudi employees may take part in assignments with the former Aramco shareholders in the United States.

Saudi Aramco provides temporary summer employment to Saudi high school and university students to introduce them to career opportunities in the enterprise. The company also offers programs for Saudi undergraduate or graduate university students in which they receive credit for course work while undertaking Saudi Aramco job assignments.

AN INTERNATIONAL WORK FORCE: Saudi Aramco stands on strong foundations laid down over more than six decades by thousands of employees from dozens of different countries. While the management and operation of the enterprise now rests largely in skilled Saudi hands, people of many nationalities helped to build the enterprise. This was especially true during such periods of rapid growth and development as the late 1940s and early 1950s and the 1970s. Saudi Aramco continues to employ expatriate professionals in fields where they often share their skills in a development capacity with up-and-coming Saudi specialists.

In the early years, and up until the last days of World War II, it was mainly men from other Arabian Peninsula and Arabian Gulf countries and from the United States and India who formed the major part of the non-Saudi work force. In fact Indians, and later Pakistanis, who had provided much of the trained manpower in the Gulf for many years before the coming of Aramco, have always made up a significant portion of the company's work force.

The period of great expansion that started in 1945 called for a rapid buildup in the company's construction manpower, and a large pool of trained craftsmen was discovered in Eritrea, across the Red Sea from Jiddah. There many hundreds of Italian colonists and former soldiers of Mussolini's defeated armies were marking time as virtual prisoners of war and welcomed the chance to take productive jobs in Dhahran and Ras Tanura. By the end of 1945 Aramco had almost 1,600 Italian employees, and although their numbers declined sharply after the period of postwar construction and expansion came to an end in the early 1950s, Italians continued to form a small part of Aramco's manpower until the 1970s.

The next group to enter the Aramco work force in considerable numbers were the nationals of other Arab countries. Many hundreds of workers from Aden and Sudan, both craftsmen and clerical and professional personnel, were employed during the late 1940s and early 1950s, and Sudan has continued to be a source of manpower for the company. Late in 1948, following the Arab-Israeli war and the expulsion or flight of many hundreds of thousands of Arabs from their homes in Palestine, King 'Abd al-'Aziz asked Aramco to make a special effort to provide jobs for as many of them as possible, and the next five years saw the addition to the work force of many hundreds of Palestinians, some of whom remained with the company for decades. Other Arab nationalities that have contributed in smaller numbers to Aramco's manpower have included Egyptians, Jordanians, Lebanese, and Syrians.

With the unprecedented demands for large numbers of additional employees to carry out the expansion of the 1970s Aramco had to look even farther afield for trained manpower and found new sources for recruitment in Britain, Ireland, and Asia, including the Philippines and Sri Lanka. The Aramco work force reached a peak of just over 60,000 employees in 1981.

Saudi and expatriate professionals work together in areas of highly advanced technology.

Saudi Aramco utilizes a mentor program to help develop specialists in a variety of fields.

Saudi Aramco's program of interest-free loans encourages Saudi Arab employees to own their own homes.

More than 100 elementary and secondary schools, serving more than 51,000 children, have been built by Saudi Aramco and integrated into the Saudi Arab Government system.

Innovative architectural design distinguishes a new girls school in Doha, built under Saudi Aramco's agreement with the government.

The fact that the company had recruited or trained such a large body of skilled employees, many of them with years of experience in the kingdom, was a major factor in the Saudi Government's decision to entrust to the enterprise the implemention of projects that have been crucial to the nation's economic welfare.

NON-OIL ACTIVITIES: On its way to becoming the world's largest oil-producing enterprise, and even before it became involved in large-scale projects such as the Master Gas System, Saudi Aramco's forerunner, Aramco, was far from limiting its activities to those connected directly with oil operations — the usual mix of exploration, drilling, production, and shipping. From the earliest days the company was involved in a very wide range of other kinds of activities, sometimes out of necessity and sometimes as a matter of choice.

When oil exploration began in Saudi Arabia in 1933 the basic requirements of a modern industrial operation were almost nonexistent. The company, perforce, had to enter into a wide variety of non-oil-connected activities until such time as the local economy could provide the goods and services needed to support its oil exploration and development operations. In addition there were a large number of other activities that the company undertook. This was because they were required under Saudi laws and regulations or because the company desired to be a good citizen and pursue a policy of enlightened self-interest — that is, it believed that whatever might help develop and enhance the local community in eastern Saudi Arabia would ultimately also benefit the corporation. Many (perhaps a majority) of the non-oil-related activities that Aramco engaged in were prompted by a combination of these different reasons.

Because of the tremendous growth of private sector and government services in the kingdom over the years, Saudi Aramco no longer pursues such a wide range of non-oil efforts. In line with good business practices Saudi Aramco is following a program to reduce manpower and expenditures, while maintaining quality, in fields not directly related to core oil activities. However the company still provides assistance to benefit the community at large, particularly when its own operations are enhanced.

HOME OWNERSHIP: An important element in Saudi Aramco's effort to draw a stable work force from across the kingdom has been its Home Ownership Program, which started in 1951. Key features of the program are land acquisition and home development. Eligible Saudis acquire free ownership of a plot made available by the government and developed by the company, or receive an allowance to purchase a lot that meets set standards in the community of their choice. Participating employees also receive long-term, interest-free loans from the company to help pay for constructing or purchasing a house.

By the end of 1994 about $3.7 billion had been lent to employees and more than 36,000 homes had been built or purchased under the program. Most of these homes are near company communities in the Eastern Province. The program has not only benefited employees but has also helped the local market by creating additional job opportunities and business opportunities in the field of home construction.

SCHOOLS FOR CHILDREN: Saudi Aramco contributes to education in a program that includes the children of many of its employees. The company's school-construction program, begun in 1953, complies with Saudi Arabian Government regulations requesting certain industrial employers to build schools for the children of their workers.

Initially elementary schools were built to accommodate male students between the ages of six and 14. In 1959 this effort was expanded to include the construction of intermediate schools for grades seven through nine. In 1960 the government moved to establish public education for women for the first time and the following year Aramco began to build elementary and intermediate schools for girls. In 1984 the company began to build high schools for both boys and girls.

The boys' schools are turned over to the Ministry of Education as soon as they are completed; the girls' schools are turned over to the General Presidency of Girls' Education. The facilities become part of the government school system, which takes full responsibility for their operation, while Saudi Aramco maintains the schools. At the end of 1994 a total of 103 schools, with a capacity to accommodate more than 51,000 boys and girls, had been built and were being maintained by the company.

SAUDI ARAMCO EXHIBIT: One of the most striking educational facilities built by the company is the Saudi Aramco Exhibit in Dhahran. A "House of Learning" devoted to telling the story of the petroleum industry in Saudi Arabia and highlighting some of the important contributions of Arab and Muslim scholars to the world's technological heritage, the exhibit opened in 1987. Standing in rocky terrain and reflecting traditional Islamic architecture in modern terms, it welcomes tens of thousands of Saudi schoolchildren and other visitors annually and — through traveling shows — carries its message throughout the kingdom and abroad.

The exhibit features state-of-the-art technology, including interactive computer games and a variety of hands-on displays. Beginning with the formation of petroleum millions of years ago, it takes visitors through the discovery of petroleum in Saudi Arabia in 1938 and then on through the crucial steps in its transportation and transformation into products vital to the world today. The exhibit's unique technological heritage section features excerpts from the works of scholars and functioning models of their achievements. To meet the needs of an international audience everything in the facility — from portions of medieval Islamic manuscripts to videos and computer games — is presented in Arabic and English.

HEALTH CARE: The labor regulations enacted by the Government of Saudi Arabia in 1942, four years after Aramco began the commercial production of oil, required employers to provide free medical care for workers and their dependents. However the company had begun medical treatment in Dhahran even earlier, in 1936, when it opened the first out-patient clinic in the region. Individuals requiring hospital care were sent to Bahrain.

In 1944 Aramco opened its first hospital facility, the Dhahran Health Center, and established a clinic in Ras Tanura, the site for its new refinery. The Dhahran Health Center, now a major hospital with a variety of advanced medical services, has undergone several expansions over the years, the latest in 1987. Smaller health centers and clinics have also been established in other company communities. The company's medical facilities provide care for more than 200,000 persons, employees and their eligible

dependents. Saudi Aramco has continued to support private medical care and services in its areas of operations. Since 1993 employees have also received treatment at contracted medical facilities near their homes in cities throughout the kingdom.

Preventive medicine efforts have also been a part of the company's medical programs. Historically, by their very nature, these have not been restricted in their application and effect to employees. The earliest and one of the most important of such efforts began in 1941 and was directed against malaria, which at the time was endemic in the Eastern Province. From 1941 to 1947 Aramco conducted an anti-malaria program at its own expense. From 1948 to 1955 the company and the Saudi Government collaborated in the campaign and in 1956 the government took over full operation of the program. These programs led to the virtual eradication of malaria in the Eastern Province. Saudi Aramco remains committed to promoting good health among its employees and their families through activities such as the Communicable Disease Control Program, health education, and occupational and environmental health services.

Young Saudi visitors to the Saudi Aramco Exhibit in Dhahran examine a globe highlighting oil reserves.

A radiologist examines ultrasound images at the Dhahran Health Center, which meets U.S. standards on hospital accreditation.

TECHNICAL ASSISTANCE: Almost from the day its geologists first set foot in the kingdom the company found itself involved in providing technical assistance in one form or another to the people of the Eastern Province. Most of the early technical assistance was provided on an informal basis — often after working hours — by various individuals and organizations within the company. By the end of World War II this activity had become a sizable function; to more formally and efficiently organize and channel it Aramco established an industrial development organization in 1946.

The group (now called the Industrial Development Division) was originally set up to seek out local entrepreneurs, giving them the technical, material, and financial assistance that would enable them to provide the goods and services needed by Aramco, its work force, and the developing local communities in the company's area of operations. The enterprise's encouragement of such contractors helped develop a large and experienced construction, transportation, and maintenance capability in the local community.

A Saudi Aramco representative, left, meets with a transformer manufacturer in Damman.

Saudi Aramco is an influential and important member of the Saudi community. Clockwise from top right: the Dhahran mosque, designed under company auspices, was completed in 1987; Saudi Aramco's Home Ownership Program, established in 1951, has enabled Saudi employees to build their own homes in neighborhoods such as Doha, near Dhahran; the company has worked with the government to build more than 100 boys and girls schools since 1953; Saudi Aramco's mobile libraries visit schools throughout the Eastern Province; and the Saudi Aramco Exhibit in Dhahran offers insights into the oil industry, science, and technology. Maintaining the quality of the environment is of special interest to the company. Representative of this commitment are, clockwise from top left: air-quality monitoring; in-depth studies of fragile marine environments; and a broad-based program of oil spill prevention and protection, as reflected in an oil spill drill near Ras Tanura.

Helping local farmers establish profitable vegetable farms was a major goal of Aramco's agricultural assistance program.

SCECO's 230-kilovolt transmission system has brought electricity to all of the towns and villages, and many of the farms, of the Eastern Province.

With a capacity of 2,400 megawatts, the Qurayyah power plant provides more than one-fourth of SCECO-EAST's power generation capacity.

As Saudi banks and government agencies began to provide many industrial development services the level of company assistance diminished. But Saudi Aramco continues to provide technical and business advice to Saudi manufacturers wishing to do business with the company and remains committed to dealing with businesses in the kingdom whenever possible. The company spends hundreds of millions of dollars annually to buy materials, most of which are purchased from Saudi suppliers, including local manufacturers. Between 1990 and 1994 the company purchased about $1.1 billion worth of materials a year and Saudi manufacturers and vendors supplied nearly 90 percent of the total. Capital projects and operations contracts between the company and hundreds of Saudi-owned and joint-venture businesses also have a major economic impact. At mid-decade the outstanding value of these contracts totaled some $2.7 billion.

AGRICULTURE: Agriculture was another field of activity not directly related to oil operations that Aramco became involved with very early in its history. The agricultural assistance efforts grew out of the company's cooperation with the Saudi Government in developing farming in the area of al-Kharj, 80 kilometers southeast of Riyadh, between 1938 and 1954.

To assist farmers in the Eastern Province oases of al-Hasa and Qatif, the company established an agricultural assistance organization in 1956. This group developed a 120-hectare demonstration farm on land provided by the Saudi Government in al-Hasa oasis in 1978 to show farmers how modern agricultural techniques and labor-saving machinery could turn salt-damaged soil into profitable farmland. The farm, which proved a great success, was turned over to the Ministry of Agriculture and Water in 1993 and the agricultural assistance organization — its mission now in the hands of the government and private parties — ceased to exist.

ELECTRICITY: The company also played an important role in the creation of a regional power system in the Eastern Province. The first step came in the 1950s when Aramco helped establish a number of private power companies in the local communities within its area of operations. When the rapid growth and

development of some of the major towns in the late 1950s and early 1960s required considerably larger and more sophisticated facilities, Aramco assisted the power companies in the two largest local communities, Dammam and al-Khobar, to merge and form the nucleus of a single large firm to serve the entire coastal area.

When the industrial growth of the 1970s got under way, the Saudi Consolidated Electric Company in the Eastern Province (SCECO-East) was created by a 1976 Royal Decree as the first of several major regional power grids, and Aramco was designated to manage the company for its first five years, beginning in 1977. The immediate tasks were to meet the needs of industrial development — especially as represented by the primary industries at Jubail Industrial City and Aramco itself — to electrify Eastern Province communities that had no electric power, and to upgrade the systems of those with inadequate supplies. The number of SCECO-East customers, which stood at about 90,000 in 1977, had climbed to nearly 200,000 by 1981 when Aramco's full management agreement was terminated and its support role substantially reduced. By the early 1990s SCECO-East had extended its grid to cover the territory's most remote villages, more than doubled its customers to over 400,000, and was even sending large amounts of power to Riyadh in central Arabia.

A NEW ERA: The march of electricity across the Eastern Province is just one of the factors that has changed the face of the company's home-base region over the years. These changes, reflected in the mirror windows of an office tower on a busy avenue at rush hour, present an image that stands in dramatic contrast to what existed even 15 years ago. A number of the old, traditional markets still function, but supermarkets, shopping malls, and fast-food restaurants are now the rule rather than the exception. Where dhows, the traditional lateen-sailed ships of the Gulf, once provided the only alternative to flying the 25 kilometers to the island of Bahrain, today the King Fahd Causeway carries cars over the water on soaring strips of concrete. And, as elsewhere in the world, high technology is a necessity — from mobile telephones to computers in businesses, at home, and in the school.

The changes that have taken place in the

Eastern Province, the area along the Arabian Gulf that was the birthplace of the nation's oil industry, represent in microcosm what has happened throughout the kingdom. What is unique today, however, is that the latest generation of boys and girls has known no world other than this new one which, to a great extent, was ushered in by their grandfathers who helped to find oil near Dhahran in 1938. This generation's fathers and mothers are now the "old-timers," for they can remember the times when changes were still novel, if not fantastic. In a sense the latest generation, the youngsters coming of age in the mid-1990s, have seen the future and it is here.

It is a future spelled out in the industrial complexes of Yanbu' and Jubail. Just being built in 1980, the cities now are flourishing. Constructed on the base provided by the country's rich hydrocarbon resources and nurtured by the Saudi Government, they provide evidence — in new factories, houses, and schools — of the kingdom's drive to diversify its economy, to expand job opportunities, to transfer technology, and to achieve a balance of development across the peninsula.

More prosaically, air conditioning remains perhaps the most pervasive symbol of the changes that have occurred since the coming of the oil industry. In modern office buildings, central air keeps workers oblivious to soaring summer temperatures — until they venture outdoors to climb into air-conditioned cars and drive to air-conditioned markets and homes. On the sidewalks of al-Khobar and Dammam in the Eastern Province hundreds of humming "window units," many manufactured in Saudi Arabia, still protrude over street-front shops and drip lukewarm condensation on the heads of unwary shoppers on sultry summer nights.

August is the month of peak loads on the electricity grid. It is a month when temperatures in towns along the Arabian Gulf's western shore frequently top 43°C (110°F) and the relative humidity hovers near 100 percent. But electricity and air conditioning, made possible by the region's oil wealth, have done for stretches of eastern Saudi Arabia what air conditioning and oil money did for the Texas Gulf coast around Houston, transforming a muggy coastal scrubland into a sprawling, modern urban conglomeration where it is possible to work and live not only comfortably, but also pleasurably.

AS IT USED TO BE: Although the winter rains along this coast provided better grazing for the sheep, goats, and camels of the Bedouins than some other parts of the peninsula, it was not rainfall that gave the region its modest prosperity and sustained man since earliest times. Rather, groundwater from distant mountains, welling up clear and plentiful from artesian springs or 'ayns in two major oases, nurtured broad date gardens, while the turquoise, shoal-filled waters of the Gulf provided fish and pearls and carried the commerce of distant shores.

Hardy Bedouins, farmers painstakingly tending groves of palms and the silver webs of irrigation channels, fishermen, divers, sailors, townsmen, merchants, and traders supplying the simple needs of these and of others in the far hinterland — these were the people who dwelt, surviving, occasionally thriving, in a harsh and demanding land. They lived, as one writer said, "lives of unselfconscious, pervasive religiosity, the name of God constantly on their lips; His intercessions sought; His interventions praised."

Their pleasures, when work was done, were the pleasures of the family and the home, playing with their children, talking with friends and neighbors over endless cups of coffee. Children imitated their parents and at an early age began to share their work. Women, secluded from men outside their family by religion as well as custom, were honored in their role as mothers. Polygamy, a practice now fading, was then widespread. Men in this land felt affection and loyalty to family and to tribe; their respect was for devoutness, age, learning, good manners, generosity, and strong, fair leadership. Hospitality extended to strangers and even, under an elaborate code, to enemies.

Al-Hasa, about 65 kilometers inland from the nearest point on the coast, is the larger of the two principal oases in the Eastern Province. In fact, with some 12,100 hectares under cultivation, it is one of the largest anywhere. Hundreds of thousands of palm trees grow there, watered by more than 50 natural springs and several hundred dug wells.

Hofuf is the largest of al-Hasa's dozens of villages and towns. After recapturing Hofuf from the Ottoman Turks in 1913, 'Abd al-'Aziz Al Sa'ud kept the name al-Hasa Province, a Turkish usage, for the whole coastal region and retained Hofuf as the capital. It remained so until 1956 when the administrative center was moved to the growing port city of Dammam

At one time, Jubail sent out as many as 200 pearling boats, but the trade declined in the early 1930s.

Coffeepot-making is a traditional craft in al-Hasa oasis.

Remains of a fortress on Tarut Island — here seen in an aerial photo of the late 1940s — are believed to rest atop evidence of millenia old civilizations.

On the shore of the Arabian Gulf, Najmah, right, is representative of Saudi Aramco's residential communities. Self-sufficient, with office buildings nearby, these communities offer nearly all the amenities of modern living. This page, top: Residents read their mail at Dhahran's al-Mujamma Building, a community center that houses the local mail center, travel office, barber shop and laundry service. Center: Two of the variety of housing styles found in Saudi Aramco communities. Facing page, clockwise from center: a Saudi Aramco community supermarket, library, and an elementary school for expatriate students. Such schools offer classes from kindergarten through the ninth grade. Bottom: the Dhahran Headquarters complex. The twin towers of the central administration building are at left; the EXPEC and Engineering buildings are at right.

Jubail was a coastal village when oilmen arrived in 1933; five decades later it had become the site of a vast industrial complex.

A seawater desalination plant is among the many installations that have turned Jubail into a thriving industrial city.

The market area, or suq, of al-Khobar in 1946 consisted of a few shop stalls along an unsurfaced street.

Today, a melange of fast-food restaurants and sophisticated shops draws visitors from throughout the Eastern Province to al-Khobar.

and the region was given its present name, the Eastern Province.

Just north of Dammam and some 130 kilometers from Hofuf is Qatif, the second of two major oases. Smaller than al-Hasa, but also comprising many villages and natural springs, the Qatif oasis is five kilometers across at its widest point. Its date gardens once stretched for kilometers along the shore of Tarut Bay. But in many locations today, as in al-Hasa, rows of plastic-covered greenhouses featuring hydroponic agriculture have taken the place of the palmeries. In the bay, across the shallows from the large town of Qatif, is the island of Tarut, once accessible only by boat at high tide and by wading at low tide, but now joined to the mainland by a causeway.

On the desert coast of Arabia plentiful water and safe anchorage have made Qatif and Tarut trading centers since at least the third millennium B.C. They probably served — as nearby Bahrain Island did — not only as a source of pearls and dates, but also as a link between the ancient civilizations of Mesopotamia and the Indus Valley. Persian Zoroastrians, Jews, and Nestorian Christians all lived and worshipped in Qatif and on Tarut before the coming of Islam. The Turks and perhaps the Portuguese had garrisons or outposts in the area and either built forts or added to ones they found already existing. It was on Tarut, in 1915, that the British Government, which had established protectorates elsewhere in the Gulf to help safeguard its sea lanes to India, signed a treaty recognizing the authority of 'Abd al-'Aziz Al Sa'ud as the independent "Ruler of Najd, El Hassa, Qatif and Jubail, and the towns and ports belonging to them," that is to say, most of central and eastern Arabia.

Jubail, about 65 kilometers farther up the coast from Qatif, with only two small springs, was not a large or, as far as anyone knows, an ancient oasis. Its importance was as a port of entry for goods destined for Riyadh (lying as it did at the beginning of one of the principal trails across the Dahna sands) and to Buraydah and 'Unayzah and other towns of northern Najd, and it had an advantage over Qatif and al-'Uqayr in its deeper waters which allowed larger vessels to unload cargoes there. The pearling fleet once based there and at Tarut never recovered from the world depression of the early 1930s and later competition from Japanese cultured pearls.

Down the coast some 95 kilometers south of Dammam is the port of al-'Uqayr, once counted as a long day's journey by camel from al-Hasa and the most favored port of entry to Riyadh and the interior by way of Hofuf. Since the development of ports farther up the coast with deeper water and more modern facilities, and with the completion of roads and a railroad from Dammam, the importance of al-'Uqayr has declined and its harbor today is little used.

Except for its people eastern Arabia had few riches, although compared with much of the world at that time the people of the region were well enough off. But before the discovery of oil brought new jobs for the young and increased income for government social services, this was not an easy place to live. Schools and hospitals had yet to be built. Smallpox, cholera, and malaria were endemic. There were no paved roads and few cars to use them if there had been. There were few radios, no telephones, no electricity except from the odd generator, and of course no air conditioning.

Viewed from the air today much of the coastal Eastern Province appears laced with the sinews of the twentieth century: four- and six-lane highways, a railroad, an endless maze of pipelines of all sizes carrying oil, gas, natural gas liquids and water, irrigation canals, chain-link fences, telephone lines, and electric transmission cables. Along the sweeping crescents of shore where desert and gulf merge, the physical manifestations of change loom up in concrete and steel. Today even the meanings of names have changed across this landscape. Places known to generations of Bedouins or seafarers as hills, plains, wells, capes, and islands are printed on maps and become oil and gas fields, pumping stations, industrial plant sites, and new communities.

LIFE AT SAUDI ARAMCO: Saudi Aramco's exploration teams have covered the kingdom from the borders with Iraq and Jordan in the north to the far reaches of the Rub' al-Khali some 2,100 kilometers to the south. The bulk of the company's producing and processing activities and the multitude of diverse and technologically complex installations and facilities to support them are located in the Eastern Province, near the main concentration of the kingdom's oil fields. But crude production has now expanded to central Arabia, and the company's refining, gas-processing, and shipping activities extend across the kingdom to the Red Sea. Furthermore, Saudi Aramco's worldwide marketing activities, its linkups with oil refining and marketing companies in other parts of the globe, and its large, advanced tanker fleet have made it a truly international oil enterprise.

Dhahran, principally an administrative center, is the oldest and largest of the company's four communities for families in the Eastern Province and has grown up over the Dammam field, where oil was discovered in commercial quantities in 1938. Ras Tanura, on a narrow headland about 65 kilometers north of Dhahran and not far from the Qatif oasis, is the site of a major refinery and a marine terminal. Abqaiq, built over the oil field of the same name some 65 kilometers southwest of Dhahran, is a producing and processing center.

The smallest family community is 'Udhailiyah, located atop the central portion of the Ghawar field, about 115 kilometers south of Abqaiq and 32 kilometers from al-Hasa oasis. The newest company community, for bachelors, is Tanajib on the Gulf coast, 150 kilometers north of Ras Tanura. Tanajib is a major operations and maintenance complex associated with offshore oil and gas production facilities. Saudi Aramco also has employees based in many different locations to support its operations across the kingdom, with the largest concentrations on the west coast at Yanbu', Rabigh, and Jiddah, and in Riyadh.

The Eastern Province has always been the work and home site for most of the company's employees. In the beginning Dhahran was a rustic camp, an all-male outpost which provided mainly Saudi Arabian and American oilmen with the bare essentials: workshops and offices in which to do the jobs they were there to do, and basic board and bed. Today Dhahran and the other Eastern Province communities look much like modern suburban towns in the American Southwest — with a few noticeable differences. They feature trim, one- and two-story houses surrounded by lawns set off with hedges, trees, and flowers. Many people bicycle to work and many go home for lunch. Each community has recreational facilities including tennis courts, swimming pools, a bowling alley, a baseball diamond, and a golf course, along with other amenities of modern life.

A number of the children of the pioneering expatriate families who worked and lived here have returned to become second-generation employees of the company, and there are even a few third-generation expatriate workers with Saudi Aramco.

The biggest difference between these oil communities and small towns elsewhere is their international character. A new arrival might find that living in the apartment or house next door, or working in the office across the hallway, is a Saudi or an American — or a Canadian, an Englishman, or Scot, an Indian or Pakistani, a Lebanese, a Palestinian, or perhaps a Dutchman or a Filipino. Saudi Aramco is the largest employer in the Eastern Province and, outside the Saudi Government, in the kingdom as a whole.

Most Saudi employees live in the long-established towns closest to the various oil operations centers, most in their own homes built with company loans. Their children, both boys and girls, attend schools administered by the Saudi Government, many of which were built and are maintained by Saudi Aramco. For the children of expatriate employees living in the four oil communities in the Eastern Province, the company operates schools following the American system.

Amid this diversity of people, different cultures flow together. For example, Saudis and foreigners learn to enjoy each other's cooking and exchange friendly greetings in each other's language. Sometimes the cultures meet unchanged in mutual respect, as when non-Muslims refrain from eating or smoking in public during the fast of Ramadan or conservative Saudis accept the fact that Western women go unveiled.

When they are not at work, employees and their families in the communities keep busy in a seemingly endless variety of pursuits. In Dhahran there exist more than 50 social activities groups dedicated to sports, hobbies, crafts, and professional interests of all kinds. Similar groups, equally active, exist in the other communities, with interests ranging from stamp collecting to sailing, rugby, and karate, and from amateur theatricals and choral singing to ceramics, weaving, natural history, and gardening. Through the years various forms of group endeavor — annual art and garden shows, scouting jamborees, and community choruses — have become some-

Comfortable homes in garden surroundings characterize Saudi Aramco communities.

Air-conditioned buses transport expatriate youngsters to and from the Saudi Aramco community schools.

Pleasant living conditions make leisure hours at home enjoyable.

Camping enthusiasts with off-road vehicles delight in the rugged beauty of the desert.

From soccer to scuba diving, most Saudi Aramco employees and their families share a love of sports that keeps them outdoors, and on the playing fields, for many of their leisure hours. Group activities are especially popular. Bowling, golf, softball and tennis tournaments are regular events in Saudi Aramco communities, as are sailing and equestrian competitions. Marathons and half-marathons highlight the running season, while soccer, always popular in Saudi Arabia, is now the number one activity for Saudi Aramco youngsters. Company communities are provided with a wide range of recreation facilities, from swimming pools to handball courts to gymnasiums. There are also well-stocked libraries, theaters, and meeting places for the many self-directed organizations, grouping artists, stamp collectors, drama buffs, and even auto repair enthusiasts. In addition, crafts, travel, and women's groups are part of each community.

Climate

Like most of the rest of the peninsula, the Eastern Province region is an area of environmental extremes. The best known features of the climate are the intense summer heat and frequent strong winds. However the Eastern Province has experienced freezing temperatures, torrential rains, thunderstorms, and extended sandstorms which add to the contrasting extremes of this desert climate.

The year can be divided into seasons in this region. The length of the summer is greater than in more temperate zones and can be considered as lasting from May until September, with the hottest months being June through August. There is normally no rain during this period, and evaporation rates are high. Relative humidity values are low in inland areas during the summer season though coastal humidities remain quite high.

Beginning in late May and June, an extensive low-pressure area develops over the Asian continent as a result of the early summer heating of this large landmass. The counterclockwise wind circulation around this low is a permanent feature over and adjacent to the continent of Asia during the entire summer. The eastern coast of Arabia lies on the southwest edge of this low-pressure zone and winds blow predominantly from a north-northwest direction. Increased winds during June and occasionally early July are known as *shamals,* the Arabic word for northerly wind. Sustained winds for two- to three-day periods at speeds of 40-50 kilometers per hour create blowing dust and limited visibility. Shamals with gusts of 95 kilometers per hour lift sand-sized particles a meter or more above the desert surface and dust fills the air. By mid-July there is no longer an appreciable pressure difference between Asia and Arabia, and the still predominantly northwest winds are weaker than in the early summer. August is normally the calmest month, though the hottest.

The autumn months of October and November are a transitional period in which temperatures fall and relative humidity begins to rise. The Asian low-pressure area weakens, and by November the first winter storms begin to affect the area. These move eastward across the Mediterranean and the Middle East. As the centers of these storms approach the Arabian Gulf, the leading boundary of the front frequently creates southeast or south winds. These are then usually followed by northwesterly winds as the front passes. Wind speeds and directions associated with these storms thus show wide variations in frequency of occurrence. Severe tropical cyclones from the Arabian Sea weaken as they approach southeast Arabia.

The winter months of December through February are characterized by stormy periods with strong winds, some rain, thunderstorms, and blowing dust interrupted by periods of relatively mild weather. February is usually the windiest month. Freezing temperatures, though rare, have been recorded in December and January in the Eastern Province.

The winter storms diminish in frequency and intensity during the spring months of March and April, though strong thunderstorms may occur locally during this period. Temperatures again begin to rise with the approach of summer, and the humidity decreases.

The climate over the Arabian Gulf is essentially the same as that of the Eastern Province since this extremely shallow marginal sea is virtually surrounded by land. The only connection with the open ocean is through the narrow Strait of Hormuz, so that the Arabian Gulf can be considered as a restricted arm of the Indian Ocean. Offshore sea surface temperatures vary from 16°C (60°F) in winter to 32°C (90°F) in summer, while coastal bays and lagoons have an even wider temperature fluctuation range, attaining temperatures of 40°C (104°F). Salinities are high in the Gulf due to excessive evaporation from the extreme temperatures and strong winds, as well as salt addition from numerous brackish undersea springs.

thing close to local institutions.

In addition there are special, sometimes unique, opportunities which come from living in the kingdom. Some employees keep and ride Arabian horses, others collect Oriental carpets. Scuba divers sail by dhow to remote islands to fish or photograph in the teeming waters of the Gulf or Red Sea. Others indulge their penchants for geological and historical field trips; they watch birds, collect shells, and learn to identify desert plants. A few even take up such esoteric interests as falconry. Teenagers enjoy racing across dune country on motorcycles, and families with four-wheel-drive off-road vehicles camp in the desert, sometimes driving long distances by compass along the fringes of the Rub' al-Khali sands.

Foreign employees interested in seeing Saudi Arabia take advantage of short holidays to explore neighboring cities and villages. New highways throughout the kingdom make driving relatively easy, but for longer trips, such as between Dhahran and Yanbu' or vice-versa, many choose to fly. The Red Sea coast boasts coral reefs that are among the most beautiful in the world; to the northwest lie the majestic ruins of Madain Salih, a Nabatean city dating to third century B.C. or earlier; and in the far southwest are the cool green mountains of 'Asir. For longer vacations it is possible to drive north to Jordan, Syria, and Lebanon or on to Europe. An adventurous few drive in yet another direction, to picturesque and mountainous Oman and Yemen.

Closer at hand, newcomers in the oil communities may hear stories about donkey-drawn wells where now there are diesel pumps, of arrow-shaped reed fish traps on tidal flats now filled in to form the foundations for hotels and amusement parks. But if they search they can find pieces of the past: There are still potters in al-Hasa who make clay water jugs; there are basketmakers in Qatif who still braid straw into circular floor mats; and there are Bedouin women who still weave homespun wool into colorful strips of sturdy textile. More and more in modern Saudi Arabia, however, one must visit a museum — such as the one operated in Dammam by the Department of Antiquities — or travel to a cultural festival — such as the annual springtime fair and camel races at al-Janadiriyah near Riyadh — to see, touch, and taste the rich parts of the past in present-day Saudi Arabia.

DEVELOPMENTS TODAY: Life in Saudi Arabia would have changed even without the discovery of oil, but certainly at a slower pace. Change has come in little ways as well as big. Trucks, not camels, are now the "ships of the desert." Today Bedouins use tanker trucks to carry water to their herds of sheep and camels, and when they move their black tents they load them — along with their radios and cylinders of bottled cooking gas — onto gaily decorated trucks. Camels, of course, are still herded in the desert, but it is as probable today that Saudis and expatriates alike will taste their first camel milk (homogenized and pasteurized) from a carton, rather than still warm in a wooden bowl offered by a herdsman.

A new generation admires the traditional crafts of Saudi Arabia at a museum in Yanbu'.

Townsmen, when they sit together in the evening, watch color television and offer their guests soft drinks and canned juices as often as tiny cups of cardamom-flavored coffee. Children in villages ride tricycles and young men's sports clubs organize taekwondo matches. Middle-class families go for weekend drives in their air-conditioned cars and stop for ice cream cones or pastries.

The art of bargaining is still practiced in covered *suqs* which have the smell of spices and musty carpets and the glitter of tin, brass, copper, and gold. But among the goods for sale are compact discs of the latest hits from Cairo and London, battery-powered toys from Hong Kong, watches and lighters and pens from Tokyo, Geneva, and Paris. In the fish market the fish — in tempting variety — are fresh from the Gulf and the Red Sea, and stacked in the next aisle of stalls are baskets of dates and bags of rice. But there are also carrots, cucumbers, and cabbages grown on local farms, apples shipped from Lebanon and Washington State, oranges from Morocco and from Jordan, mangoes from India, and bananas from Honduras.

Today it is not uncommon to see the beast once called the "ship of the desert" defer to a faster mode of travel.

Hofuf and Qatif have spilled beyond the walls that once surrounded them, and nearby date gardens are yielding to houses and streets. Inexorably, Dhahran and its two coastal neighbors, al-Khobar and Dammam, are growing together.

Dhahran itself has spread out with new housing for families, and new suburbs have flowered around the community. Next to it, like an acropolis crowning a low limestone jabal, is the magnificent King Fahd University of Petroleum and Minerals. Below the university, toward the Gulf is Dhahran International Airport, then the town of Thuqbah. Al-

King Fahd University of Petroleum and Minerals overlooks Dhahran and the Gulf beyond.

*A*bove: Dhahran was a modest oil camp in 1938, the year Casoc, predecessor of Aramco and Saudi Aramco, discovered oil in the formation below the nearby jabal.

Below: By the mid-1990s, Dhahran had become the headquarters for Saudi Aramco's global activities; it remained a pleasant residential community for employees from around the world.

Towering office buildings abound in the bustling city of al-Khobar, a center of commerce for the Eastern Province.

Khobar is a bustling shopping center with bumper-to-bumper rush-hour traffic, flashing neon lights, and shops selling most of the consumer products of the industrialized world. It has a number of modern office buildings, several first-class hotels, and any number of good places to eat out, including those most ubiquitous of institutions, Chinese restaurants and American fast-food franchises. In the expanding residential areas pastel stuccoed houses and apartments, schools, mosques, and supermarkets flank tree-lined streets. A new corniche hugs the coast, carrying many of the cars bound toward or heading off the causeway linking Saudi Arabia to Bahrain.

The roads on the east coast of the kingdom near the conurbation of Dhahran, al-Khobar, and Dammam feature a succession of wholesale showrooms, construction storeyards, light industrial areas, and palatial gas stations with their own restaurants and mini-markets — mirroring developments elsewhere in the kingdom. On the outskirts of Dammam, the provincial capital, are a fertilizer plant; two industrial estates with dozens of factories producing items as varied as ice cream and potato chips, plastic pipes, aluminum buildings, paints, paper bags, and air conditioners; and the railroad yards and warehouses supporting the busy commercial harbor.

All government ministries have established branches here; daily newspapers are published; modern soccer stadiums draw capacity crowds; and King Faysal University, Saudi Arabia's youngest, has a new campus in the city. And more changes are coming. Sixty-three kilometers northwest of Dhahran the new King Fahd International Airport was nearing completion at mid-decade. New shopping malls are opening. New industries are on the drawing boards.

For the most part the people of al-Hasa, of Dhahran, al-Khobar, and Dammam, of Qatif, Tarut, and Jubail, like their fellow countrymen throughout the kingdom, realize that the oil and gas resources which have brought all these changes will not last forever. They understand the need to exchange these depletable sources of wealth for others which are enduring — technology, industry, education — so that their grandchildren will not experience the hardships their grandparents knew. They also realize that wealth is not only bringing material changes, but also, in countless ways, is changing the time-tested patterns of their lives.

Not many years ago the people of the region lived by a kind of natural rhythm determined by the changing seasons. The Bedouins clustered near the wells when summer was at its zenith and moved on with the first flicker of lightning on the night horizon in the fall. Seamen followed the monsoon winds into the Indian Ocean, and pearlers and fishermen each had their time. In the oases farmers gathered their dates in the heat of late summer and early autumn and planted other crops beneath the palms to harvest before the burning winds of June.

Today families plan holiday trips according to the vacation periods at their children's schools, while Saudi university students in America and Europe return to the kingdom to join their colleagues in seeking summer jobs. Merchants schedule business trips with a close eye on trade fair calendars around the world. Imported fruits and vegetables know no season, and if in the coastal towns August is still hot and humid, today it no longer really matters.

The pace of life is different now. People are concerned that traditional values will change too. The wealth they had was their faith, a code of conduct, a generosity of spirit, a foundation man could build on. Now education, travel, television, contact with foreigners — all have made Saudi Arabia a full partner in the modern world. Through technology the people of the kingdom hope to strengthen their future, but not at the cost of destroying the foundation, of losing what they had: the closeness of the family with its loyalties and responsibilities, their respect for custom and pride in the richness of their language, the sustaining strength of their religion.

People in the Eastern Province and the kingdom as a whole believe that tradition and modernity can live honorably side by side; that women will be able to take a more active part in society and in the development of their country within the context of Islam without neglecting their roles as wives and mothers; that boys and girls can go to school and learn how to flourish in the larger world and still cherish their past and esteem their elders. They believe also that man can take the time, even in the rush of twentieth-century commerce and industry, to close a shop or stop work at the amplified call of the muezzin to prayer or, pulling a truck to the side of the road at sunset, kneel in the vastness of the desert and offer thanks to God for what He has given.

Even in the remote reaches of the silent desert, men hear in their hearts the call to prayer...and they answer.

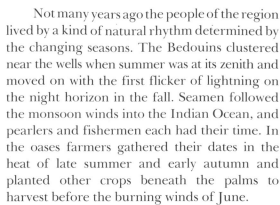

Further Reading

This selection of books for further reading about the subjects covered in this volume does not pretend to be more than an introduction to the vast number of works that have been published on the Middle East and the Islamic world. The books listed are all in English, and they have been chosen for their intrinsic worth, readability, and availability. Many have been published in paperback editions. Although most of the authors are specialists, the books are — with few exceptions — general works written for the nonspecialist. Many offer annotated bibliographies enabling the reader to pursue particular subjects in as much detail as desired.

Following common present-day practice, no attempt is made to list publishers. It is not uncommon now for books to have separate publishers for hardbound and paperback editions in America and in Britain. It is impractical to list all and unfair to list only one.

Brief descriptions of certain books are provided, especially where titles do not clearly indicate the material covered (e.g., *The Loom of History*). On the other hand, descriptions are not normally provided for books as explicitly titled as *The Middle East: A Physical, Social, and Regional Geography*. All books included are recommended, though they are not, of course, of equal merit. The selection is unabashedly weighted in favor of books on the Arabs and the Arabian Peninsula.

A good source of well-illustrated articles on a wide variety of subjects connected with the Arab and Muslim worlds is the magazine *Aramco World*, published six times a year.* Recent special issues of particular interest have dealt with the eastward spread of Islam (1991), the past and present of the Kingdom of Saudi Arabia (1991), the Silk Roads (1988), the role of the Muslim world in the Age of Discovery (1992), Arab immigrants to the United States (1986), the legacy of Islamic Spain (1993), the Arab role in science and technology (1982), and Muslim scientists' contributions to the space age (1986).

---

* Subscriptions are available to a limited number of readers, without charge, from these three addresses: Public Relations Department, Publications Division, R-2213, Admin. Bldg., Saudi Aramco, Dhahran 31311, Saudi Arabia; Public Relations, Aramco Overseas Co., Box 222, 2300 AE Leiden, the Netherlands; Public Affairs/Subscriptions, Aramco Services Co., Box 2106, Houston, Texas 77252, USA. Back issues, where in print, are available from the Houston address.

THE MIDDLE EAST AND THE ISLAMIC WORLD: SURVEYS AND COMPILATIONS

Berger, Morroe. *The Arab World Today.* Garden City, N.Y., 1962.

Brill, E. J., publisher. *First Encyclopedia of Islam, 1913-1936.* 9 vols. Leiden, 1927, 1993.

Coon, Carleton S. *Caravan: The Story of the Middle East.* Revised edition. Huntington, N.Y., 1976.

al-Faruqi, Isma'il R. and Lois Lamya'. *The Cultural Atlas of Islam.* New York and London, 1986.

Glassé, Cyril. *Concise Encyclopedia of Islam.* New York, 1989.

Hayes, John R., editor. *The Genius of Arab Civilization, Source of Renaissance.* 2nd edition. Cambridge, Mass., and London, 1983.

Held, Colbert C. *Middle East Patterns: Places, Peoples, and Politics.* Boulder, Colo., 1994.

Lewis, Bernard, editor. *The World of Islam: Faith, People, Culture.* London, 1976.

Malone, Joseph J. *The Arab Lands of Western Asia.* Englewood Cliffs, N. J., 1973.

Mansfield, Peter. *The Arabs.* 3rd edition. London and New York, 1991.

Robinson, Francis. *Atlas of the Islamic World Since 1500.* New York and Oxford, 1982.

Rogers, Michael. *The Spread of Islam,* in the series The Making of the Past. Oxford and New York, 1976.

Schacht, Joseph, and Bosworth, C. E., editors. *The Legacy of Islam.* Oxford, 1979.

Weekes, Richard V., editor. *Muslim Peoples: A World Ethnographic Survey.* Westport, Conn., 1984.

The Legacy of Islam and the handsome books edited by Hayes, Lewis, and Rogers contain excellent essays by specialists on many aspects of Middle Eastern and Islamic civilization. Berger writes as a sociologist, Coon as an anthropologist, Malone as a political scientist, and Mansfield as a journalist.

THE MIDDLE EAST AND THE ISLAMIC WORLD: HISTORY

Antonius, George. *The Arab Awakening: The Story of the Arab National Movement.* London, 1938; New York. 1965.

Atiyah, Edward. *The Arabs.* London etc., 1955; New York, 1968.

Brockelmann, Carl. *History of the Islamic Peoples.* Translated by Joel Carmichael and Moshe Perlmann. New York, 1960.

Fromkin, David. *A Peace to End All Peace: Creating the Modern Middle East 1914-1922.* New York, 1989.

Hitti, Philip K. *The Arabs, A Short History.* Princeton, 1949.

———. *History of the Arabs From the Earliest Times to the Present.* 10th edition. Boston, 1970.

Hodgson, Marshall G. S. *The Venture of Islam: Conscience and History in a World Civilization.* 3 vols. Chicago and London, 1974.

Holt, P. M.; Lambton, Ann K. S.; and Lewis, Bernard. *The Cambridge History of Islam.* 4 vols. Cambridge, 1977.

Hourani, Albert Habib. *A History of the Arab Peoples.* London and Cambridge, Mass., 1991.

Hourani, Albert; Khoury, Philip S.; and Wilson, Mary C. *The Modern Middle East: A Reader.* Berkeley, 1994.

Ibn Khaldun. *The Maqaddimah: An Introduction to History.* New York, 1958.

Kirk, George E. *A Short History of the Middle East, from the Rise of Islam to Modern Times.* 6th edition. New York, 1960.

Lapidus, Ira M. *A History of Islamic Societies.* Cambridge, 1990.

Lewis, Bernard, editor and translator. *Islam: from the Prophet Muhammad to the Capture of Constantinople,* in the series The Documentary History of Western Civilization. 2 vols. New York and London, 1974.

Mas'udi. *The Meadows of Gold: The Abbasids.* Translated by Paul Lunde and Caroline Stone. London, 1989.

Monroe, Elizabeth. *Britain's Moment in the Middle East, 1914-1956.* London, 1963.

Sinor, Denis, editor. *The Cambridge History of Early Inner Asia.* Cambridge and New York, 1990.

Toynbee, Arnold J. *The Islamic World Since the Peace Settlement*, in the series Survey of International Affairs. London, 1927.

The works of Brockelmann, Hourani, and Hodgson are scholarly and thorough. Antonius writes of the political renaissance of the Arabs in this and the last century. Hitti brings his history of the Arabs down to the establishment of the Ottoman Empire in Egypt in 1517. The books by Toynbee and Kirk are particularly valuable for their treatment of the Palestine problem.

THE RELIGION OF ISLAM

Ahmad, Kurshid, editor. *Islam: Its Meaning and Message.* Ann Arbor, 1976 and Leicester, 1980.

Ali, Abdullah Yusuf. *The Holy Qur-an: Text, Translation and Commentary.* 2 vols. Lahore, 1934; New York, 1946; and other editions.

Amin, Mohamed. *Pilgrimage to Mecca.* London, 1978.

Amin, S. M. *The Holy Journey to Mecca.* Naples, 1976.

Arberry, Arthur J. *The Koran Interpreted.* 2 vols. London and New York, 1955.

Archive Editions, publisher. *Record of the Hajj, A Documentary History of the Pilgrimage to Mecca.* 10 vols. London, 1993.

'Azzam, 'Abd-al-Rahman. *The Eternal Message of Muhammad.* Translated by Caesar A. Farah. New York, 1964.

Esin, Emel, and Doganbey, Haluk. *Mecca the Blessed, Madinah the Radiant.* London, 1963.

Esposito, John L. *Islam: The Straight Path.* New York and Oxford, 1991.

Gibb, H. A. R. *Mohammedanism: An Historical Survey.* 2nd edition. London etc., 1961.

Guellouz, Ezzedine, and Frikha, Abdelaziz. *Mecca: The Muslim Pilgrimage.* New York and London, 1979.

Hamidullah, Muhammad. *Introduction to Islam.* Paris, 1969.

Hourani, Albert *Europe and the Middle East.* Berkeley and London, 1980.

Lippman, Thomas W. *Understanding Islam: An Introduction to the Muslim World.* New York, 1990.

Maududi, Abul A'la. *Towards Understanding Islam.* Chicago and Leicester, 1980.

Nasr, Seyyed Hossein. *Ideals and Realities of Islam.* Boston, 1972 and London, 1985.

Pickthall, Mohammed Marmaduke. *The Meaning of the Glorious Qur'an: Text and Explanatory Translation.* New York, 1988.

Schimmel, Annamarie. *Islam: An Introduction.* Albany, 1992.

Von Grunebaum, G. E. *Muhammadan Festivals.* New York, 1951.

Watt, W. Montgomery. *Muhammad at Mecca.* Oxford, 1953.

———. *Muhammad at Medina.* Oxford, 1956.

Zakaria, Rafiq. *Muhammad and the Quran.* London, 1991.

There are many translations of the Quran. Each has its strong points. Those by Arberry and Pickthall are particularly good. Abdullah Ali's translation is especially useful for its explanatory notes. The newcomer to the subject is well advised, however, to start his or her study of Islam with a general introduction, such as is provided by 'Azzam or Gibb or Hamidullah. Von Grunebaum's book is a pleasant introduction to the Muslim at prayer and on pilgrimage. Esin and Doganbey are Turkish Muslims who describe the Holy Cities by word and picture.

ARCHAEOLOGY AND THE ANCIENT NEAR EAST

Al Khalifa, Shaikha Haya Ali and Rice, Michael, editors. *Bahrain through the ages, the Archaeology*. London, 1986.

Baines, John and Malek, Jaromir. *Atlas of Ancient Egypt*. Oxford and New York, 1988.

Bibby, Geoffrey. *Looking for Dilmun*. London, 1984.

Breasted, James Henry. *A History of Egypt from the Earliest Times to the Persian Conquest*. 2nd edition. New York and London, 1909 etc.

Burney, Charles. *The Ancient Near East*. Ithaca and Oxford, 1977. The title of the British edition is *From Village to Empire*.

Harding, G. Lankester. *The Antiquities of Jordan*. Revised edition. London, 1967.

James, T.G.H. *An Introduction to Ancient Egypt*. London, 1989.

Johnson, Paul. *Civilizations of the Holy Land*. London, 1979.

Potts, D.T. *The Arabian Gulf in Antiquity*. Oxford, 1990.

Rice, Michael. *The Archaeology of the Arabian Gulf*. London, 1994.

———. *Egypt's Making*. London and New York, 1991.

Roaf, Michael. *Cultural Atlas of Mesopotomia and the Ancient Near East*. Oxford and New York, 1990.

Saudi Arabia, Ministry of Education, Department of Antiquities and Museums. *An Introduction to Saudi Arabian Antiquities*. Riyadh, 1975.

Simpson, William K., et al., editors. *Ancient Egypt: Discovering Its Splendors*. Washington, 1978.

———

Bahrain through the ages, the Archaeology details discoveries relating to Bahrain and the Arabian Gulf area. Readers should also consult the series of journals, *Atlal*, published by the Saudi Arabian Dept. of Antiquities and Museums from 1977, which provide much detailed data on excavations and surveys in the kingdom.

SCIENCE, LITERATURE, AND THE VISUAL ARTS

Brend, Barbara. *Islamic Art*. London and Cambridge, Mass., 1991.

Blunt, Wilfrid. *Splendours of Islam*. London and New York, 1976.

Burckhardt, Titus. *Art of Islam: Language and Meaning*. London, 1976.

Du Ry, Carel J. *Art of Islam*. Translated by Alexis Brown. New York, 1970.

Eiland, Murray L. *Oriental Rugs: A Comprehensive Guide*. Revised edition. Boston, 1976.

Gibb, H. A. R. *Arabic Literature: An Introduction*. 2nd edition. Oxford, 1974.

Grabar, Oleg. *The Mediation of Ornament*. Princeton, 1992.

al-Hassan, Ahmad Y. and Hill, Donald R. *Islamic Technology: An Illustrated History*. New York and Cambridge, 1987.

Hawley, Walter A. *Oriental Rugs, Antique and Modern*. New York, 1913, 1970.

Haywood, John A. *Modern Arabic Literature, 1800-1970: An Introduction, with Extracts in Translation*. London, 1971.

Jayyusi, Salma Khadra, editor. *The Literature of Modern Arabia: An Anthology*. London and New York, 1988.

Johnson-Davies, Denys. *Arabic Short Stories*. London, 1983.

King, David A. *Astronomy in the Service of Islam*. 1993.

Kritzeck, James. *Anthology of Islamic Literature, from the Rise of Islam to Modern Times*. New York etc., 1964.

Lichtenstadter, Ilse. *Introduction to Classical Arabic Literature*. Boston, 1974.

Lings, Martin. *The Quranic Art of Calligraphy and Illumination*. London, 1976; Boulder, Colorado, 1978.

Nicholson, Reynold A. *A Literary History of the Arabs*. 2nd edition. Cambridge, 1969.

Rice, David Talbot. *Islamic Art.* London, 1991.

Ross, Heather Colyer. *Bedouin Jewellery in Saudi Arabia.* London, 1978.

Safadi, Yasin Hamid. *Islamic Calligraphy.* London, 1978.

Scerrato, Umberto. *Islam,* in the series Monuments of Civilization. New York, 1976.

Tahan, Malba. *The Man Who Counted: A Collection of Mathematical Adventures.* London and New York, 1993.

Topham, John, et al. *Traditional Crafts of Saudi Arabia.* London, 1981.

———

Gibb, Lichtenstadter, and Nicholson offer introductions to the literature of the Arabs, and Kritzeck and Haywood have compiled anthologies of that literature in translation. Bernard Lewis's two volumes entitled Islam and listed above under "History" are a fascinating compilation of translations from Arabic, Persian, and Turkish literature up to the fifteenth century. Brend, Burckhardt, Du Ry, and Rice survey the visual arts, with Blunt and Scerrato accenting architecture.

CLASSICS

Arabian Nights. Many editions, but the most famous translations are by Richard F. Burton and Edward William Lane.

Burkhardt, John Lewis. *Notes on the Bedouins and Wahábys.* 2 vols. London, 1831; New York and London, 1967.

———. *Travels in Arabia.* 2 vols. London, 1829; Beirut, 1972.

Burton, Richard F. *Personal Narrative of a Pilgrimage to Al-Madinah and Meccah.* 2 vols. London, 1893; New York, 1964.

Doughty, Charles M. *Travels in Arabia Deserta.* 2 vols. London, 1921, and New York, 1979.

Lawrence, T. E. *Seven Pillars of Wisdom.* London and New York, 1991.

Palgrave, William Gifford. *Narrative of a Year's Journey through Central and Eastern Arabia (1862-63).* 2 vols. London, 1865, 1969.

Pelly, Lewis. *Report on a Journey to the Wahabee Capital of Riyadh in Central Arabia.* Bombay, 1866; Cambridge and New York, 1978.

———

As the entries above indicate, a number of publishers have recently been bringing out new editions of some of the best-known older books on Arabia and the Middle East—in some cases a hundred years and more after they were first published. The books by Burton, Doughty, and Lawrence listed here are classics of the English language, quite apart from their merits as authorities on matters Arabian. None of these works is easy reading, but any effort by the reader will be well rewarded.

THE ARABIAN PENINSULA AND THE GULF

Abdullah, Muhammad Morsy. *The United Arab Emirates: A Modern History.* London and New York, 1978.

Ajmi, Nassir. *Legacy of a Lifetime: An Essay on the Transformation of Saudi Arabia.* London, 1995.

Almana, Mohammed A. *Arabia Unified: A Portrait of Ibn Saud.* London, 1980, and New Brunswick, 1985.

Azzi, Robert. *Saudi Arabian Portfolio.* Manchester, N.H., 1978.

Belgrave, Charles. *Personal Column.* London, 1960.

Beling, Willard A., editor. *King Faisal and the Modernisation of Saudi Arabia.* Boulder, Colo., and London, 1980.

Bin Sultan, Khaled. *Desert Warrior, A Personal View of the Gulf War by the Joint Forces Commander.* London, 1995.

Bogari, Hamza. *The Sheltered Quarter, A Tale of a Boyhood in Mecca.* Austin, Texas, 1991.

Brent, Peter. *Far Arabia: Explorers of the Myth.* London, 1977.

Buchan, James. *Jeddah: Old and New*. London, 1991.

De Gaury, Gerald. *Faisal, King of Saudi Arabia*. London, 1966.

Dickson, H. R. P. *Kuwait and Her Neighbours*. London, 1956.

Facey, William. *The Story of the Eastern Province*. London, 1994.

———. *Riyadh, The Old City*. London, 1992.

Freeth, Zahra, and Winstone, H.V.F. *Explorers of Arabia: From the Renaissance to the End of the Victorian Era*. London etc., 1978.

Guise, Anthony. *Riyadh*. London, 1988.

Al-Hariri-Rifai, Wahbi and Mokhless. *The Heritage of the Kingdom of Saudi Arabia*. Washington, 1990.

Hawley, Donald. *The Trucial States*. London, 1970.

Hill, Ann and Daryl. *The Sultanate of Oman: A Heritage*. London and New York, 1977.

Hopwood, Derek, editor. *The Arabian Peninsula: Society and Politics*. London, 1972.

Ingrams, Harold. *Arabia and the Isles*. 3rd edition. London, 1972.

Kelly, J. B. *Britain and the Persian Gulf, 1795-1880*. Oxford, 1968.

Kiernan, R. H. *The Unveiling of Arabia: The Story of Arabian Travel and Discovery*. London etc., 1937; New York, 1975.

Long, David E. *The Persian Gulf: An Introduction to its Peoples, Politics and Economics*. Revised edition. Boulder, Colo., 1978.

Mauger, Thierry. *Undiscovered Asir*. London, 1993.

Monroe, Elizabeth. *Philby of Arabia*. London, 1973.

Pesce, Angelo. *Jiddah: Portrait of an Arabian City*. London, 1974.

Peterson, J. E. *Oman in the Twentieth Century: Political Foundations of an Emerging State*. London and New York, 1978.

Philby, H. StJ. B. *Arabia of the Wahhabis*. London, 1928, 1977.

———. *Arabian Jubilee*. London, 1952.

———. *The Empty Quarter: Being a Description of the Great South Desert of Arabia Known as Rub' al-Khali*. London, 1933.

Rihani, Ameen. *Maker of Modern Arabia*. Westport, Conn., 1983.

al-Saud, Noura bint Muhammad; and al-'Anqari, al-Jawharah Muhammad. *Abha, Bilad Asir: Southwestern Region of the Kingdom of Saudi Arabia*. Riyadh, 1989.

Sheean, Vincent. *Faisal: The King and His Kingdom*. Tavistock, England, 1975.

Stacey International. *The Kingdom of Saudi Arabia*. Revised edition. London, 1993.

Troeller, Gary. *The Birth of Saudi Arabia: Britain and the Rise of the House of Sa'ud*. London, 1976.

Twitchell, K. S. *Saudi Arabia*. 3rd edition. Princeton, 1958.

Wilson, Arnold T. *The Persian Gulf: An Historical Sketch from the Earliest Times to the Beginning of the Twentieth Century*. Oxford, 1928; London, 1954.

Winder, R. Bayly. *Saudi Arabia in the Nineteenth Century*. New York etc., 1965.

Beling's book is a compilation of papers on Saudi Arabia. Belgrave's autobiographical work is about the author's years as an advisor to the ruler of Bahrain, while Kelly and Winder collect detail for the historian. Ingrams' *Arabia and the Isles* is about the area known as Hadhramaut. Brent, Freeth, and Kiernan write of the development of knowledge of Arabia by European explorers. The books by Azzi, Buchan and Guise consist for the most part of beautiful color photographs. Facey's book on the Eastern Province is a photographic report with much descriptive and historical text as well. Philby's historical works are sound, and his records of exploration are valuable compendiums of information. Rihani's book is a first-hand account of the Saudi State in the 1920s by an American of Arab descent who was with King 'Abd al-'Aziz at the conference of al-'Uqayr in 1922.

SYRIA, LEBANON, AND IRAQ

Fisk, Robert. *Pity the Nation: Lebanon at War.* London, 1990.

Hitti, Philip K. *History of Syria, Including Lebanon and Palestine.* 2nd edition. London and New York, 1957.

Hourani, A. H. *Syria and Lebanon: A Political Essay.* London etc., 1946.

Khadduri, Majid. *Independent Iraq, 1932-1958: A Study in Iraqi Politics.* 2nd edition. London etc., 1960.

———. *Republican 'Iraq: A Study in 'Iraqi Politics since the Revolution of 1958.* London etc., 1969.

Lloyd, Seton. *Twin Rivers: A Brief History of Iraq from the Earliest Times to the Present Day.* 3rd edition. London etc., 1961.

Longrigg, Stephen Hemsley. *Four Centuries of Modern Iraq.* Oxford, 1925; Beirut, 1968.

———. *Syria and Lebanon under French Mandate.* London etc., 1958.

Penrose, Edith and E. F. *Iraq: International Relations and National Development,* in the series Nations of the Modern World. London and Boulder, Colo., 1978.

Salibi, Kamal S. *The Modern History of Lebanon.* London, 1965; Westport, Conn., 1976.

Seale, Patrick. *The Struggle for Syria: A Study of Postwar Arab Politics, 1945-1958.* London etc., 1965.

Thesiger, Wilfred. *The Marsh Arabs.* London, 1985.

————————————

Thesiger's book is about the inhabitants of the marshes of southern Iraq.

PALESTINE/ISRAEL

Ashrawi, Hanan. *This Side of Peace.* New York, 1995.

Barbour, Nevill. *Nisi Dominus: A Survey of the Palestine Controversy.* London etc., 1946. Published in New York in 1947 as *Palestine: Star or Crescent?*

Chacour, Elias with Jensen, Mary E. *We Belong to the Land: The Story of a Palestinian Israeli Who Lives for Peace and Reconciliation.* New York, 1992.

Doughty, D. and El Aydi, M. *Gaza: Legacy of Occupation.* West Hartford, Conn., 1995

Friedman, Thomas. *From Beirut to Jerusalem.* New York, 1990.

Hirst, David. *The Gun and the Olive Branch: The Roots of Violence in the Middle East.* New York and London, 1977.

Husseini, Hassan Jamal. *Return to Jerusalem.* London, 1989.

Quandt, William B. *Decade of Decisions: American Policy Toward the Arab-Israeli Conflict, 1967-1976.* Berkeley and Los Angeles, 1977.

Rodinson, Maxime. *Israel and the Arabs.* Translated by Michael Perl. New York, 1968.

Sayegh, Fayez A. *The Arab-Israeli Conflict.* New York, 1956.

————————————

Virtually all books about the Middle East in this century discuss the partition of Palestine and the creation of the State of Israel. There are, in addition, a growing number of books about individual aspects of these events, few avoiding the tendentious. However, those recommended here are good introductions to the most pressing current problem of the Middle East.

IRAN AND TURKEY

Black, Cyril E. and Brown, Carl L., editors. *Modernization in the Middle East: The Ottoman Empire and Its Afro-Asian Successors.* Princeton, 1992.

Kinross, Lord. *The Ottoman Centuries: The Rise and Fall of the Turkish Empire.* New York, 1977.

Lenczowski, George. *Russia and the West in Iran, 1918-1948: A Study in Big Power Rivalry*. Ithaca, N.Y., and London, 1949.

Lewis, Bernard. *The Emergence of Modern Turkey*. 2nd edition. London etc., 1968.

Mottahedeh, Roy. *The Mantle of the Prophet: Learning and Power in Modern Iran*. New York and London, 1986.

Muller, Herbert J. *The Loom of History*. London and New York, 1966.

Wilber, Donald N. *Iran, Past and Present*. 8th edition, Princeton, 1976.

Muller's book is considered a minor classic on civilization in Asia Minor from Troy to Ataturk.

EGYPT, NORTH AFRICA, AND SPAIN

Burckhardt, Titus. *Moorish Culture in Spain*. Translated by Alisa Jaffa. London, 1972.

Knapp, Wilfrid, editor. *North West Africa: A Political and Economic Survey*. 3rd edition. Oxford, 1977. The first two editions (1959 and 1962) were edited by Nevill Barbour.

Samsó, Julio. *Islamic Astronomy and Medieval Spain*. 1994.

Sordo, Enrique, and Swaan, Wim. *Moorish Spain: Cordoba, Seville, Granada*. New York, 1963.

Vatikiotis, P. J. *The Modern History of Egypt*. London, 1969.

Waterbury, John. *Egypt: Burdens of the Past, Options for the Future*. Bloomington, Ind., and London, 1978.

MIDDLE EAST AND WORLD OIL

Barger, Thomas C. *Energy Policies of the World: Arab States of the Persian Gulf*. Newark, Dela., 1975.

Longrigg, Stephen Hemsley. *Oil in the Middle East: Its Discovery and Development*. 3rd edition. London etc., 1968.

Philby, H. StJ. B. *Arabian Oil Ventures*. Washington, 1964.

PennWell Publishing Co., publisher. *International Petroleum Encyclopedia*. Tulsa, Okla., 1994.

Sarkis, Nicolas, publisher. *Arab Oil and Gas Directory*. Paris, 1994.

Stegner, Wallace. *Discovery!* Beirut, 1971.

Stocking, George W. *Middle East Oil: A Study in Political and Economic Controversy*. Nashville, Tenn., 1970.

Tughendhat, Christopher. *Oil: The Biggest Business*. London, 1968.

Turner, Louis. *Oil Companies in the International System*. London, 1978.

Yergin, Daniel. *The Prize: The Epic Quest for Oil, Money and Power*. New York, 1991.

Longrigg, Stocking, and Tughendhat present surveys, each with its own strength. Longrigg provides the most detail, and Tughendhat is easiest to read and the widest ranging. *Arabian Oil Ventures* contains Philby's story of the negotiation of what became the Aramco concession and of Major Frank Holmes's earlier concession covering much the same area. Stegner tells the story of Aramco's pioneering days. The books published by PennWell and Sarkis are solid reference works.

BEDOUINS, OASIS DWELLERS, AND SAILORS

Cole, Donald Powell. *Nomads of the Nomads: The Al Murrah Bedouin of the Empty Quarter*. Arlington Heights, 1975.

Dickson, H. R. P. *The Arab of the Desert: A Glimpse into Badawin Life in Kuwait and Sau'di Arabia*. London, 1949.

Hawkins, Clifford W. *The Dhow: An Illustrated History of the Dhow and Its World*, in the series Nautical Historical Record. Lymington, Hants., 1977.

Keohane, Alan. *Bedouin, Nomads of the Desert*. London, 1994.

Lancaster, William. *The Rwala Bedouin Today*. Cambridge, 1981.

da Silveira, Humberto. *Bedu*. Lausanne, 1994.

Thesiger, Wilfred. *Arabian Sands*. London etc., 1985. Braille edition, 1991.

Villiers, Alan. *Sons of Sinbad*. New York, 1940.

Weir, Shelagh, editor. *The Bedouin: Aspects of the Material Culture of the Bedouin of Jordan*. London, 1976.

Thesiger's work is already considered a classic. It describes the Bedouins of southern Arabia with great perception. Dickson is rich in material on the people of eastern Arabia. Villiers gives a popular account of the life of the Arab sailor, while Hawkins describes and illustrates the ships he sails.

NATURAL HISTORY

Baron, Stanley. *The Desert Locust*. London, 1972.

Basson, Philip W.; Burchard, John E.; Hardy, John T.; and Price, Andrew R. G. *Biotopes of the Western Arabian Gulf: Marine Life and Environments of Saudi Arabia*. Dhahran, 1977.

Benson, S. Vere. *Birds of Lebanon and the Jordan Area*. London and New York, 1970.

Bundy, G.; Connor, R. J.; and Harrison, C. J. O. *Birds of the Eastern Province of Saudi Arabia*. London, 1989. [published in association with Saudi Aramco]

Cloudsley-Thompson, J. L., and Chadwick, M. J. *Life in Deserts*. London, 1964.

Collenette, Sheila. *An Illustrated Guide to the Flowers of Saudi Arabia*. London, 1985.

Cornes, M. D. and Cornes, C. D. *The Wild Flowering Plants of Bahrain*. London, 1989.

Gasperetti, J. "Snakes of Arabia", *Fauna of Saudi Arabia*, vol. 9 (1988), pp. 169-450.

Harrison, David L. *The Mammals of Arabia*. 2nd ed. Sevenoaks, 1991.

Hollom, P. A. D.; Porter, R. F.; Christensen, S.; and Willis, Ian. *Birds of the Middle East and North Africa*. Calton, 1988.

Mandaville, James P. *Flora of Eastern Saudi Arabia*. London and New York, 1990.

McKinnon, Michael. *Arabia: Sand, Sea, Sky*. London, 1990.

——— and Vine, Peter. *Tides of War*. London, 1991.

Migahid, Ahmad Mohammad. *Flora of Saudi Arabia*. 3rd edition. Riyadh, 1989.

Mountfort, Guy. *Portrait of a Desert*. London, 1965.

Nelson, Bryan. *Azraq: Desert Oasis*. London, 1973.

Stacey International, publisher. *The Wildlife of Saudi Arabia and Its Neighbors*. London, 1990.

Walker, D. H. and Pittaway, A. R. *Insects of Eastern Arabia*. London, 1987.

Western, A. R. *The Flora of the United Arab Emirates: An Introduction*. n.p. 1989.

Readers should refer to the journal series, *Fauna of Saudi Arabia*, copublished by the National Commission for Wildlife Conservation and Development, Riyadh. The book on marine life in the Arabian Gulf by Basson and others is a well-illustrated standard work that will appeal to the general reader as well as the professional. Mandaville contributed the material on plants found in this book; the many illustrations in Migahid's flora will aid the nonspecialist seeking names of Saudi Arabian plants.

Index

I

J